Underpinning and Retention

Underpinning and Retention

Edited by

S. THORBURN, OBE
Director
Thorburn Limited

Visiting Professor
University of Strathclyde

and

G.S. LITTLEJOHN
Professor of Civil Engineering
University of Bradford

CRC Press
Taylor & Francis Group
Boca Raton London New York

CRC Press is an imprint of the
Taylor & Francis Group, an **informa** business
A TAYLOR & FRANCIS BOOK

CRC Press
Taylor & Francis Group
6000 Broken Sound Parkway NW, Suite 300
Boca Raton, FL 33487-2742

First issued in paperback 2019

© 1993 by Taylor & Francis Group, LLC
CRC Press is an imprint of Taylor & Francis Group, an Informa business

No claim to original U.S. Government works

ISBN-13: 978-0-7514-0094-6 (hbk)
ISBN-13: 978-0-367-86595-5 (pbk)

Typeset in 10/12 pt Times by Thomson Press (India) Ltd., New Delhi

A catalogue record for this book is available from the British Library

Library of Congress Cataloging-in-Publication data available

**Visit the Taylor & Francis Web site at
http://www.taylorandfrancis.com**

**and the CRC Press Web site at
http://www.crcpress.com**

Preface

The purpose of this book is to introduce the principles associated with the design and construction of underpinning and retention systems. The art and practice of underpinning developed from the requirements of strengthening and supporting structures built in former times, and with the modern trend towards deep excavations, particularly in urban areas, a retention system is often an essential element of the underpinning solution. Both systems seek to limit movements by support being introduced into structures and materials, in various states of stress, and to control safely the interaction between the existing and new works. In this regard, planning authorities generally strive to maintain the historical character of towns and cities, and the juxtaposition of old and new has highlighted the problems of maintaining the equilibrium of old buildings and avoiding damage to brittle fabrics.

Incomplete knowledge concerning the history and condition of structures to be supported, combined with a wide variety of foundation types and ground conditions, demands considerable engineering judgement, risk assessment and practical innovation when evolving underpinning designs. As a consequence, underpinning is generally handled by specialist consultants and contractors. With this in mind, case histories are included to demonstrate some of the difficulties encountered and the safe practical solutions which have been employed.

Competence in the design and construction of underpinning requires an understanding of materials science, combined with an awareness of the distributions of strain and stress unique to a structure and its particular ground support conditions. The paths of load transfer, both primary and secondary, must also be fully defined within any structure when underpinned. Dependence on theoretical skills without adequate experience of the latent weaknesses in earlier forms of construction can be a recipe for malperformance of underpinning designs.

Where a building suddenly displays movement, it is important to establish the cause and to assess whether the movement can be arrested solely by underpinning. Structural strengthening may be an essential requirement for a safe solution. For underpinning adjacent to a deep excavation, it is important to appreciate that the behaviour of the retention system is influenced by the ground type, method of construction, sequence of excavation and rigidity of the system of support. In the overall field of underpinning and retention, specialist firms now use a wide range of modern techniques which have been developed to provide support to all kinds of civil engineering structures. In this book, however, simple methods of excavation and underpinning have not been neglected since they can still be effective and economical in appropriate circumstances.

In the first chapter, the emphasis is on proper and adequate methods of investigating and appraising the structure, foundations and supporting ground, as a prelude to selection of the best method of underpinning. The legal aspects involving right of support are also discussed in this chapter, which concludes with an historical account of underpinning for comparison with the more modern techniques described later. Much of the art and practice of underpinning was developed without the benefit of current knowledge of the physical sciences. Subsequent chapters describe different methods of underpinning and retention ranging from simple methods of excavation and traditional piling, to ground improvement techniques such as chemical grouting and freezing, and soil anchorages and nailing. An adequate knowledge of theoretical soil mechanics has been assumed to enable the text to concentrate on the practice of underpinning and retention.

It is hoped that this assemblage of knowledge and experience will provide a useful work of reference for the construction industry.

S.T.
G.S.L.

Contributors

Mr H. Bradbury
Director Quality and Training, Roger Bullivant Ltd, Walton Road, Drakelow, Burton-on-Trent, Staffordshire DE15 9UA, UK

Dr D.A. Bruce
Technical Director, Nicholson Construction of America, PO Box 308, Bridgeville, PA 15017, USA

Mr K.W. Cole
Project Director, Ove Arup and Partners, 13 Fitzroy Street, London W1P 6BQ, UK

Mr J.S. Harris
Geotechnical Consultant, Specialist in Ground Freezing, 5 Foxhill Road, Burton Joyce, Nottingham NG14 5DB, UK

Mr J.F. Hutchison
Chief Engineer, Henry Boot Scotland Ltd, Garrowhill, Baillieston, Glasgow G69 6EY, UK

Mr P. Light
Associate Director, Trollope & Colls Construction Ltd, Mitcham House, 681 Mitcham Road, Croyden, Surrey CR9 3AP, UK

Professor G.S. Littlejohn
Head of Department of Civil Engineering, University of Bradford, Bradford, West Yorkshire BD7 1DP, UK

Dr F. Lizzi
Via de Nardis 7, 80127, Naples, Italy

Mr J.F.S. Pryke
Director, Pynford South Ltd, Warlies Park House, Upshire, Waltham Abbey, Essex EN9 3SL, UK

Mr S. Thorburn
Director, Thorburn plc, 243 West George Street, Glasgow G2 4QE, UK. Visiting Professor, University of Strathclyde, George Street, Glasgow G1 1XW, UK.

Contents

6 The Bullivant systems   199
H. BRADBURY

7 Ground freezing   220
J.S. HARRIS

11 In-situ earth reinforcing by soil nailing 340

D.A. BRUCE

1 Introduction

S. THORBURN

L'objet de la construction est d'executer toutes les parties d'un ouvrage projeté avec toute le solidité et la perfection dont elles sont susceptibles, en y employant les matériaux les plus convenables mis en tourne avec art et économie. RONDELET, *L'Art de bâtir*, Vol. 1, p. 8.

1.1 General description

The separate actions of shoring and underpinning have been described in the literature, notably in the classic treatise by Cecil Haden Stock in 1882. Stock found that little of the practice of shoring and underpinning had been committed to print in a form suitable for proper study of the subject and he collated and presented the works of various authorities then engaged in that practice. Two discerning comments were made by Stock; firstly, that direct involvement with site works was the only way to gain proper knowledge and adequate practical ability of underpinning, and, secondly, that theoretical knowledge *per se* should not be allowed to interfere with the application of sound techniques derived from long experience. Regardless of the fact that Stock was dealing with ruinous and dangerous building structures constructed before the turn of the century and although considerable technological progress has been made over the past two decades, these statements are still relevant today in the context of underpinning. Much benefit can be obtained from Stock's comprehensive descriptions both of earlier buildings and of the practice of shoring and underpinning at that time. Reference may also be made to Prentis and White (1950), Hunter (1952) and Tomlinson (1978) for useful information on earlier and recent forms of shoring respectively.

Shoring is generally used to provide temporary support to structures while the underpinning works are being executed. The interaction between shoring and underpinning should be appreciated, and great care must be taken during the final phase of the operations involving the removal of the temporary shoring and the acceptance of all structural loads by the underpinning. The temporary relief of load and/or the temporary restraint provided by shoring demands careful assessment and shoring must be judiciously positioned and stressed to support a structure without damage, while its basal support is strengthened or removed and replaced. Successful underpinning requires a

1

knowledge of the state of balance of a building and its foundations, and of the ground conditions. Paths of load transfer, both primary and secondary, need to be fully investigated within any structure to be underpinned, as do the probable concentrations of stress in the building while in its passive condition.

Changes to the state of balance and to the pattern of load distribution within the structure take place during all phases of the underpinning operations, and it is important to identify the mechanisms of load distribution and load sharing. An awareness and knowledge of the effects of age on the durability and performance of the materials and of the fabric of structures is essential. Great care must be taken during the final phase involving removal of all temporary support, since elastic and permanent strains will be experienced.

This series of individual operations places great demands on the skills of the engineer and the contractor, and a thorough study of the mechanics of a structure is a prerequisite to the successful outcome of the complete sequence of shoring and underpinning operations.

The transfer of load from a structure to its underpinning components must be carefully executed and the mechanism of load distribution must be identified and controlled to an extent commensurate either with the simplicity of the operation or with its complexity and the need to restrict movements.

Foundations need to be underpinned either when a structure is being distorted by foundation movements or because proposed new works are likely to cause structural malperformance or failure if the foundations are not improved. If investigations show that movements are continuing or that the risk of fresh movements is unacceptably high then underpinning should be considered.

It is important to establish the cause of settlement in a building that suddenly displays movement and, in particular, to know whether the event can be arrested by underpinning alone. Underpinning may be unnecessary if the event is initiated by a short-term phenomenon and the movements do not disturb significantly the natural state of balance of a building structure.

Some forms of underpinning involve excavations for the installation of deeper or wider foundations. The excavations will remove support from part of the foundation while the work is in progress, and care must be taken to ensure that the structure remains safe. The structure and existing foundations should be able to arch safely over the excavations. If this is not the case because the wall is too weak or too fragmented at foundation level, for example poorly jointed random rubble stone masonry, then additional work should be carried out to strengthen the foundations before the underpinning commences. Structural strengthening measures, in conjunction with underpinning, may be an essential requirement to arrest movements.

Special care should be taken when constructing traditional underpinning

segments at corners, short return walls, or beneath—or partly beneath—
existing piers or isolated foundations. At such points building elements above
cannot arch over and additional support is often required.

If only parts of a foundation are to be underpinned then the engineer
should be satisfied that any movements between those parts being underpinned and the remainder will be acceptable. The success of partial
underpinning, where properly designed and executed, has been demonstrated
by experience.

Three main categories of structure and classes of underpinning can be
identified.

Categories of structure: (i) Ancient—greater than 150 years since
                                          completion
                                     (ii) Recent—50 to 150 years since completion
                                   (iii) Modern—less than 50 years
Classes of underpinning: (i) Conversion works
                                     (ii) Protection works
                                   (iii) Remedial works

A knowledge of building construction as previously practised is of great
assistance to the engineer preparing designs for each of the three classes of
underpinning. The determination of the condition of the fabric and foundations of ancient and recent structures can be a major and daunting task
for the engineer, particularly where the means of access and working space
for investigations are restricted. The underpinning of ancient building
structures presents hazards in the form of deterioration of the condition
of the materials of loadbearing walls, pillars and buttresses, which are often
of composite construction. Historical records reveal that in medieval times
too great a reliance was placed on the supportive ability of essentially rubble
masonry contained by relatively thin ashlar facing stones. Sound ashlar
masonry construction should not be expected as a common provision and
careful exploratory work should be carried out to ascertain the real nature
and condition of all loadbearing components which are important to the
safe and successful execution of underpinning works. The loss of the beautiful
spire of Chichester Cathedral in 1861 may be attributed to the belief by
those responsible for the repair work that the stone filling of the pillars, upon
which the 83 m spire relied for support, would continue to provide safe
support while the ashlar facing stones were being repaired. Tertiary crinoidal
limestone quarried at the Isle of Wight had been used for the ashlar facing to
the pillars, but the internal rubble stone filling had been cemented with
chalk–lime mortar. It was alleged by the workmen who carried out the repair
work that dry mortar dust poured occasionally from the joints in the ashlar
facing. It is possible that the chalk–lime mortar lost moisture prior to setting
due to the hygroscopic nature of the limestone blocks, in addition to the
loss of its cementitious properties with age deterioration. The outer shell of

ashlar stone, being stronger and stiffer, probably carried a significant proportion of the total load in addition to its containment function. The repairs must have adversely affected the distribution of stress within the pillars, and also perhaps caused some loss of fines from the rubble stone filling, to the extent that the highly stressed shell of facing stones failed. Deformation of some or all of the pillars must have altered seriously the natural state of balance of the high tower structure and caused the disastrous failure of an historical heritage. That this incident is not singular is attested by the failure of the towers of Winchester Cathedral in the 12th century; of Gloucester Cathedral in 1160; of Worcester Cathedral in 1222; of Ely Cathedral in 1322; and of Norwich Cathedral in 1361. The possibility of the state of balance of an ancient structure being altered during underpinning work should be anticipated and recognized in the design of the shoring. The following incident reflects this possibility and demonstrates that the alterations may result in portions of the structure apparently remote from the loci of the underpinning works being subjected to a new pattern of loading. In 1841, during repairs to the tower of the Church of St Mary, Stafford, a sudden fracturing of one of the pillars of the chancel occurred with a distinct noise. The chancel provided no direct support to the tower structure. The pillar split from top to bottom due to some change in the natural pattern of load sharing caused by the repairs to the tower.

Buttresses present a similar problem where rough stones were used for the main body of the buttresses and dressed stones for the facing. The dressed stones generally were more compactly laid than the rough interior stones and inequality of stress is a likely result. It is also possible that the outer portions of major buttresses having large projections carry much less load than the portions nearest the walls which they support. A further inequality is experienced from the random combination of soft and hard stones, and it may be safely concluded that the composite loadbearing elements of ancient structures are unevenly stressed and may be highly stressed at critical locations.

It should not be assumed that structural movements are due to foundation settlement alone, since ancient structures founded on bedrock can be subjected to significant differential movements due to poor construction of the loadbearing elements. The practice of allowing parts of the original foundations of ancient building structures to remain during conversion or reconstruction and incorporating them into new foundation systems has always been considered inadvisable but could not always be avoided by earlier builders. The possibility of this circumstance should always be borne in mind.

Arches provide another hazard in the form of lateral components of load derived from thrust at the springers. Loss of restraint in the vicinity of springers must be avoided, and care must be taken in case the spandrel walls of arches contain loose stone filling which could result in failure if the natural restraint is removed. The loose nature of stone infilling between the spandrel

Figure 1.1 Mature growth of tree root in bridge structure.

walls of a bridge structure is evidenced by the extreme root growth (Figure 1.1) which remained undetected until repairs commenced. It seems obvious that the shoring for an arch should be capable of supporting not only the self-weight of the masonry, but also the loads transmitted to the arch, before any underpinning involving removal of material is attempted. It is advisable to introduce suitable grouts under carefully controlled pressures through the outer shells of the stone facings of all loadbearing elements of ancient structures, and to consolidate any loose stone filling hidden behind the outer facing stones. This injection of grout should be carried out prior to the construction of the major shoring works to ensure that the local concentrations of stress induced by the often high shoring loads do not cause local damage or even result in harmful movements as the shoring accepts its temporary supportive role. In critical situations, or where some uncertainty exists, it is preferable to use only those construction techniques which minimize stress relaxation and loss of restraint; which maintain essentially the natural state of balance and pattern of loading; and which supplement the natural support conditions.

The hazard provided by inaccurate or poor workmanship in ancient buildings should also be borne in mind, and the following account of the collapse of the Long Room at Custom House, London, in 1825 provides a good illustration of this risk. The site on exposure was found to comprise a confusion of irregular old walls, ancient quays and sewers, and to contain the debris of former building activities. It would appear that extensive site clearance to provide a proper working surface was not carried out, and a new foundation of timber piles was driven through the mass of debris and obstructions. The building had two storeys of vaults formed by a series of stone piers and brick arches. Each pier was supposed to bear upon nine timber piles in a square pattern. Great difficulty was experienced in driving the piles because of the numerous obstructions, and the quantity of debris prevented accuracy both in positioning the piles and verifying their positions later. As a result of the total lack of clarity in respect of the true positions of pile groups, the outer row of piles of one of the foundations was mistaken for the middle row of the nine-pile group and the foundation was constructed with only half of the piles supporting it. The load transmitted eccentrically to this particular pile group by a stone pier eventually caused structural failure, since the debris outside the pile group was very loose and incapable of preventing the rotational, translational, and vertical movements of the pier. This single pier moved sideways off the eccentric pile group and sank about 1.5 metres, causing major movements and imbalance of the vault structure culminating in the total collapse of the building. The assumption should never be made that any structure is accurately located over its foundations.

Stress concentrations should be anticipated in any form of structure, and the following simple incident exemplifies the ease with which this hazard can be created even in an ancient building. The mortar bed for a ceremonial stone was prepared by a mason and took the form of four pats of strong mortar on the foundation stone of the ornamental base of a pillar. In order to assist the dignitary invited to lay the ceremonial stone, the mason mixed and spread a soft mortar of high workability over the foundation stone between the four stiff pats of strong mortar. The dignitary spread the soft mortar with a great flourish but the stiff pats supported effectively the ornamental base stone and the pillar itself. About one year after application of the full load on the pillar, major cracks developed in the base stone which was constrained to span between the hard pats of strong mortar, and shoring and repairs had to be hastily executed. This classic example of stress concentration should be a lesson to designers of modern buildings and a warning to all who design and build masonry structures that a ceremonial stone should not be a critical component of a structure. The ambience of a ceremony is not always conducive to good workmanship because of the danger of distraction.

Recent buildings constructed during the nineteenth century to accommodate the demands of the Industrial Revolution are not without faults despite improvements in building technology in that era. Forms of composite masonry construction used in recent buildings were similar to those used in ancient buildings, where rough stones in the interiors of masonry elements were contained by dressed facing stones. Extensive use of walls was made for recent structures and crosswall construction provided the means of stability against lateral loads. The relatively fast rates of construction sought during

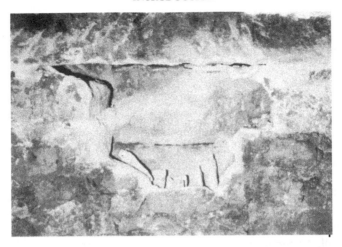

Figure 1.2 Photograph of interior of masonry wall showing wedge-shaped stones between inner and outer leaves.

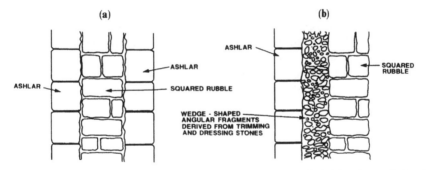

Figure 1.3 Diagrammatic presentation of construction of masonry walls in (a) ancient period; and (b) recent period.

the recent period introduced a new hazard, since the walls could be built faster by forming outer leaves of dressed stone, and using irregular wedge-shaped stones, which were the product of the trimming and dressing operations by the mason, for the interiors of the walls (Figure 1.2). The inequalities caused by the simultaneous use of rough and dressed stones in ancient times were made more acute by the greater use of this poorer material in the heart of walls in recent structures (Figure 1.3). Portland cement mortar was used as well as lime mortar during the recent period to bind the stone filling, but the nature of the stone fragments and the lack of control over the grouting of the stone filling produced walls of variable quality and strength.

Deterioration has taken place in the condition of walls built during the recent period although some buildings are only a century old. It should not be assumed that the techniques of construction used for recent structures were greatly superior to those of ancient structures, and walls may require to be grouted before the commencement of shoring and underpinning. The wedging action of sharp irregular stones within the heart of a wall due to vertical load must be prevented by the bonding action of the mortar, and its failure as a cementitious agent will be accompanied by an outward thrust on the thin shell of dressed stones. Fortunately, crosswall construction and the smaller rooms required for accommodation during the recent period provided a greater spread of load, and the masonry walls were relatively narrow. The outer stone facings were capable of safely supporting the loads applied at that time, although the stone filling could be of relatively poor quality. The ability of these walls to sustain modern high loads or shoring and underpinning loads without distress should not be assumed, however. The lateral restraint of walls at junctions is dependent on the number of keystones provided over the height of the wall. Experience has shown that the keystones were often few in number, and over the past 100 to 150 years the natural deterioration in the condition of the keystones under tensile stresses has resulted in brittle failure. The interval of time between the complete fracturing of all keystones connecting masonry crosswalls, and facades and failure of the load-bearing walls themselves is relatively short, and the effect can be dramatic. Attempts have been made to repair keystones which have failed, but the task is difficult, and often the situation is irretrievable since the degree of outward movement can be such that progressive movement due to load eccentricity is usually a continuous and gradual process which often culminates with acceleration of

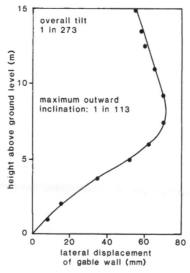

Figure 1.4   Outward movements of masonry wall.

movement. Measurements taken of the pattern of outward movement of an ashlar masonry wall of a prestigious building are given in Figure 1.4 and show the extent and form of curvature of this external wall subsequent to failure of all internal keystones.

Recent buildings can possess very heavy ornamental stone parapets at roof level, and the removal of this load at such high elevations during structural strengthening and underpinning greatly assists the maintenance of stability of facades, particularly in situations where the keystones to transverse walls have fractured. Deficiencies in the fabrics of recent buildings also result from poor workmanship, building inaccuracies, insufficient bonding to transverse walls, and age deterioration.

The problem of mutual gables or common walls arose during the recent period, and walls built in juxtaposition may require each other to maintain their stability. Often there was no servitude of support and the use of adjacent walls for the purposes of achieving structural stability was clandestine. Situations have arisen where the demolition of recent properties has led to progressive tilt and translational movements to such an extent that demolition became a necessity. The demolition of old property of the recent period within an urban situation should not be carried out without due care. Shoring and propping should be judiciously located to support adjacent buildings before and during the demolition and site clearance works, and certainly before any underpinning commences. The legal situation must be clearly understood by all parties involved with the development of urban sites. Clear definitions of responsibilities should be made at the outset and the criteria for the support works should be carefully and explicitly defined. Basements beneath buildings of the recent period may have been formed or extended after completion of a building and the degree of support afforded to adjacent property by the later construction can be marginal.

Brickwork was extensively used during the recent period, and was bedded with mortars comprising either lime, portland cement or combinations of both cementitious agents. The quality control of brick production has varied, and the changes in the strengths of mortar as well as bricks can present problems during underpinning. In damp conditions, lime mortar can deteriorate to a considerable extent with age. Damp conditions can also result in sulphate attack on mortar, the source of the sulphates being the bricks themselves. Situations have been found where portland cement mortar had deteriorated in strength to such an extent that a scraping action readily removed the mortar between the bricks. The practice, derived from ancient times, of forming a shell of facing stones and infilling with rough stone was also used during the recent period for brick walls and piers having large dimensions. Thick walls and piers of recent structures were shaped by an outer skin of brickwork and loose bricks were hand-placed within the brick shell. The loose bricks were bedded on soft mortar and the surface of each succeeding or alternate layer of bricks was grouted with a liquid mortar. The workmanship is known to have varied widely and loose brick filling with little or no mortar has been found. The need to build quickly on occasions created unfortunate legacies which can present a hazard to underpinning. Grouting of all loose brickwork of this kind is recommended prior to the execution of underpinning works. Fortunately, many brick walls of crosswall construction forming buildings of the recent period were of thicknesses equivalent to only one or one-and-a-half bricks, and such walls may be expected to provide fewer difficulties to underpinning.

Modern structures built within the last 50 years were subject to Building Regulations and Codes of Practice, and their form of construction can often be assessed reasonably easily from a search of plans which are extant and from a thorough structural survey. Modern structures differ from those of ancient and recent times by virtue of the greater use of stronger cementitious materials with the objectives of economy of materials and slender structural elements. Higher brick crushing strengths have been achieved through strict control of the composition, kilning and curing procedures. The compressive strengths of bricks have increased by about 50% over the past 50 years, and portland cement mortars have much greater strengths, elastic moduli, and resistance to deterioration. The use of monolithic forms of concrete construction and higher-strength bricks and mortars has changed the response of structures to ground movements. New building materials have also been introduced over the past 20 years which, in conjunction with much stronger bonding agents, have reduced the flexibility of structures and produced brittle behaviour. Ancient and recent masonry and brick structures bonded with relatively weak lime mortars often could accommodate significant ground movements by virtue of small slip strains along the bedding joints and minute tensile strains or microcracks in the vertical joints, in the manner of a discrete material. The total displacements permitted by the extensive development of these small strains produced essentially flexible behaviour of a complete structure. The relative brittleness of modern ceramic or cementitious materials reduces the global flexibility and a few wide cracks can develop rather than numerous microcracks at close intervals. Considerable use has been made in modern structures of composite materials such as brick and blockwork and timber panels with brickwork facades. Cavity wall construction was introduced in modern times and this separation of the inner and outer brick leaves brought a greater dependence upon wall ties for adequate structural behaviour. Internal leaves of concrete blockwork and external leaves of brickwork have introduced inequalities which should be taken into account in underpinning works. The use of precast concrete cladding panels for modern buildings has provided structures which behave in an essentially rigid monolithic manner, and load transfers due to inequalities of ground support are a distinct possibility depending on the strengths of the mechanical connections between the panels. The mechanical levelling devices at the bases of precast concrete panels can cause stress concentrations despite the introduction of low-water-content cement mortar rammed into the gap beneath the panels after completion of the positioning and levelling operations. Prestressing and post-tensioning of beams in modern structures has introduced considerable strain energies into these structural elements and great care should be taken during conversion or remedial works. The diaphragm action of reinforced concrete floors can be of benefit in respect of providing restraint against lateral movements, provided proper wall-to-floor connections were formed. Steel structures of recent and modern times possess considerable ductility and strength, and shoring and underpinning of most types of steel structure are generally simple in concept and construction and present fewer hazards than ancient structures.

The previous descriptive passages of ancient, recent and modern structures are intended to provide a background against which an appreciation of the hazards and problems of underpinning can be developed. Reference should, however, be made to technical literature on the manufacture and properties of engineering materials which are likely to be encountered during underpinning works, since this book cannot cover the subject of materials and their uses in a sufficiently comprehensive manner. The design and construction of shoring and underpinning of all types of structures should err on the conservative side, since uncertainties always exist regardless of the thoroughness of any archive searches and of non-destructive investigations. The objectives of underpinning now require definition before solutions and techniques are presented for the guidance of the reader.

1.1.1 Conversion works

Alteration, improvement, and renewal of old properties may be prompted by the need to preserve national assets of historical importance; to take commercial advantage where existing buildings are essentially suitable for change of use; and often to minimize expenditure on housing provisions.

The architect must ensure that the character of the building after renovation and refurbishment conforms with the present setting or recognizes apparent change in the local environment and avoids incongruity in the modern setting. The planning authorities often insist, therefore, that the existing masonry or brickwork facades of buildings should be preserved although the interior structure of the building may be demolished. These conversion works present a considerable challenge to the engineer who must design his shoring and underpinning works to safeguard the important facades, and construct new support works within working spaces which are very restricted. Figure 1.5 shows the extensive nature of shoring to a masonry facade which had to be provided together with the bored pile foundations which were constructed to support new internal colums.

The engineer must develop a proper understanding of the nature of the building and its relationship to the existing environment, particularly from the aspects of stability and the groundwater regime. Changes in the local

Figure 1.5 Shoring and underpinning using bored piles.

environment due to development by conversion can affect adversely both the states of balance of existing buildings and the local groundwater table. Recourse to law can affect redress, but the damage to property will already have happened, with possible serious commercial consequences. A knowledge of building construction as formerly practised is thus of great benefit to the engineer preparing designs for renovation and refurbishment, and an awareness of the effects of age on the materials and fabric of old buildings is also of advantage.

Adequate information on the foundations of old buildings requiring conversion is often difficult to obtain, particularly where the means of access and working space for investigations are restricted. A judgement must often be made regarding the condition and stability of old building foundations based on a minimal amount of direct information. In these circumstances it is essential to examine archivistic records and study the available history of building performance in the locality, including careful and comprehensive observations of the condition of the building to be refurbished. The condition of the immediately adjacent property must also be examined in case of structural interaction between the buildings, such as reliance upon support from an adjacent wall which may be clandestine. Since investigations of the foundations of old buildings are often restricted and judgements may require to be based on minimal evidence, it is important to be reasonably conservative, and, more important, to employ construction methods which permit adaptation. The ability to contend with unforeseen variations in ground conditions because of a measure of versatility of the chosen construction technique is invaluable during the period of strengthening or of replacing old foundations. Deficiencies of support can also be encountered during conversion works due to age deterioration, errors in the earlier construction, and poor workmanship. Knowledgeable site supervision during the period of the underpinning works is essential.

The relief of stress caused by removal of ground from within a building undergoing conversion in order to provide a basement structure can result in the development of significant elastic and permanent strains, and the effect of these ground displacements on the temporary shoring and permanent underpinning and new construction works should be carefully evaluated. The legal implications of inadequate temporary support provisions during conversion works can be serious and tortuous.

Ground anchorage may be used as an alternative to internal shoring to provide lateral restraint to the perimeter walls of basement structures. Anchoring often involves drilling through existing walls of uncertain dimensions and condition, and the choice of the methods of drilling and anchoring must be selected carefully. The method of drilling must not cause loss of ground or disturbance to the state of balance of existing buildings, and grouting of anchors should not cause lateral ground displacements or heave.

1.1.2 Protection works

Protection works to structures may be required in the situations described below:

(a) Construction of new buildings in sufficiently close proximity to existing buildings to warrant the provision of underpinning works under the existing properties in order to limit ground movements and protect the existing buildings.

(b) The construction of new basements within existing buildings.

(c) The construction of new buried structures, such as services tunnels and pipelines in close proximity to existing buildings.

(d) The growth of trees in clay soils causing adverse ground displacements due to moisture movements.

(e) The construction of new manufacturing facilities which contain mechanical sources of harmful vibrations to the extent that the ground support beneath existing buildings would be affected.

(f) Groundwater lowering.

The construction of new buildings within an urban environment can involve large excavations below the foundations of existing buildings. Shoring and underpinning of existing buildings as a protective measure is often required as shown in Figure 1.5 before the construction of new buildings can commence. Figure 1.6 shows both the magnitudes of vertical ground surface

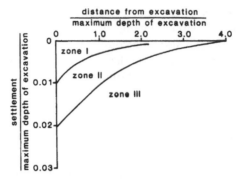

Figure 1.6 Ground movements near braced excavations (Peck, 1969).

displacements and their distribution as a function of distance from the faces of braced excavations in various types of soil. The displacements and distances are presented in dimensionless form as fractions of the depths of the excavations, and the consolidation settlements which occurred within the construction period are included. This graphical plot was interpreted by Peck (1969) from a relatively small number of observations but it permits a rough evaluation of the vertical ground displacements which may be experienced by existing buildings adjacent to braced excavations for new buildings. Lateral ground displacements will also be experienced adjacent to braced excavations for new buildings and their magnitudes will depend on the design and construction of the retention systems (see Chapter 11).

The construction of deep trenches for services, sewers and pipelines adjacent to existing buildings presents the possibility of significant movements in ground adjacent to the deep trenches. The Transport and Road Research Laboratory have made observations at a number of sites where trenches were excavated to depths of between three and five metres in London Clay (Chard and Symons, 1982). The distribution of lateral surface movements towards the trenches measured at various stages is presented as a dimensionless plot in Figure 1.7.

There is generally no need for temporary support when piling methods of underpinning are employed. Conventional piles for underpinning are either formed in-situ in bored holes, driven by vibratory hammers, or jacked. Bored cast in-situ piles are the most common form and have a minimum diameter of 300 mm. A more usual size is 450 mm diameter. Piles that are jacked or vibrated may be of a smaller size and need not be circular.

Table 1.1 shows a simple classification of piles used for underpinning works. Small diameter piles may be installed close to or beneath the loads to be supported, and very small diameter minipiles are frequently drilled through existing foundations or the bases of thick walls. Although the bearing capacities of individual minipiles are small, these piles have proved effective

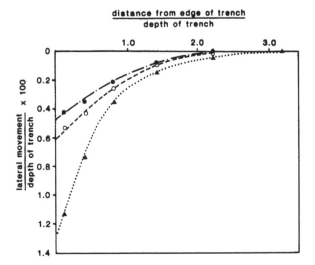

Figure 1.7   Ground movements near trench excavations (Chard and Symons, 1982). ●, 3 days after completion of excavation; ○, 1 day after completion of backfilling; ▲, $7\frac{1}{2}$ weeks after completion of backfilling. No further change after 80 weeks.

Table 1.1   Classification of piles used for underpinning works

Types of pile					
Ground removal			Ground displacement		
Micro	Mini	Small	Micro	Mini	Small
Methods of construction or installation					
(i) Rotary	(i) Flight auger	(i) Flight auger	(i) Driven and pushed tubular steel sections, open and closed ended	(i) Driven and pushed tubular steel and concrete sections	(i) Driven and pushed tubular steel sections, open ended
(ii) Rotary percussive	(ii) Rotary	(ii) Percussive (clay cutter and bailer)		(ii) Driven and pushed steel sections (H-piles)	
	(iii) Rotary percussive				

Notes   (1) Piles having lateral dimensions less than 600 mm but more than 300 mm are described as small piles
(2) Piles having lateral dimensions less than 300 mm but more than 75 mm are described as minipiles
(3) Piles having lateral dimensions less than 75 mm are described as micropiles

in difficult situations and increasingly cost-effective as improved drilling and driving equipment has been developed.

The installation of conventional piling as part of the foundation works for new buildings can cause harmful vibrations, and it should not be assumed that the cable percussion boring methods used in the construction of small-diameter bored piles will not cause adverse effects on existing structures. These piles are formed using a tripod rig which raises and lowers the cutting tools by means of a cable. The temporary steel casings which are used to support the pile bores are driven into position by a percussive technique, and the vibrations induced by the driving of the casings have been known to cause damage to adjacent buildings. In medium-dense to dense sands and gravels it is often difficult to advance the temporary steel casings ahead of the bases of the pile bores, and there is always the danger that the boring tools will be used to loosen the granular soils below the leading edges of the steel casings. The loosening of the granular soils permits the temporary casings to be driven with relative ease, but excess volumes of soil have been removed and loosening of the strata takes place. If a sufficient number of piles are installed adjacent to an existing building using this technique then significant ground movements can be experienced by the building and damage can result.

The design of piles should recognize the elastic response of the pile–soil or pile–rock system and no criteria for movement at the head of the pile should be specified which are less than the elastic response of the system to the applied load. Theoretical considerations indicate that linear-elastic analyses of small-diameter piles having length-to-diameter ratios greater than 20 give reasonable predictions of performance provided the service conditions do not

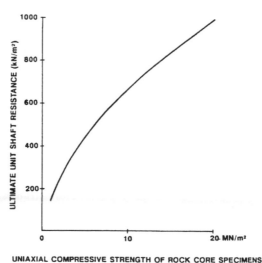

Figure 1.8 Relationship between ultimate values of unit shaft resistance in rock sockets and uniaxial compressive strength of rock core specimens.

impose stresses exceeding 50% of the limiting stresses from the aspect of soil failure.

Conventional piles founded in sedimentary rocks are often designed on the basis that their service loads can be supported by side or peripheral shear in rock sockets and using factors of safety of about one-and-a-half to two. A relationship between ultimate values of unit shaft resistance in rock sockets and uniaxial compressive strengths of rock core specimens is presented in Figure 1.8. These values may be considered lower bound but do not take into account the adverse effects of smear.

Conventional piles founded on hard igneous and metamorphic rocks are designed mainly on the basis of the service loads being supported by base resistance and using factors of safety of about three. It is difficult to penetrate such hard rocks using conventional piling equipment.

Recognition of the necessity for simplicity in routine designs and of the assumptions which must be made concerning soil stiffnesses and structural interaction imposes constraints on the use of more rigorous mathematical approaches. Randolph (1980) has developed methods of analysis of individual piles and pile groups using approximate, but compact solutions for the elastic behaviour of piles. The approximations have been formed from a study of rigorous numerical solutions.

With regard to pile testing procedures, it is important to differentiate between integrity testing and load testing. *Integrity testing* involves the measurement of a property of the body of the pile which can be related to its soundness but does not provide any assurance of the capability of the pile to support safely the service loads. *Load testing* involves the measurement of the response of a pile–soil or a pile–rock system to loads applied to the head of the pile, but does not provide any assurance that the construction of the pile conforms with the requirements of the specification. Proper supervision of piling works by site personnel having adequate and relevant experience is essential to the sound construction of piles.

The installation of steel sheet piling can cause vertical ground displacements due to the effects of vibration. Clough and Chameau (1980) monitored the work of several contractors who were confronted with damage claims from owners of nearby buildings during the construction of new stormwater culverts. The construction works involved the vibratory driving of steel sheet piles, and cracks appeared in both adjacent streets and buildings. The vibratory equipment operated at about 19 Hz and accelerations of as much as $0.4\,g$ were measured close to the sheet piling. The vertical ground displacements resulting from the installation of the steel sheet piles are shown in Figure 1.9 for two sites underlain by granular materials in loose and medium dense states of compaction.

The underpinning used for protection works must be founded on suitable bearing strata which will not be affected by the construction operations to be carried out adjacent to the structures to be protected. The protective

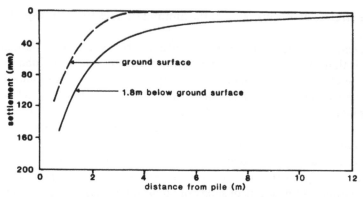

Figure 1.9 Ground movements due to vibratory driving of steel sheet piles.

underpinning must be capable of resisting the vertical and lateral displacements which can result from stress relief, ground vibrations, and stresses imposed by the new construction. Reasonable predictions of the magnitudes of ground displacements caused by stress relief or stress increases can be made by means of finite element analyses using appropriate moduli obtained either from in-situ tests or from reliable data derived from the retrospective analyses of suitable case histories. The quality and quantity of data must be adequate to permit reasonable predictions to be made.

1.1.3 Remedial works

Recent modern buildings which should have continued to perform satisfactorily for some years have often suddenly undergone movements and displayed structural defects which required immediate attention. The determination of the reasons for the change of behaviour, an assessment of the residual strength and stiffness of the building structure and the selection of suitable economical remedial measures can involve much time and expenditure, regardless of the relative importance of the building. The timing and extent of the remedial underpinning works to foundations are important aspects of the process of assessment and care must be taken that the problem has been clearly defined and appropriate remedial measures taken.

The legal implications of over-expenditure on over-conservative solutions confront all engineers preparing designs for remedial works, and care must be taken that improvement of a structure is not achieved at the expense of a party being sued for damages. The expense incurred in addition to that necessary to remedy the situation may be contested successfully by the defender. Remedial works may involve structural strengthening as well as underpinning, particularly if the structural displacements take the form of significant tilt or lateral movements. Masonry structures are particularly prone to separation of facades from internal walls due to fracturing of the keystones, and structural ties are often required in addition to underpinning. The underpinning must bear on suitable ground which is capable of accepting the loads with the minimum of movement. Sound rock, heavily over-consolidated clays, such as glacial soils, and compact sands and gravels are suitable bearing materials for simple conventional underpinning and the best means of transferring the loads to these materials without causing harmful ground disturbance should be employed. Loose sands and silts should not be used as bearing materials for simple underpinning, since experience has shown them to be most unsuitable. Recognition must be given in the design of underpinning to the non-linear stress–strain behaviour of soils and the fundamental response of soils which requires displacement of the underpinning in order to mobilize soil resistance. Those techniques which involve minimal displacements to achieve maximum resistance should be employed; particular care must be taken with techniques involving methods of excavation since the underpinning is relatively free of stress after construction, and building movements take place as the loads are transferred from the temporary shoring to the simple conventional underpinning and the ground beneath the underpinning accepts the direct transfer of load.

Before a building is underpinned by simple conventional methods it will have consolidated the ground directly beneath its foundations. Excavation will remove some of this consolidated ground and the underpinning will apply load to less-consolidated ground. It is important to ensure that the 75 mm thick layer of low-water content cement mortar formed on top of each underpinning segment is rammed into the gap and allowed to achieve adequate strength before load is imposed.

Piles may be pushed into the ground by means of a hydraulic jack placed between the heads of segments of the piles and the undersides of the footings of the building being underpinned. The vertical displacement required to mobilize the necessary pile resistance is automatically mobilized by this method of underpinning, provided the strata within which the piles are thrust will behave in an essentially elastic manner.

Simple methods of constructing remedial underpinning can be appropriate even in difficult circumstances and Figure 1.10 shows the physical land features which existed at a particular site in 1846 and in 1982. In the nineteenth century the site was traversed by an old canal adjacent to a dam. The modern development of the site in the twentieth century covered the earlier features, and a modern three-storey building was constructed in the position shown on the site plans. The building was subjected to significant differential movements due to the consolidation settlement of the poor-quality fill materials filling and concealing the existence of the old canal cutting. The degree of damage and continuance of building movements necessitated underpinning, which was successfully accomplished by simple methods of excavation at minimal cost.

Grouting of relatively permeable granular soils can be an effective means of consolidating and underpinning the foundations of structures where these rest

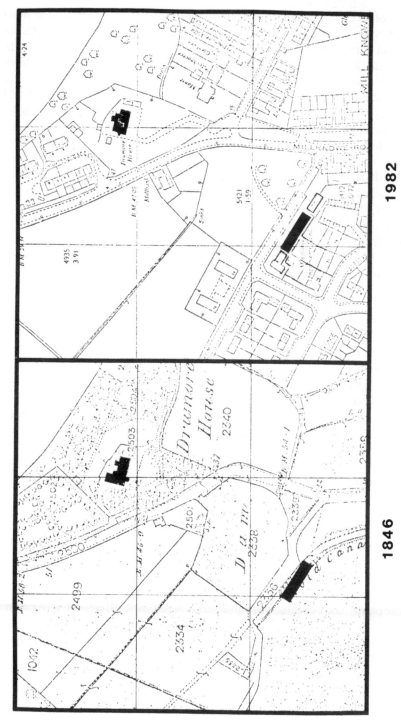

Figure 1.10   Ordnance Survey sheets, 1846 and 1982.

1982

1846

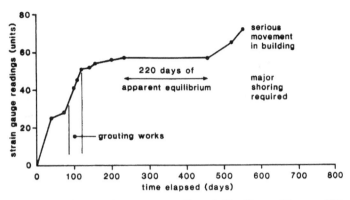

Figure 1.11 Structural movements at Lanarkshire House, Glasgow, UK.

Figure 1.12 Shoring at Lanarkshire House.

directly on granular deposits. Care should be taken, however, that the grouts comprise materials which have an adequate life in service. Even using long-life chemical grouts does not necessarily ensure permanent stability in situations where the ground disturbance due to construction operations within an adjacent site is excessive. Demec strain gauge measurements across a major structural crack, favourably orientated with respect to any major and critical building movements, revealed sudden and serious outward movements of the corner and gable of a high and heavy masonry building. A long period of equilibrium had been experienced after grouting of loose medium to fine sands disturbed by bored piling operations within an adjacent site. The cement and chemical-based grouts maintained a period of equilibrium of 220 days (Figure 1.11) but this temporary state of balance suddenly changed and the heavy shoring shown on Figure 1.12 had to be erected in a very short period of time—only two days—to arrest the movements. It would appear that the grouting works effectively consolidated the loose sands beneath the old masonry foundations, but that the state of balance was marginal and depended on the lateral restraint afforded by a few keystones and the diaphragm action of heavy timber floors supported by the masonry facade of the building.

1.2 Investigatory works

It is important to distinguish between site investigation works and ground investigation works, since each are separate but equally important phases of the investigations required to provide proper and adequate information for the design and construction of underpinning works. The former embraces the comprehensive investigation of a site, including past use and environmental constraint, and much of this information can be obtained from a desk study and a search of local archives. The latter is an exploratory and geotechnical investigation of the ground conditions to determine the geological structure and the characteristics of the superficial and solid deposits. Both investigatory phases should be carried out in accordance with BS 5930:1981, and a useful guide to investigation procedures and equipment is provided in CIRIA *Special Publication 25 (PSA Civil Engineering Technical Guide 35)*, 1983.

All conversions of building structures and protective and remedial works must be preceded by adequate investigations of the ground conditions and a realistic assessment of the state of the foundations, materials and fabrics of the buildings. Underpinning works present more than just technical risk for the engineer, and uncertainty as to the real character of ground and old foundations occasionally results in unfortunate events involving damage to adjacent property, as well as to the building structure being renovated and refurbished. Also, excessive expenditure on remedial works due to the adoption of over-conservative solutions may be contested by parties involved with construction on adjoining sites or even by the owners of the properties being remedied.

A site investigation involves the acquisition of information on some or all of the matters listed below.

   (i) Historical use of the site
  (ii) Mineral support conditions
 (iii) Geology of the district
 (iv) Hydrogeology
  (v) Seismicity
 (vi) Existing services
(vii) Condition of adjacent buildings—their structural form and performance
(viii) Physical features within the site and the relationship of the site to the environment within the district
 (ix) Effect of the development on the environment in physical rather than aesthetic terms.

A few exploratory boreholes and/or trial pits may be sunk as part of the site investigation phase. The information gained from a site investigation is used to determine the extent and the content of the ground investigation.

A ground investigation is a geotechnical investigation of the ground conditions required to determine the geological structure and the characteristics of the superficial and solid deposits. The existence of groundwater and its behaviour are determined at this phase of the investigation by the installation of piezometers in boreholes. Seasonal variations in the elevation of the groundwater table should be considered and, if necessary, measurements must be made over an adequate period of time. Care must be taken to ensure that sufficient information about the ground conditions and groundwater table is gathered to permit an adequate definition of the site.

It is important to identify any buried features associated with the historical use of sites, and proper and extensive search of old records, plans, and memoirs should be made. Bitter experience has established the need for thoroughness of search, particularly in districts known to have experienced historical development.

Made ground is the natural consequence of human activity. In large industrial cities, extensive areas of land surface are artificial, and have resulted from the deposition of a wide variety of materials to elevate low-lying ground and to backfill old stone quarries and clay pits. Fill materials often comprise boiler ash, steelworks slag, coarse discard from former mineral workings, chemical waste, demolition debris and excavation spoil consisting of mixtures of sand, silt and clay. Household refuse and landfill gases can also be found within landfill sites. Ancient watercourses have been culverted throughout the past few centuries and now exist as buried features. All such historical hidden features can remain unknown until exposed by the unwary.

It is important to realize that even the sinking of site investigation boreholes can cause significant ground disturbance. The interception and release of artesian and sub-artesian groundwater can create problems due to the formation of artificial vertical free-flow channels, unless the boreholes are sealed. Overbreak or draw-off can also occur during the sinking of site investigation boreholes through water-bearing non-cohesive strata. Figure 1.13 shows the dramatic effect of the removal of excess volumes of sand on the relative densities of sand deposits caused by the sinking of two 150 mm diameter boreholes to depths of about 20 metres by cable percussion drilling methods. The variations in the cone penetration resistances are a clear indication of the removal of excess volumes of sand, causing loosening of the thick sand layer C. It is of interest that loosening of the relatively thin sand layers A and B also occurred.

It is generally appreciated that lack of definition of variations in ground conditions can result in completely misleading predictions of the performances of underpinning works. Concomitantly, proper methods of sampling and field testing are essential if the characteristics of superficial and solid deposits are to be determined with acceptable accuracy for rigorous analyses of foundation behaviour.

In cohesive soils having soft to firm consistencies, continuous piston sampling is now recognized as the best method of recovering samples in a reasonably undisturbed condition for testing. The U100 open-drive sampler causes serious disturbance to the fabric of clay soils, although its use for recovery of samples of stiff to hard clays will probably continue because of the

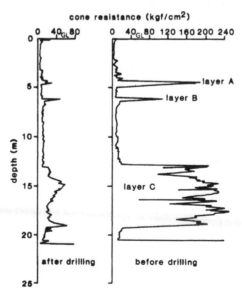

Figure 1.13 Overbreak caused by sinking a borehole in sand.

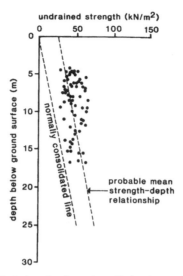

Figure 1.14 Variations in shear strength due to sample disturbance.

ability of the sampling equipment to withstand the hard driving stresses in such heavily over-consolidated soils. Figure 1.14 shows the wide scatter of undrained strengths derived from routine triaxial compression tests on U100 samples of a soft to firm late-glacial marine deposit.

In situations where there is a wide variation of results from routine testing which is suspected to result mainly from sampling disturbance, it is recommended that the probable undrained strength of the clay soil be assessed from the stress history of the soil. Overconsolidation related to general stress history can result from groundwater table movements, soil erosion, glaciation, chemical weathering, cementation, and secondary compression. Overconsolidation related to local stress history is the result of events either natural, and involving desiccation of near-surface layers, or artificial, and related to past loading caused by historical use of the site.

Cone penetration tests (CPT) and self-boring or push-in pressuremeter tests are currently used to determine the undrained strengths of clay soils, and greater use may be expected to be made of in-situ testing to obviate the effects of sampling disturbance on laboratory tests.

With regard to fine-grained non-cohesive soils, information on the in-situ condition of sand deposits can be obtained from the Standard Penetration Test (SPT) as well as the CPT.

However, good drilling and cleaning techniques and careful execution of the SPT are essential in order to ensure that typical values of penetration resistance are obtained from this form of testing. Figure 1.15 compares the penetration resistances obtained from good-quality ground investigation work (Curve 2) with poor-quality work (Curve 1) at the same site. The latter

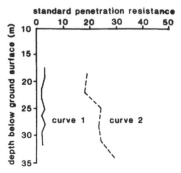

Figure 1.15 Comparisons of high- and low-quality SPT results.

curve reflects the serious disturbance caused by boring operations and completely misrepresents the state of compaction of the sand deposits. These tests were carried out below the water-table in the same uniformly graded fine sand deposit at the same site and within distances of about 15 metres from each other.

A qualitative assessment of the strength of rock strata can be made from visual examination of rock cores and from the rock quality designation (RQD). Compressive strengths obtained from uniaxial compression should be compared with the subjective assessment of strengths as defined by the Geological Society Engineering Group Working Party Report (The Logging of Rock Cores for Engineering Purposes, *Q.J. Eng. Geol.* **3**, 1970, 14) and subsequently referred to in BS 5930, Code of Practice for Site Investigations.

In order to assess intact rock strengths point load tests may be carried out on pieces of rock core. Point load tests permit strength measurements to be carried out on small portions of rock cores. The test gives a Point Load Index which takes account of each individual sample size and is generally normalized to a standard 50 mm sample size. A number of authors have given values for the correlation between Point Load Index (I_s) and the uniaxial compressive strength (UCS) (Broch and Franklin, 1972; Bieniawski 1975; Carter and Sneddon, 1977). Point Load tests should be carried out axially in order to avoid simply measuring bedding plane separation failures, and also because axial testing involves a loading direction which is more representative of the essentially vertical stresses imposed by foundations.

Considerable use has also been made of the standard penetration test as a means of assessing the strength and stiffness of rocks. An extensive study has been made by Stroud (1974) concerning the standard penetration test (SPT) in insensitive clays and soft rocks. Stroud demonstrated that the SPT can be used to estimate the properties of clays in situ, and extended the correlations for stiff fissured London Clay to a wide variety of clays and weak rocks.

Stroud emphasized that for engineering design purposes the mass shear strength must be determined which takes into account the weakening effect of

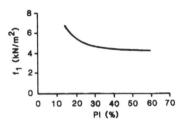

Figure 1.16 Stroud relationship between SPT results and mass shear strength

the system of discontinuities in stiff clays, and indicated that the mass shear strength of fissured London Clay may be only one-quarter to one-half of the shear strength of the intact material. To measure the mass shear strength of a fissured clay, a large enough volume of material must be tested to properly reflect the weakening effect of the system of discontinuities. The empirical relationship in Figure 1.16 was suggested by Stroud and gives the mass shear strength as a product of the standard penetration resistance (N value) and f_1. The factor f_1 was considered to be a constant, essentially independent of depth and the spacing discontinuities, although it could apparently vary from site to site according to the characteristics of the materials.

It is important that in the very large number of tests carried out by Stroud the standard splitspoon sampler was used. In hard materials where the full penetration of 450 mm was not achieved, the tests were stopped at 100 blows and the actual penetrations noted. The number of blows for 300 mm penetration was then obtained by extrapolation. Cole and Stroud (1976) provided the scale of strengths and N values for weak rocks shown in Figure 1.17.

The existence of a groundwater regime and the seasonal behaviour of the groundwater table must be defined during the ground investigation stage of the work. It is particularly important that a proper definition should be made of the groundwater regime in the locality of a building requiring under-pinning. Information on the behaviour of the groundwater table is required both to aid the selection of the most suitable form of underpinning and to ensure that no adverse change is made to the groundwater regime during and after construction of the underpinning which would cause damage to adjacent property. Diversion of normal paths of groundwater flow caused by any artificial barriers created by underpinning could result in flooding of basement property formerly enjoying relatively dry conditions. Artificial lowering of the groundwater table to facilitate underpinning operations, or reductions in the water table derived from environmental changes due to the construction of underpinning, can cause significant increases in effective vertical soil stresses which may result in the settlement of adjoining buildings. It should be a requirement of the investigation works to establish the real groundwater conditions within a site in relation to sources, piezometric heads, hydraulic

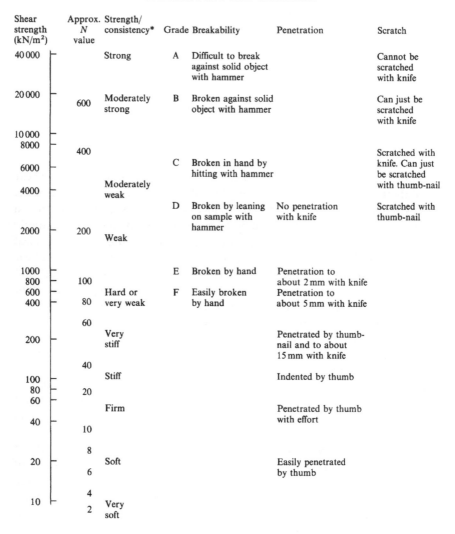

Shear strength (kN/m^2)	Approx. N value	Strength/ consistency*	Grade	Breakability	Penetration	Scratch
40 000		Strong	A	Difficult to break against solid object with hammer		Cannot be scratched with knife
20 000	600	Moderately strong	B	Broken against solid object with hammer		Can just be scratched with knife
10 000						
8000	400					Scratched with knife. Can just be scratched with thumb-nail
6000			C	Broken in hand by hitting with hammer		
4000		Moderately weak				
			D	Broken by leaning on sample with hammer	No penetration with knife	Scratched with thumb-nail
2000	200	Weak				
1000	100		E	Broken by hand	Penetration to about 2 mm with knife	
800						
600	80	Hard or very weak	F	Easily broken by hand	Penetration to about 5 mm with knife	
400						
	60					
200		Very stiff			Penetrated by thumb-nail and to about 15 mm with knife	
	40					
100	20	Stiff			Indented by thumb	
80						
60		Firm			Penetrated by thumb with effort	
40	10					
	8					
20	6	Soft			Easily penetrated by thumb	
	4					
10	2	Very soft				

*Geol. Soc. Working Party Report (1970) and BS8004 (1986) except that the designation 'hard' for soil materials has been given a separate identity, which is analogous to very weak for materials classifiable as rock.

Figure 1.17 Cole and Stroud relationship for strength of weak rocks. Note that grades and shear strengths for rock refer to intact specimens. The N value, however, is an in-situ test and includes some effect of discontinuities. For cohesive soils the correlation between N values and in-situ strength assumed is that given by Stroud (1974) for clays of low plasticity.

gradients, and the influence of climatic change. The real situation can rarely be determined during the relatively short period of time afforded by boring operations, and open-type piezometers should be installed in boreholes selected because of their advantageous location within a site in respect of lateral and vertical variations in the ground conditions. The long periods of time (as much as three months) to attain equilibrium of water levels in piezometers installed in low permeability soils emphasize the impossibility of measuring the elevations of a groundwater table during the relatively short times involved in sinking boreholes.

Before selecting and designing underpinning for conversion, protective or remedial works, it is important to carefully assess the mineral support conditions under a site by reference to such bodies as the National Coal Board and the British Geological Survey. The current state of the art is defined in CIRIA Special Publication 32 and the ICE publication, *Ground Subsidence* (ICE, London 1977). The ground displacements caused by mining subsidence are unrelated to the stresses imposed by foundations unless the bases are unknowingly founded immediately above voids caused by mineral extraction. Mining subsidence can adversely affect underpinning which by its very nature is securely integrated with the ground being subjected to large movements.

An over-emphasis on the definition of the ground conditions to the detriment of acquisition of knowledge on the nature and condition of all building structures affected by underpinning works should be avoided. In order to properly assess each particular situation, careful investigations must be made of the building to be underpinned and the environment as well as the ground. The necessity for dilapidation surveys involving photographic records and subsequent monitoring of building performance are paramount.

The legal situation must be clearly understood by all parties to the underpinning contract. Clear definitions of responsibilities should be made at the outset and the criteria for the support works carefully and explicitly defined.

Nuisance caused by ground and airborne vibrations to the residents of existing buildings must also be assessed to avoid expensive litigation and contractual delays. The selection of the most suitable type of underpinning may be determined by the need to minimize disturbance to adjacent buildings caused by construction operations. Cost need not dictate selection.

1.3 Serviceability limits

As stated in the Preface, it is important to differentiate between damage to the primary support elements of a structure and damage to cladding, partitions, and finishes. Ground movements affect visual appearance as well as function and serviceability, but it is essential to recognize the relative unimportance of purely aesthetic considerations. If the investigatory works establish that the ground movements have ceased and that the state of balance of the structure is adequate, then underpinning is generally unnecessary. Classifications of visible damage to building structures in relation to widths of structural cracks vary considerably and there is the danger that the indiscriminate use of these partly subjective criteria will result in decisions to underpin when it is unnecessary. The decision to underpin a structure should only be taken when there is clear evidence either that movements will continue for an unacceptable period of time, or that the apparent structural equilibrium is only temporary and harmful movements may be experienced in future times. The relationship between serviceability and amount of visible damage is not simple and the structural engineer must make a decision based on an assessment of the particular circumstances. It should be appreciated that slight damage may be

Table 1.2

Degree of damage	Description of typical damage* (ease of repair is italicized)	Approximate crack width (mm)
	Hairline cracks of less than about 0.1 mm width are classed as negligible.	≯ 0.1**
1. Very slight	*Fine cracks which can easily be treated during normal decoration.* Perhaps isolated slight fracturing in building. Cracks in external brickwork visible on close inspection.	≯ 1.0**
2. Slight	*Cracks easily filled. Redecoration probably required.* Several slight fractures showing inside of building. Cracks are visible externally and *some re-pointing may be required externally to ensure weathertightness.* Doors and windows may stick slightly.	≯ 5.0**
3. Moderate	*The cracks require some opening up and can be patched by a mason. Recurrent cracks can be masked by suitable linings. Repointing of external brickwork and possibly a small amount of brickwork to be replaced.* Doors and windows sticking. Service pipes may fracture. Weathertightness often impaired.	5 to 15** or a number of cracks ≥ 3.0
4. Severe	*Extensive repair work involving breaking-out and replacing sections of walls, especially over doors and windows.* Window and door frames distorted, floor sloping noticeably. Walls leaning or bulging noticeably, some loss of bearing in beams. Service pipes disrupted.	15 to 25** but also depends on number of cracks
5. Very severe	*This requires a major repair job involving partial or complete rebuilding.* Beams lose bearing, walls lean badly and require shoring. Windows broken with distortion. Danger of instability.	usually > 25** but depends on number of cracks

*It must be emphasized that in assessing the degree of damage account must be taken of the location in the building or structure that it occurs.
**Crack width is one factor in assessing degree of damage and should not be used on its own as a direct measure of it.

unacceptable for a hospital structure in contrast to an industrial building where moderate damage is acceptable since it probably would not affect serviceability or function. Table 1.2 was presented by Jennings and Kerrich (1962) and was intended to act as a guide to ease of repair of brickwork and masonry rather than as a direct measure of the degree of damage and of the necessity for underpinning.

In contrast, Table 1.3 presents a classification of damage to walls of buildings in relation to their use in service. This table was developed by Pynford and was based upon thirty years of experience of the amount of damage which prompted property owners to express concern and their legal advisers to seek assistance from engineering specialists. Damage should not be related only to the widths of cracks, and any proper assessment of damage should take into account the means by which a structure is supported (i.e. frame or shear wall); its state of balance; the nature of the cracking (i.e. tensile or shear or a combination of both); and whether ground movements may be expected to continue. Differential movements can cause cracking and separation of walls into units which are then capable of articulation without experiencing failure provided the structure is capable of maintaining its state of balance. Unfortunately, experience has shown that once cracking develops, from whatever source, it is probable that movements due to other sources will be concentrated at these lines of weakness. Cracking that is initiated by one cause and is initially negligible may become excessive and unacceptable when other movements are superimposed. In such circumstances, underpinning in conjunction with structural strengthening and the introduction of movement joints may be necessary measures.

Table 1.3

Crack width (mm)	Degree of damage			Effect on structure and building use
	Dwelling	Commercial or public	Industrial	
≯ 0.1	Insignificant	Insignificant	Insignificant	None
0.1 to 0.3	Very slight	Very slight	Insignificant	None
0.3 to 1	Slight	Slight	Very slight	Aesthetic only. Accelerated weathering to external features.
1 to 2	Slight to moderate	Slight to moderate	Very slight	
2 to 5	Moderate	Moderate	Slight	The serviceability of the building will be affected, and towards the upper bound, stability may also be at risk.
5 to 15	Moderate to severe	Moderate to severe	Moderate	
15 to 25	Severe to very severe	Moderate to severe	Moderate to severe	
> 25	Very severe to dangerous	Severe to dangerous	Severe to dangerous	Increasing risk of structure becoming dangerous.

1.4 Litigation

The legal aspects of underpinning involving right of support are a complex subject, of which the proper treatment is the prerogative of the legal profession. It may be helpful, however, to present the fundamental principles of the laws of Scotland and England as they may be applied by the Courts and Arbiters of these countries, in the hope that some general pointers may emerge.

Scots law has its foundations in Roman or Civil Law, in contrast to English law, which arose as a consequence of the Norman Conquest and has its roots in Common Law and Equity. In general, the rules observed in both Scots and English law are similar but, as we shall see, not identical.

The legal systems comprising International Law may be placed in three basic categories. Some are derived from the Civil or Roman Law, and others are based on the Common Law, which evolved from the actions of English courts and was conveyed by communication and emigration from England. Civil Law is the basis of the law of Europe and of many former European colonies; Common Law is the basis of the law of most of the United States. Islamic Law is the basis of the law for most of the Middle East and those non-Middle Eastern countries which base their legal systems upon Shiriah law.

In both Scots and English law, every owner of land has a natural right, as an incident of ownership, the right to prevent such use of the neighbouring land as will withdraw support which the neighbouring land naturally affords to his land. In the natural state of land, one part of it receives support from the other, upper from lower strata, and soil from adjacent soil, and, therefore, if one piece of land is conveyed so as to be divided in point of title from another contiguous to it, or (as in the case of mines) below it, the right to support passes with the land as an essential incident to the land itself.

When the natural condition of the surface is changed and pressure upon it has been artificially increased by the erection of buildings, this is a different matter. The right of support to buildings from adjacent land or the right to have buildings supported by other buildings are questions of an acquired right and no longer of a natural right of property. Such a servitude (in Scotland) or easement (in England) may be created by express or implied grant. In England, it may also be acquired by prescription (*Dalton* v. *Angus*) 1881 (6 Appeal Cases at page 740) but this question has not been settled in Scotland. It is, therefore, the position in England that the owner of a new building has no common law right of lateral support for it until the prescriptive period of twenty years elapses and then on the day when the period does elapse the owner acquires an almost absolute right of support. Thus the English doctrine of prescription logically implies acquiescence on the part of the adjoining owner and the power to interrupt prescription running. This means, as Lord Penzance observed, that the owner of the adjoining soil may for twenty years 'with perfect legality dig that soil away and allow his neighbour's house, if supported by it, to fall in ruins to the ground'. It was also held in *Ray* v. *Fairway Motors (Barnstaple) Ltd* (1968) 20 Property and Compensation Reports at page 261, that the owner of the supporting land did not even owe a duty of care in negligence to the owner of the new building and that in the light of *Dalton* v. *Angus* only the House of Lords could introduce such a duty. This, however, would not appear to be the law of Scotland as a common law right of Action in Nuisance may be available. The modern law of Scotland relating to nuisance is directly derived from the Civil Law maxim 'use your own property so as not to injure that of your neighbour'. The judicial approach to nuisance has been expressed thus:

> ... the proper angle of approach to a case of alleged nuisance is rather from the standpoint of the victim of the loss or inconvenience than from the standpoint of the alleged offender; and that if any person so uses his property as to occasion serious disturbance or substantial inconvenience to his neighbour or material damage to his neighbour's property, it is in the general case irrelevant as a defence for the defender to plead merely that he was making a normal and familiar use of his own property. The balance in all such cases has to be held between the freedom of a proprietor to use his property as he pleases, and the duty of a proprietor not to inflict material loss or inconvenience on adjoining proprietors or adjoining property; and in every case the answer depends on considerations of fact and of degree .... the critical question is whether what he was exposed to was *plus quam tolerabile* when due weight has been given to all the surrounding circumstances of the offensive conduct and its effects ... any type of use which in the sense indicated above subjects adjoining proprietors to substantial annoyance, or causes material damage to their property, is *prima facie* not a 'reasonable use'. (Per Lord President Cooper in *Watt* v. *Jamieson* 1954 SLT at page 57.)

In the leading Scottish Case of *Lord Advocate* v. *Reo Stakis Organisation Limited* 1982 SLT page 144, we are provided with important authority on the Law of the Right of Support especially with regard to the lateral support of land to buildings. As a result of this case, it appears the Law of Scotland is that the owner of a building may recover damages for loss of support on the following three grounds:

(1) infringement of the right of support
(2) nuisance
(3) negligence.

It was held in that case that the owner of adjacent land within which excavation work is taking place was under a duty under the Law of Nuisance not to deprive an adjoining building, belonging to his neighbour, of lateral support. This duty and the right of action exist as soon as the new building is constructed and do not accrue after twenty years' prescriptive enjoyment of lateral support as is the case in England.

The boundaries of nuisance as expounded in English law are uncertain but historically breach of support (including a prescriptive right of support of buildings) was a breach of the tort of nuisance and is still treated as such by the standard English texts on tort. This appears to conflict with the opinion of the Court in *Stakis* in which it was observed

> ... it is no doubt the case that in some instances an occupier whose property suffers such damage (subsidence caused by building operations) may find himself with a remedy *both* under the Law of Nuisance and upon the basis of infringement of an acquired right of support but it is by no means unusual to find that more than one right of action is available upon the same set of facts... the law of support in relation to buildings is part of the law of heritable right. The Law of Nuisance, on the other hand, is part of the Law of Neighbourhood.

It would, therefore, appear that the decision in *Stakis* cannot be reconciled with English decisions in *Dalton* v. *Angus* and *Ray* v. *Fairway Motors (Barnstaple) Ltd.*

In *Stakis*, structural damage had been caused to the pursuer's building in the centre of Glasgow, allegedly as a result of piling operations carried on in connection with the building of a hotel situated across a lane from the damaged building. The old buildings on the site had been demolished and the construction of a new hotel commenced. Structural cracking then appeared at the portion of the pursuer's building nearest to the building site of the new hotel. The representatives of the developer took steps to prevent further subsidence and damage but without success. Thereafter, the pursuer raised a court action in which he sought to recover the cost of his own remedial works. Claims based on nuisance and negligence were made against the developer and their piling contractors, and claims based on negligence against the consulting engineers. At first instance, the judge declined to reject the pursuer's plea of a right of action based on nuisance (1980 SLT 237) and the decision was upheld on appeal (1982 SLT 140). The pursuer had pleaded that 'piling involving boring by cable percussion method is hazardous in that it gives rise to a high risk of subsidence in nearby ground because of the draw-off of the soil ... resulting in lack of support to the buildings on the ground so affected'. Neither the Court of first instance nor the Appeal Court considered whether the conduct complained of was deliberate, negligent or otherwise. The Court observed that 'the Law of Nuisance applies without exception to provide a remedy for any relevant damage suffered by a neighbouring occupier as a result of any type of use of adjoining subjects by the occupiers thereof' and applied the above-quoted passage from WATT which was described at page 143 as 'the best of the most recent descriptions of our Law of Nuisance'. Subsequent to this decision the view has been expressed that the Court has overstated the rights in nuisance available to a neighbour affected by building operations and a further case is awaited with keen interest.

In brief, the effect of *Stakis* appears to be that although the Judge at first instance appeared to think that the Law of Scotland and England do not diverge, it seems clear that *Stakis* impliedly rejects *Dalton* v. *Angus* and the doctrine of prescriptive acquisition of rights of lateral support of land to buildings.

What is clear, however, in both the Scots law of delict and the English law of tort is that if an engineer or contractor negligently carries out piling, blasting or other excavating operations which cause damage, an action in delict or tort based on the ordinary principles of the reasonable man, duty of care and negligence will arise independent of any right of action which may be available due to loss of support or nuisance.

In Scotland, questions involving the support afforded by one building in favour of another, adjoining or discontiguous, are less frequent and such questions are sometimes dealt with by other branches of the law of property. For example, the law relating to flatted houses known as the Law of the Tenement gives rise to a species of right differing from common property which arises among the owners of subjects possessed in separate portions but still united by their 'common interest'. In the absence of any specific provisions in the Title Deeds to the properties, this common interest is the source of common law rules in connection with the roof, walls, etc. For example, the gables are common to the owner of each flat so far as they bound his property, but he and the other owners in the tenement have cross-rights of common interest to prevent injury to the stability of the building. As adjoining houses, the law of mutual gables is that most usually involved.

In both Scotland and England it appears that a servitude or easement giving right of support by one building to the other may be constituted by express or implied grant and in England by prescription: The absence of such an easement means that in English law if a man pulls down his house and thereby deprives his neighbour's house of the support it has been enjoying, and his neighbour's house is thereby damaged, his neighbour has no cause of action although he must take care to interfere as little as possible with the adjoining house. He is not called upon to take active steps for its protection. Again, in the light of *Stakis* it is likely that a pursuer will sue upon the ground of nuisance which is so favourable as compared with breach of a supposed right of support. Whether the action of a neighbour is founded upon nuisance, negligence or breach of a right of support, his remedy can be interdict, including interim interdict (which if granted would halt the building or excavating operations) and damages for loss sustained or simply damages.

In conclusion, it is perhaps worthy of note that every new subsidence event causing damage grounds a new action for reparation although damages have been awarded for earlier events. It is difficult, in many instances, to state categorically that a building requires the support of adjacent land and the matter is further complicated when the subsidence of a building follows at a significant interval of time after the construction operation which caused the ground movements particularly when doubts may be expressed of the state of repair of the building which is damaged.

Debate over contractual responsibilities often arises in situations in which major works are required to retain buildings and roads adjacent to a site under development. The prime responsibilities of the Engineer and the Main Contractor are separate but related, as described in the following paragraphs.

'Permanent Works' means the permanent works to be constructed, completed, and maintained in accordance with the Contract. 'Temporary Works' means all temporary works of every kind required in or about the construction, completion, and maintenance of the Works. 'Works' means the Permanent Works together with Temporary Works.

The Engineer is responsible for the design and supervision of the Permanent Works under contract to the Client. The Engineer has a contractual duty to protect the Client by satisfying himself as to the reasonableness of the proposals for the Temporary Works prepared by the Contractor, and to be in attendance during the critical phases of the temporary works.

The Contractor is responsible for all aspects of the Temporary Works under contract to the Client. The Contractor has a contractual duty to identify those items of the Temporary Works that require to be given special consideration (e.g. in situations where there are Third Party implications), and which should be brought to the attention of the Engineer by submission of proposals for the Temporary Works.

The preceding commentary has attempted to explain the possible interpretations of the law in connection with the damage to buildings caused by unsuccessful underpinning operations, but it is important that legal advice should be sought before arriving at a conclusion in any particular set of circumstances.

1.5 Historical background

As a generalization it may be stated that the underpinning of structures in the days of growth of ancient empires depended on the practical skills passed from master to apprentice. Knowledge of soil behaviour would depend on experience within particular localities, with reliance on knowledge of foundation behaviour rather than on an understanding of the reasons for the behaviour. There was no science of foundations.

The application of scientific thought to foundation engineering began in the seventeenth and eighteenth centuries with the gradual development of theoretical concepts from that time. The use of the physical sciences to attain an understanding of why soils behaved in a certain manner under applied loads opened the pathway to enlightenment and gave foundation engineering the opportunity of becoming a science.

The development of the theories of elasticity and plasticity during these exploratory years provided the mathematical tools for engineers to analyse a wide variety of foundation problems. Recognition must always be given to the strains which occur in soils as stress fields are modified by excavation and underpinning.

The year 1925 was a milestone in the development of the science of foundation engineering with the publication of the book *Erdbaumechanik* by Karl Terzaghi. Terzaghi demonstrated that the mechanical behaviour of soils depended on the fluid pressure in the pore spaces between the discrete mineral particles. His theoretical concept of effective stress controlling changes in volume or strength of soils provided a major advance in the science of soil mechanics and greatly advanced the science of foundation engineering.

Terzaghi also introduced the philosophical thought that indiscriminate application of theory based on poor physical models was inappropriate for practical solutions, and was of the conviction that the development of the science of foundation engineering required careful field observations of real behaviour. In the opinion of Terzaghi the difference between theoretical and real behaviour could only be ascertained by field experience. In every branch of applied mechanics the researcher or theoretician considers the behaviour of an ideal material, and Terzaghi emphasized that it was necessary to be aware that theory must be combined with a thorough knowledge of the physical characteristics of real soils with the additional awareness of the difference between the behaviour of soils in the laboratory and in the field.

It is important to recognize that soils are quite different from structural materials, such as concrete, being discrete particles with the void spaces between the particles containing a liquid. Volume change during shear is an important characteristic of soils and the dependence of mechanical behaviour on effective stress acting between the discrete particles is unique in the sciences of material behaviour. The economical design and safe execution of underpinning requires a sound theoretical knowledge of foundation engineering and a wide experience of real behaviour.

The contributions made by the construction industry from the aspects of the development of plant, techniques, and materials have been significant over the past thirty years and many major underpinning problems have been solved by practical innovation rather than application of theory.

Ingenuity of method and skill of execution is not the prerogative of engineers of the present century and the following account by G.L. Taylor, RIBA, of the methods used in underpinning the Long Storehouse at H.M. Dockyard, Chatham, demonstrates earlier skills and innovative ability.

The Storehouse was a five-storey masonry structure about 165 m long and 16 m wide. The structure displayed serious defects in its walls and floors and the arches of the exterior walls had been damaged by settlement of the piers supporting the arches. The walls leant outwards about 250 mm and settlement of the piers of the order of 225 mm had taken place. Shoring was erected to support the masonry walls and the timber floors were propped to enable them to carry the weight of the fittings and equipment for ships. The propping

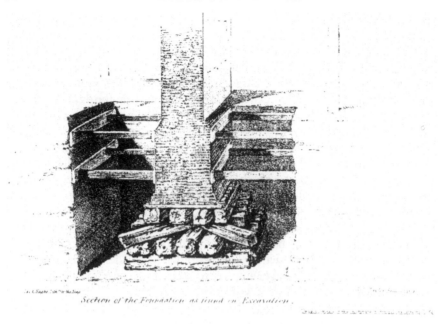

Section of the Foundation as found in Excavation.

Figure 1.18   Underpinning in 1848. Existing foundation.

successfully prevented settlement of the floors but the heavy masonry external walls continued to settle. The differential movements caused severe upward bending (hogging) of the timber floors and the floor girders lost contact at their ends with the walls. It was considered that the masonry walls of the Storehouse were supported by timber piles or on timber sleepers, but the exact nature of the foundations was unknown.

The damage to the building became so serious that the decision was taken to underpin the heavy masonry walls. Investigatory works effected by simple excavation techniques revealed that the walls had been built upon oak timbers founded at depths of about 4.5 metres, as shown in Figure 1.18. The timbers were in poor condition due to natural decay and the several tiers of longitudinal and transverse timbers were being compressed differentially under the weight of the heavy masonry walls.

The initial proposal for the underpinning comprised removal of the timbers and the construction of a thick mass concrete foundation on solid ground of adequate bearing capacity, leaving a gap of about one metre between the underside of the old masonry wall and the top of the new concrete foundation. The initial proposal was that the gap of one metre was to be carefully filled by a bricklayer making sure that the new brickwork was tightly built into the intervening space.

The innovatory solution which was adopted made use of a strong iron frame as shown in Figure 1.19. The ground was excavated in short 1.8 metre

Sections showing the Method of underpinning with Concrete, the Store Houses, Chatham Dock Yard.

Figure 1.19   Underpinning in 1848. Underpinning operations.

lengths beneath the heavy masonry walls to a depth of 600 mm below the wall to facilitate the extraction of the oak timbers. The excavations were planked and strutted and the bases of the excavations were found to be sound hard chalk. Shuttering was erected within the trench excavation and concrete poured by means of troughs. This mass concrete was poured to within 300 mm of the bottom of the wall which was found to be relatively smooth and planar. Large slates were bedded on the upper surface of the mass concrete before it set and hardened. An iron frame was placed on these slates and positioned as shown in Figure 1.19. The frame was 300 mm high, and measured 1.2 × 2.1 metres, the latter dimension being the width of the base of the masonry wall. Finely ground lime and sharp river gravel in the proportions of one of lime to six of gravel were mixed with hot water, poured into the iron frame, and the cementitious mix squeezed by turning screws until it was in intimate contact with the underside of the old wall. The movable rams within the iron frame were slackened when the first batch of lime concrete had set and succeeding batches of concrete squeezed into position until the entire width of a wall had been underpinned. The heavy masonry walls were shored before the excavation and underpinning works commenced and the entire operation was most successful.

Towards the river the ground conditions worsened and the excavations had to be sunk to depths of about 6 metres and these exposed 3 metres of oak timber laid in a crosswise fashion. Timber piles were found under the walls of a portion of the storehouse and the same form of underpinning was used with equal success.

The philosophy of underpinning is concisely and appropriately contained in the words of Rondelet: 'The aim of construction work is to carry it out with all of the strength and quality required and using reliable materials handled with skill and economy'.

References and further reading

1. Attewell, P.B. and Taylor, R.K. (eds.) (1984) *Ground Movements and their Effects on Structures*. Surrey University Press, Glasgow and London.
2. Bieniawski, Z.T. (1975) The Point Load Test in geotechnical practice. *Eng. Geol. J.* 9, 1–11.
3. Broch, E. and Franklin, J.A. (1972) The Point Load Strength Test. *Int. J. Rock Mech. Ming. Sci. Geomech. Abstr.* 669–696.
4. BS 5930: 1981. Code of Practice for Site Investigation. British Standards Institution, London.
5. Carter, P.G. and Sneddon, M. (1977) Comparison of Schmidt Hammer, Point Load and Unconfined Compression Test in Carboniferous strata. *Proc. Conf. on Rock Engineering*, University of Newcastle on Tyne.
6. Chard, B.M. and Symons, I.F. (1982) Trial trench excavation in London Clay: a ground movement study at Bracknell. *TRRL Report LR 1051*.
7. Clough, G.W. and Chameau, J-L. (1980) Measured effects of vibratory sheet pile driving. *J. Geotech. Eng. Div. ASCE*, October.
8. Cole, K.W. and Stroud, M.A. (1977) Rock socket piles at Coventry Point, Market Way, Coventry. *Piles in Weak Rock, Proc. ICE Symp., ICE*, London.
9. Fleming, W.G.K., Weltman, A.J., Randolph, M.F., and Elson, W.K. (1985) *Piling Engineering*. Surrey University Press, Glasgow and London.
10. Geological Society Engineering Group Working Party (1970) Report on the logging of rock cores for engineering purposes. *Q. J. Eng. Geol.* 3, 14.
11. Healy, P.R. and Head, J.M. (1984) Construction over abandoned mine workings. *CIRIA Special Publication 32*, CIRIA, London.
12. Hunter, L.E. (1952) Underpinning and strengthening of structures. *Contractors Record and Municipal Engineering*.
13. Institution of Civil Engineers (1977) *Ground Subsidence*. ICE, London.
14. Jennings, J.E. and Kerrich, J.E. (1962) The heaving of buildings and the associated economic consequences. *Civil Engineering in South Africa*, 5(5) 112.
15. Peck, R.B. (1969) Deep excavations and tunnelling in soft ground. State-of-the-Art Report. *Proc. 7th Int. Conf. on Soil Mechanics and Foundation Engineering*, Mexico City. 225–290.
16. Prentis, E.A. and White, L. (1950) *Underpinning—Its Practice and Applications*. Columbia University Press, New York.
17. Randolph, M.F. (1980) PIGLET: A computer program for the analysis and design of pile groups under general loading conditions. *Cambridge University Engineering Department Research Report*, Soils TR91.
18. Reasonable neighbourhood; the province and analysis of private nuisance in Scots Law. Parts I and II. *J. Law Soc. Scotland*, Dec. 1982, Jan. 1983.
19. Royal Institute of British Architects, (1836–1867) extracts, *Proc. R. Inst. Br. Architects*, London.
20. Stock, C.H. (1902) *A Treatise on Shoring and Underpinning*. Batsford, London.
21. Stroud, M.A. (1974) The Standard Penetration Test in insensitive clays and soft rocks. *Proc. Eur. Symp. on Penetration Testing*, Stockholm.
22. Tomlinson, M.J. (1978) *Foundation Design and Construction*. Pitman, London.
23. Walker, D.M. (1969) *The Scottish Legal System*. 3rd edn. (revised), W. Green, Edinburgh, p. 67.
24. Weltman, A.J. and Head, J.M. (1983) Site Investigation Manual. *CIRIA Special Publication 25*.

2 Traditional methods of support

J.F. HUTCHISON

2.1 Introduction

'Traditional methods of support' is a broad title for this chapter and it describes the more commonly used techniques of excavation, shoring and underpinning. These methods have been successful for many years and can today still be effective and economical in certain circumstances. Since underpinning is usually a combination of temporary and permanent works, and the former are generally the responsibility of the contractor and the latter that of the engineer, close co-operation between these two parties is essential at every stage of construction. The simple methods which will be described in the chapter make it easier for this co-operation to take place and, since underpinning is by its nature a complex subject where methods can differ from job to job, the techniques will be couched in general terms.

Due to the modern trend towards refurbishment of existing buildings and other structures, and the development of 'gap' sites in cities, access is almost always very cramped and limited. While the development of the mini-excavator has enabled mechanical plant to be used in many restricted sites, there are many more where hand digging is the only solution to the access problem—as in the situation where the existing structure requires extensive internal propping and shoring. The use of the hand dig technique can also have other benefits; for example, it will enable the engineer to monitor the work step by step where the ground investigation has not, of necessity, been as detailed as he would have wished. It is of paramount importance during excavation for underpinning that the trenches and pits are adequately supported to prevent ground movement. A careful technique such as hand excavation will enable the contractor to control his temporary works and will permit time for modification should this prove necessary.

Where a building is sensitive to minor movements, whether it is the structure being underpinned or an adjacent structure or structures, vibrations will be kept to a minimum by the use of the 'shovel and barrow'. This technique was successfully used in the construction of the Museum of Modern Art in Brussels. Due to environmental and other pressures, the decision was taken to build the Museum underground in a pedestrian square, bounded by elegant 18th century buildings which house the Museum of Classical Art. The construction called for an excavation below existing ground level of up to 8 m,

41

and the first thoughts of supporting this excavation involved a contiguous piled wall. However, space was very limited and the required drilling rig would have had great difficulty in operating successfully. It was also thought that vibrations from the rig would set up unacceptable movements in the adjacent structures.

The solution was to construct a non-contiguous retaining wall of 2 m-diameter shafts. The shafts were hand dug to a depth of 25 m and built up from hand-placed precast segments reinforced with steel hoops. This enabled the bulk excavation to be carried out, and with movements in the existing structures being closely monitored and the excavation carefully phased, maximum recorded displacement was of the order of 15 mm—considered to be acceptable for the magnitude of the excavation.

The previous paragraphs have described situations where hand digging was considered to be advantageous. However, the method has disadvantages too, perhaps the main one being the slow rate of progress which could expose the structure being underpinned to a longer period of risk than more expeditious techniques. Where, too, the excavation is required to be taken to a stratum which requires a depth of excavation of more than say 6 m, the costs of hand digging could prove to be unacceptable. It therefore requires careful consideration before deciding to use these techniques.

Shoring is a vast and varied subject, and requires detailed treatment on its own. However, since shoring and underpinning are frequently complementary to each other, the opportunity will be taken to describe, in general terms, the more commonly used methods of shoring. There will also be descriptions of underpinning techniques that employ hand excavations.

2.2 Shoring

Where a building or structure is in poor condition due to settlement, and underpinning has to be carried out to limit or to arrest resulting movements, external shoring will probably be required. If, however, the building has soundly constructed loadbearing walls in good condition, internal ties may be used which would not restrict the access round the building. These ties would normally be supplemented by internal bracing.

The main shoring members can be of timber, steel or scaffolding. Where timber is used, swelling and shrinking will take place and provision should be made, in the shape of hardwood wedges, to allow for any adjustment which will be required. In external shores, changes in the ground will have an effect; for example, freezing and thawing, softening of soils due to heavy rain, and nearby excavation during the underpinning operation causing ground movement. Much of the effect of the first two can be obviated by careful attention to drainage and weather protection, and ground movements can be limited by a good support system to the excavation. Where the structure is

Figure 2.1 Example of raking shore using scaffold tube and fittings.

situated close to a heavily trafficked road, or if it houses heavy machinery, allowance for the effects of vibration should be made in the design of the shoring system. Some thought should also be given to the consequences of accidental impact and steps should be taken to protect the system from these as far as is practicable. Where space permits and loading is not too severe, scaffolding can be most effective in forming shores, flying shores or ties (Figures 2.1–2.3). This method also has the advantage of being less sensitive to weather effects and it is easily adjusted.

No matter which shoring system is used, it is essential that regular checking is done during the underpinning operation so that adjustments and modifications can be carried out without delay, where these are necessary.

The most common forms of shores are

(1) raking shores,
(2) flying shores, and
(3) needles and dead shores.

2.2.1 Raking shores

These are used where a building or structure requires external support, and each raking shore consists of one or more members set at an angle to the

Figure 2.2   Example of raking shore showing treatment at base of shore.

Figure 2.3   Example of flying shore used in 'gap' site.

building. The angle should be no more than 75° to the horizontal and the top of each shore should terminate at a short needle set into the fabric of the building just below each floor level. The tendency of the shore to move upwards is resisted by a cleat fixed to a wall plate set immediately above the needle.

The load from the shores is transferred to the ground by means of grillages on sole plates and, as in all forms of temporary supports, the safe bearing capacity of the soil should be determined and a suitable soleplate designed so that this is not exceeded. Where possible the soleplate should be kept clear of the edge of the underpinning excavation. This distance will depend on the nature of the soil, but should be equal to at least half the depth of the excavation. If, due to lack of room, this is not possible, the excavation supports should be designed to resist any additional surcharge load imposed by the shoring system (Figure 2.4).

2.2.2   Flying shores

Where raking shores are likely to cause an obstruction to underpinning operations and where a 'gap' site is being developed, flying shores can be used. Generally these are used where the distance between the opposing buildings is no more than about 10 m. This limitation is for practical reasons.

Flying shores should be needled as already described in section 2.2.1 and should be taken from the floor level of one building to that of the other if these are on the same horizontal plane. Where this is not so, a stiff vertical member

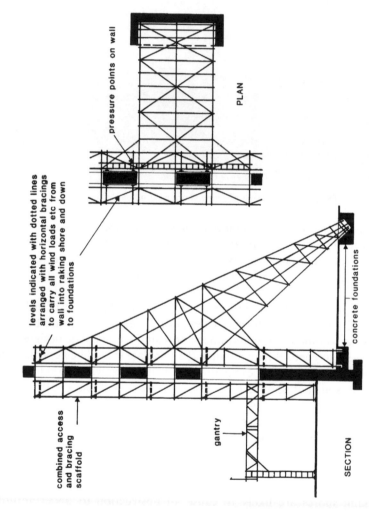

Figure 2.4 Arrangement of two-way raking shore using scaffolding. Some bracing has been omitted for clarity.

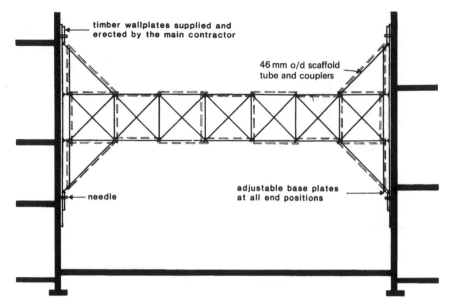

Figure 2.5 Typical arrangement of flying shore.

should be introduced to distribute the load. An example of flying shores using scaffold tube is shown in Figures 2.5 and 2.6. Flying shores are provided to prevent bulging or tilting of walls and will not carry loads imposed by self-weight of walls and floors.

2.2.3 Needles and dead shores

These are used to support walls while underpinning operations are carried out and consist of vertical members spaced on either side of the underpinning

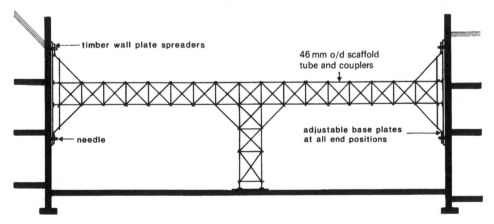

Figure 2.6 Arrangement of flying shore with centre support tower. Some bracing has been omitted for clarity.

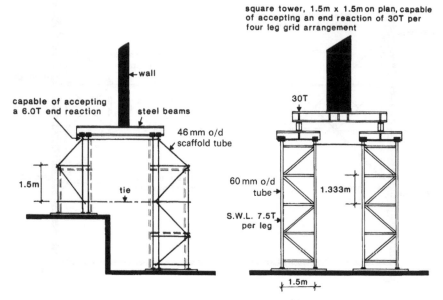

Figure 2.7 (*Left*) dead shore constructed in tube and fittings. (*Right*) dead shore constructed in Triframe 30 components.

excavation. The vertical members in turn support needles spanning between them. These needles should, as far as possible, be placed near to the points of maximum load. Before carrying out this operation, all openings and recesses in the wall should be adequately braced and the holes for the needles should be the minimum size necessary. Horizontal beams may also be used at right angles to the needles on top of the dead shores if it is desirable to limit the number of shores used (Figure 2.7).

The above brief descriptions are of shoring systems which are commonly used and, as mentioned before, careful monitoring is essential. It is also important that a thorough survey of the structure or building is carried out before shoring and underpinning operations are commenced. A record of all cracks, movements and defects, should be compiled. This should be agreed among all interested parties and the record should be continuously updated until all are satisfied that movements have ceased.

2.3 Simple underpinning

The simplest form of underpinning can be used where ground conditions are good, and consists of excavating for and constructing a series of columns, or legs, beneath the wall to be underpinned. The length of these legs is usually determined by the condition of the existing structure, its composition, and the nature of the ground supporting the underpinning. BS 8004 recommends that this length should be between 1 m and 1.4 m for brick and/or stone walls in good condition.

It is good practice to construct the legs on a 'hit or miss' basis up to a maximum of six in a group. This will, of course, depend on the length of the wall or elevation to be underpinned, and Tomlinson (1982) suggests that the sum of the legs being constructed should not exceed one-quarter of the total length of the wall. Where the structure shows some signs of distress this should be reduced to one-fifth or one-sixth of the total length. Careful attention should be paid to the position of openings and piers above the underpinning so that due allowance is made for increased loads in local parts of the structure. Each group of legs should be completed before the next group is commenced (Figure 2.8).

When the legs are constructed of concrete, this should be started immediately after the excavation is completed. If this is not possible, the last few millimetres of the excavation should be left and should be removed immediately before concreting commences. One other method is to protect the solum of the excavation with a 50 mm layer of blinding concrete.

Each leg should be constructed to within 75 mm of the underside of the old foundation or wall and the top should be carefully levelled off to receive the final pinning. The leading edge of each leg should incorporate a vertical groove or chase so that the adjacent leg, when constructed, can be keyed into it and the total length of underpinning can act as a unit. After a suitable period (to allow the concrete to set and shrink) the final pinning should be carried out using a fairly dry concrete mixture with a maximum aggregate size of 10 mm. 'Fairly dry' means that the mixture has the minimum amount of water added so that it will form a ball when squeezed in the hand. This concrete should be well rammed into the gap between the new leg and the old foundation. This final gap can also be filled with a dry mortar mix or brickwork in cement mortar. Where a wide foundation is being underpinned, this should be constructed in steps from back to front. At all times during the construction of this type of underpinning, the sides and ends of the excavations should be supported by temporary timbering or sheeting. Some common methods of doing this will be described later in the chapter.

Where the underpinning is being carried out because of settlement of the existing foundation, or where the safe ground bearing capacity is likely to be exceeded during the construction of the legs, the sections of the wall on either side of the leg being constructed should be supported at of near ground level

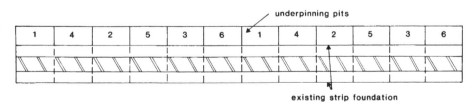

Figure 2.8 Strip foundation. Simple underpinning construction. Sequence of legs.

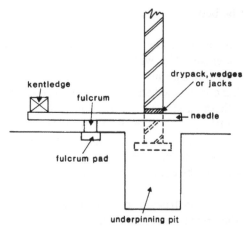

Figure 2.9   Cantilevered needles for access restricted to one side of wall.

by needles as already described. Bedding of the wall on the needles should be done with dry pack and in such a way that overstressing of the wall is avoided. It is good practice to use wedges or jacks to support the wall on the needles.

There will be circumstances where it is not possible to gain access to the inside of a building or structure; for example, where the building is still occupied or its contents cannot be disturbed. On these occasions, a system of cantilevered needles can be constructed with the needles being supported on a fulcrum and the counterbalance being supplied by kentledge. This system is shown in Figure 2.9.

It will sometimes be necessary to construct king posts to support one end of the needle; for example, where the ground conditions make it imperative to carry the load to a deeper layer of soil. A typical sequence of operations for this exercise in illustrated in Figure 2.10.

If ground conditions are particularly difficult, or where the water table is higher than the excavation level, also where the legs have to be taken much lower than the existing foundation, a system of pier foundations, with beams spanning between them to carry the wall loads, can be used. This system is only practical where there is a layer of ground of sufficient bearing capacity to carry the heavy loads applied by the piers. The beams spanning between the piers can be constructed in a number of ways, and some of these are described in *Foundation Design and Construction* (Tomlinson, 1982).

Mention has been made of the importance of support to the excavation where underpinning operations are being carried out and since this forms an integral part of the construction, some of the more commonly used methods are described hereafter. As already noted, a careful examination and assessment of the structural soundness of the building to be underpinned should be made, and this will have considerable influence on the type of excavation support to be used.

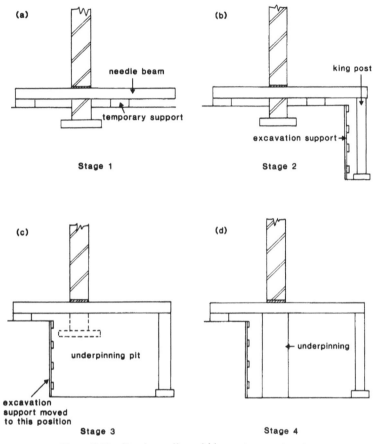

(a)

needle beam

temporary support

Stage 1

(b)

king post

excavation support →

Stage 2

(c)

underpinning pit

excavation
support moved
to this position

Stage 3

(d)

← underpinning

Stage 4

Figure 2.10 Simple needle and kingpost support system.

Traditional timbering methods are applicable in shallower excavations where slight vibrations only can be tolerated. The amount and type of timbering used will also, of course, depend on the type of ground and on the following considerations:

(1) Size and depth of excavation
(2) Variations in soil conditions, e.g. pockets of sand, fissures, or results of previous soil disturbances
(3) Amount of ground water, if any
(4) Surface drainage conditions
(5) Weather and moisture conditions—soil strength can break down through heavy rain or frost, clay can shrink in drying out
(6) Routes and depths of existing underground services, e.g. gas, electricity, telephone cables, etc.
(7) Possible draw down of material through prolonged pumping

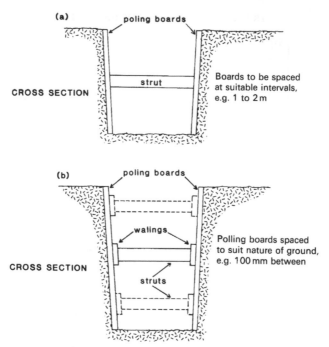

(a) poling boards

CROSS SECTION

strut

Boards to be spaced at suitable intervals, e.g. 1 to 2 m

(b) poling boards

walings

CROSS SECTION

struts

Polling boards spaced to suit nature of ground, e.g. 100 mm between

Figure 2.11 Open sheeting. (*a*) Two poling boards and a strut for use in good ground. (*b*) Frame—any pair of walings on opposite sides together with strut. Alternative arrangements shown dotted. 'Frame' may also include poling boards.

ISOMETRIC PROJECTION

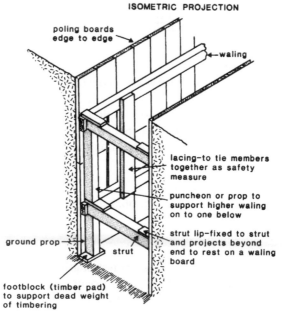

poling boards edge to edge

waling

lacing-to tie members together as safety measure

puncheon or prop to support higher waling on to one below

strut lip-fixed to strut and projects beyond end to rest on a waling board

ground prop

strut

footblock (timber pad) to support dead weight of timbering

Figure 2.12 Close sheeting. Typical example using vertical poling boards; alternatively, horizontal sheeting could be used.

(8) The need to remove supports progressively and replace, as the underpinning operation proceeds
(9) Whether the excavation can be treated as a double-sided or single-sided
support.

If the soil is good firm clay and it is possible to have double-sided support,
then pairs of vertical boards can be used spaced at 1 to 2 metres centres
along the length of the excavation and strutted apart across the width of
the excavation. This method is termed open sheeting, and alternative ways
of doing this are illustrated in Figure 2.11. If the ground to be excavated is
of poor quality, e.g. soft clay, sand and gravel, etc., then close sheeting
(Figure 2.12) can be used. This can be done either with close poling boards
(Figure 2.13) or, if the ground is so bad that it will not stand long enough
to permit the placing of the poling boards, then the use of runners will
have to be considered. These consist of long vertical timbers at least 50 mm
thick and with their lower end chisel-shaped. These are driven downwards in
advance of the excavation. Sketches of both methods are shown in

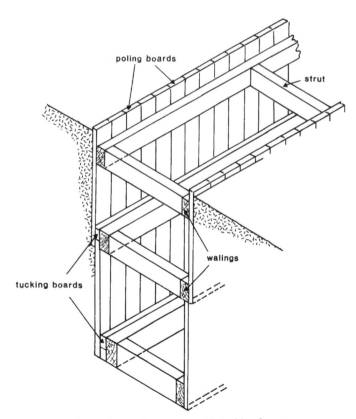

Figure 2.13 Close poling with tucking frames.

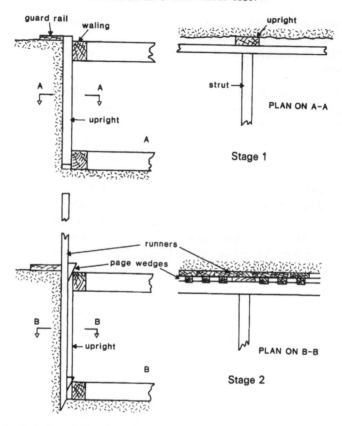

Figure 2.14 Typical methods of sheeting with runners. Stage 1: uprights, walings and struts ready for pitching runners. Stage 2: runners pitched ready for excavation.

Figures 2.14 and 2.15. The above methods may also be used in a single-sided support application where the struts are replaced by raking shores taken down to a thrust-block at the foot of the excavation. Where the excavation is deeper than say 6 m, and vibration is still a consideration, then one method of support would be to use soldier piles of 'H' section located in pre-bored holes with horizontal boards placed and wedged between them as the excavation proceeds. This is illustrated in Figure 2.16.

The control of groundwater during excavation and underpinning operations requires expert soil advice so that the correct method can be adopted and the draw down of fine material from adjacent structures, or indeed, the structure to be underpinned, is avoided. Of equal importance to excavation support is the backfilling operation after the underpinning has been completed. This should be done carefully and with the proper compaction to minimize any future settlement of the replaced material.

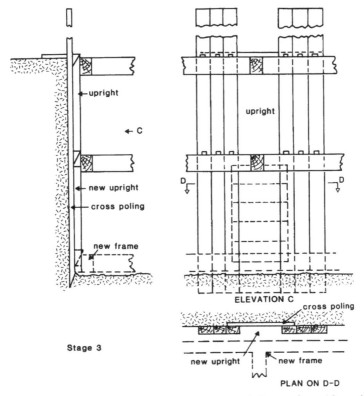

upright

← C

new upright

cross poling

new frame

Stage 3

upright

ELEVATION C

cross poling

new upright

new frame

PLAN ON D-D

Figure 2.15 Stage 3: runners driven and trench excavated ready for next frame (shown dotted).

2.4 Case study

To illustrate the use of simple excavation methods, the following is a description of a contract which—due to restricted access and the need for careful monitoring during the construction of the foundation—required the use of hand digging techniques.

'Browns of the Mound' was a local name for an eighteenth-century eight-storey block of flats situated at the top of the Mound in the city of Edinburgh. The Mound itself has an interesting history, being originally known as the 'earthen mound', and is even referred to as such in old maps. It was man-made, and was built of the excavated material arising from the construction of the New Town to the North of Princes Street, Edinburgh's main thoroughfare. Its purpose was to form a link from the New Town to the Old Town across what was then the ill-drained Nor'Loch. The situation of 'Browns' is therefore a prime one: the building, together with its neighbouring property, stands high above Princes Street and forms an important part of the Edinburgh skyline.

ELEVATION

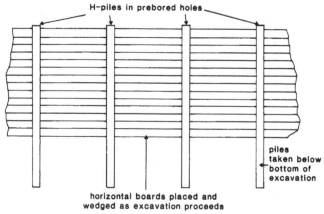

horizontal boards placed and
wedged as excavation proceeds

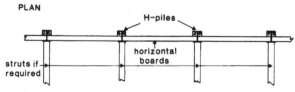

Figure 2.16 Support of excavation by H-pile soldiers and horizontal boards.

The contract consisted of the conversion of the existing flats to modern executive suites, together with office accommodation and a restaurant, all with upgraded services. Due to the building's prominent position, the need to preserve and restore, where necessary, the north and east elevations was obvious, and the decision was taken to build within the existing facades.

The block stands on the slope of the rock outcrop at the top of the Mound and has two halves divided by a thick masonry crosswall. The rear half of the building is in fact built over Lady Stair's Museum which was to remain undisturbed. The decision was taken by the architects and consulting engineers (J & W Johnston & Partners and Cundall Johnston & Partners respectively) to upgrade the floors, replacing joists and flooring as necessary.

The front or north half, however, presented a different problem. When the developer took over the structure the north and east elevations were showing signs of movement outwards from the rest of the building and emergency shoring measures were immediately put in hand. These consisted of steel channel sections placed horizontally on the outer face of the masonry at each floor level. These were tied back through window openings to timber trusses again placed horizontally against the inner face of the walls. The timber trusses gained their end support from heavy steel angle sections bolted to the masonry crosswalls and were carried on vertical timber legs on the existing floors.

Figure 2.17 View of timber truss support to east elevation.

The new reinforced concrete structure was designed to be built within the north and east elevations, the masonry being tied into the new floors at each level. It was therefore important to maintain the temporary shoring system ahead of the construction of the new frame, and each set of trusses was left in place together with the timber floor supporting them until the new floor had been constructed at the level below. This involved substantial temporary propping of the existing floors from ground floor level and imposed further restrictions on an already difficult access. The truss arrangement is shown in Figure 2.17.

The rock level below the existing building, established by exploratory pits and boreholes, dipped quite steeply from back to front and from south-west to north-east corners. The rock was overlain by a stiff clay of varying thickness. During the excavation for ground floor beams and column footings, it was discovered that the heavy masonry crosswall dividing the north and south halves of the building was not (as had been expected) built on the rock, but on

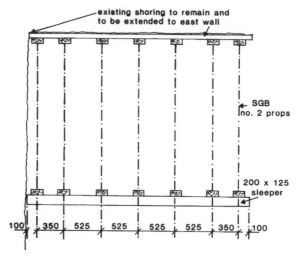

Figure 2.18 Plan of shoring arrangement.

Notes: (1) Existing props to remain until S.G.B. props are in position
(2) All uprights to be 150 × 75 and to be driven to rock
(3) Sleepers to be staggered and make-up pieces to be inserted as shown on elevation
(4) Excavation for sleeper wall to be kept neat and all voids behind sleepers to be packed with soil well compacted.

the overlying clay layer. Since the level of the underside of the new ground beams was some 1 m to 1.5 m below the underside of the cross-wall, immediate steps had to be taken to prevent loss of ground from under the wall.

To allow time for the engineer to redesign the foundations to incorporate underpinning of the crosswall, a temporary timber support to the exposed clay face to the excavation was erected; see Figures 2.18, 2.19 and 2.21. Construction then proceeded on the remaining beams and column footings, and some of the columns were also built. This enabled a more positive support system to be implemented using the foot of one of the permanent columns as a thrust-block; Figure 2.20 illustrates the actual method used.

The permanent underpinning of the crosswall was achieved by using a vertical reinforced concrete wall cantilevered from the ground floor slab in the manner of a basement wall. This was constructed on a 'hit and miss' basis, the temporary propping system being removed section by section as the permanent structure was completed.

Although not a large underpinning job, the contract described serves to illustrate some of the points made at the beginning of the chapter, viz.:

(1) Because of restricted access and the need for careful excavation, hand digging was decided upon.
(2) Close co-operation between the contractor and the engineer enabled measures to be put in hand immediately the unexpected situation arose.

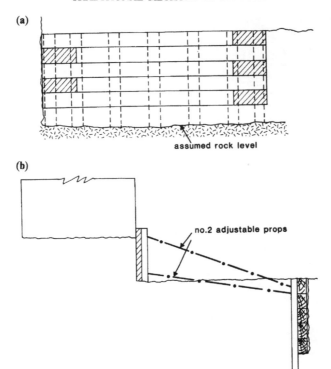

Figure 2.19 (*a*) Elevation of sleeper wall—make-up pieces shown hatched. (*b*) Section showing props—props to be wedged securely to prevent movement in any direction.

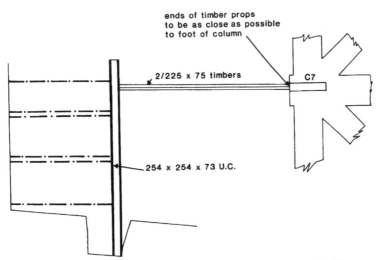

Figure 2.20 Alternative to propping arrangement, Figure 2.19(*b*).

Figure 2.21 Temporary support of clay face under main crosswall.

(3) The existing structure was closely observed during the whole construction and propping was frequently checked.

The outcome was a successful structure in which no movement took place during construction nor has any taken place since construction was completed.

Acknowledgement

Figures 2.1, 2.2 and 2.3 were provided by Mr A.S. White, SGB Building Equipment Division, Mitcham, Surrey.

References

1. BS 8004: 1986 Foundations British Standards Institution, London.
2. Tomlinson, M.J. (1982) *Foundation Design and Construction*. Pitman, London.

3 Conventional piles in underpinning

K.W. COLE

This chapter concerns itself with the use of conventional piles as under-pinning. As already described (Chapter 1) the objective of underpinning is to create new foundations on to which the existing load may be wholly or partially transferred without harmful movements. The need to use piles comes about when

(a) the normal procedure of underpinning by excavating and forming mass concrete foundations (see Chapter 2) is not possible, or is uneconomical either because of the depths to which the excavations would to be taken, or because of problems caused by groundwater; or
(b) the loads on the ground from the existing structure are so great that the large dimensions of the excavations necessary for normal underpinning would cause the structure (or other nearby structures) to settle or move sideways by unacceptable amounts.

Even though piles may be the 'correct solution' to the design problem, there are many ways in which piles can present problems during and subsequent to installation. This may possibly make them less cost-effective than more laborious conventional methods.

Problems with piles mainly arise through two causes.

(i) The pile type is not easily constructed in the prevailing ground conditions. In extreme cases the pile type selected may be entirely inappropriate; for example a bored pile where the underlying strata are waterlogged fine sands and gravels.
(ii) The pile is not constructed properly.

The Construction Industry Research and Information Association (CIRIA), in association with the Department of the Environment, commissioned and published in the period 1977 to 1980 a series of guidebooks on various subjects concerned with piling, of which reports PG2, PG3, PG4, PG8 and PG9 deal especially with 'problems' during the installation of piles and with methods of detecting problems. Other guidebooks in the series are mentioned in the appropriate parts of this chapter.

3.1 Connecting piles to the structure

The cardinal rule in underpinning is that whatever method is used, the structure being underpinned should not be irreparably damaged by the underpinning process, as otherwise the whole point of underpinning is lost. In practical terms this often means that if installing the underpinning will be damaging, a prior stage of underpinning will be necessary.

This is illustrated in Figure 3.1 (a) which shows a temporary 'needle' beam inserted to support the load while the existing footing is undermined to insert the new spreader beam which distributes the load to the new pile foundations. As illustrated in Figures 3.1 (a) and (d), pockets can be made to allow the insertion of jacks, or the entire structural arrangement can be made suitable for lifting the existing structure to a higher elevation. Figures 3.1 (b) and (c) illustrate how the new underpinning piles can be connected to the existing foundations either by being constructed in holes drilled through them or pockets cut out of them, or by transferring the load through a 'corset' structure.

In all cases symmetry of the new pile positions about the existing loads is desirable, to prevent large bending moments and consequent high stresses being imposed on the existing structure.

Where possible piles should be placed close to the lines of action of the loads so as to minimize the amount of connection structure. In many cases the headroom required by conventional drilling and driving equipment will be greater than is available immediately adjacent to the loads to be supported, and structural arrangements of varying degrees of complexity, such as those illustrated in Figures 3.1 (e), (f) and (g), may be necessary.

Figure 3.1 (a) Spreader beams as temporary and permanent underpinning. 1, structure load; 2, spreader beam to support structure load while new pile cap is built under existing; 3, pockets for jacks if required; 4, existing footing or pile cap; 5, new pile cap also acts as spreader beam; 6, existing piles; 7, new piles.
(b) Underpinning piles installed through existing foundation. 8, structure load; 9, piles; 10, existing piles; 11, pile cap drilled through for new pile; 12, pile cap broken out for new pile.
(c) Corset structure to take load from existing foundation. 13, structure load; 14, existing footing or pile cap; 15, foundation reinforced to act as corset and to transmit load to new piles; 16, existing piles; 17, new piles.
(d) Structure arranged for lifting to higher elevation. 18, structure load; 19, existing foundations 20, new corset structure using prestressing to 'clamp' existing foundation lifted by jacks; 21, jacks and packing; 22, new pile cap; 23, new piles.
(e) Using extended bearing beam to overcome headroom difficulty. 24, no headroom problems when outside; 25, new bearing beam; 26, new piles; 27, headroom required for piling equipment; 28, piles and bearing beams installed at intervals.
(f) Using 'balancing' beam when headroom is inadequate for piling equipment. 29, headroom inadequate; 30, new 'balancing' beam; 31, new piles installed outside structure; 32, compression pile; 33, tension pile.
(g) Using a combination of steeply raked piles and prestressing when access to structure is obstructed. 34, new piles to provide reaction installed at same rake (θ); 35, flat jacks if required; 36, reaction beams at intervals along structure; 37, working clearance for piling equipment; 38, longitudinal pile cap; 39, prestressing cables; 40, piles installed at a rake (θ) at allow working clearance on existing structure.

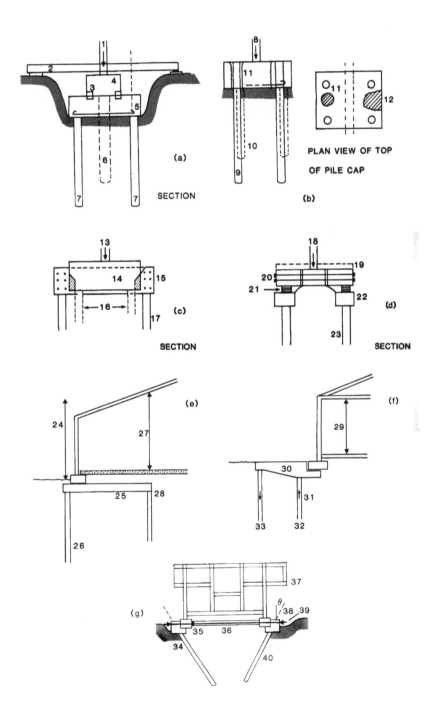

(a)

SECTION

(b)

PLAN VIEW OF TOP
OF PILE CAP

(c)

SECTION

(d)

SECTION

(e)

(f)

(g)

The alternatives if headroom is less than that required by conventional equipment, are as follows.

(i) Adopt a method which installs the piles in short sections. For driven piles this can be done using short pile lengths of either precast concrete or steel H-section or tubular piles, and a low head driving frame requiring about 5 m of headroom. Joints are expensive, vulnerable to damage if not properly aligned, and time-consuming to complete, so the cost of using short sections will be greater than for unrestricted conventional piling. One pile type, the Franki Mega pile, is designed specifically to be installed in short lengths using a hydraulic jack to thrust the pile into the ground against the reaction of the existing structure. Details are given in Tomlinson (1987).

(ii) Construct bored piles using very low headroom tripod rigs and percussion boring equipment. The rate of construction is likely to be slow, and the process may produce large amounts of mud slurry of which it is difficult to dispose. The minipiling process described in Chapter 4 offers less disruptive and more economical methods, particularly in circumstances where the piles have to be installed from basements or such places where access is confined.

3.2 Conventional piles

Conventional piles are those which in usual circumstances are constructed to support new structures but can be adapted to the role of underpinning; non-conventional piles are described in Chapter 4, and many are especially suitable for underpinning, particularly in circumstances where the space available to construct piles is restricted.

The conventional piles considered in this chapter have the following principal characteristics.

(a) Types of piles. Conventional piles are available from a wide range of types, all falling within the categories of displacement types or non-displacement types as shown in Figure 3.2. Displacement piles are forced or driven into the ground by vibratory or thrust techniques, whereas for non-displacement piles the ground is (to a large extent) excavated and the pile placed or formed within the excavation. Piles driven by percussion (self acting or drop hammer) are unlikely to be suitable in the majority of cases due to the headroom requirements, noise and vibration.

Detailed descriptions of the principal bearing pile types are given in 'A review of pile bearing types', (DoE/CIRIA, 1977), and in Tomlinson (1987).

(b) Materials. Piles used for underpinning consist generally of steel and

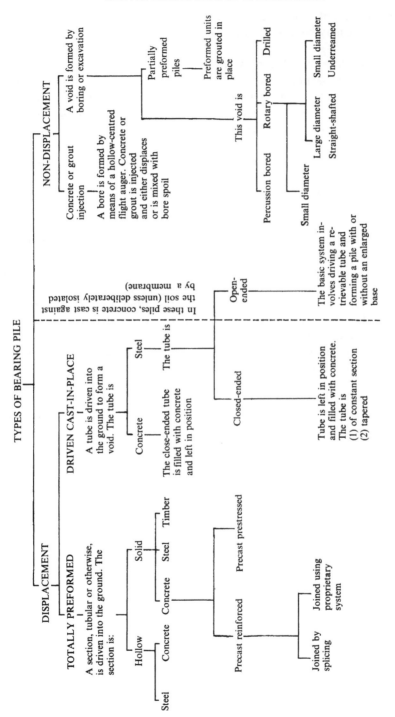

Figure 3.2 Classification of bearing pile types. After Weltman and Little (1977).

concrete or grout (sand–cement mix) reinforced with steel. The steel reinforcement may be of rods, structural sections, wires or tendons.

(c) Shapes. Piles installed by vibratory devices or by hydraulic jacking techniques have sections that combine the greatest possible stiffness with lightness, so that they do not deform unduly while being handled and installed, and yet are light in weight for economy in handling and use of materials. Bored piles are cylindrical in shape, being formed either by percussion boring or by rotating machinery, and often have concentric enlargements of the excavations called 'bells' or 'bulbs'. A cylindrical excavation, with or without cylindrical steel 'lining' tube to support the ground, is inherently the most stable shape likely to remain open, and in good condition to receive the plastic concrete.

(d) Sizes. The two dimensions which characterize the size of a pile are its diameter (or side dimension, if square) and length. The diameter often quoted is nominal, meaning that a cylindrical pile should nowhere be less than the diameter stated, or for an octagonal precast pile it is the diameter of the equivalent circular pile. Piling contractors often offer a range of sizes, and care should be taken to establish the actual size as constructed, rather than to rely on the nominal size described.

A pile is considered to be of 'large' size if it has a nominal diameter (a) greater than 750 mm for a bored pile; and (b) greater than 600 mm nominal diameter (or width) for a driven pile. A pile is considered to be 'small' if it has a nominal diameter or width of (a) less than 300 mm for a bored pile; or (b) less than 150 mm for a driven pile. There is no special term for piles with intermediate dimensions. Piles with lengths greater than 20 times their nominal diameter or longer than 20 m are considered to be 'long'; piles of such length exposed above the ground surface usually require support against buckling. For piles largely beneath the ground buckling is unlikely to be a problem unless the ground is very weak.

The above characteristics are used to describe the piles illustrated in Figure 3.3: (a) shows small-diameter, driven, precast reinforced concrete or tubular steel piles; (b) driven universal bearing piles; and (c) small-diameter, bored, cast-in-place, reinforced concrete piles. Large-diameter bored reinforced concrete piles may be used in certain situations to form pier foundations.

3.3 Designing piles

The process of designing piles for underpinning consists of

  (i) gathering together and understanding information about the site, the existing structures and the ground conditions;

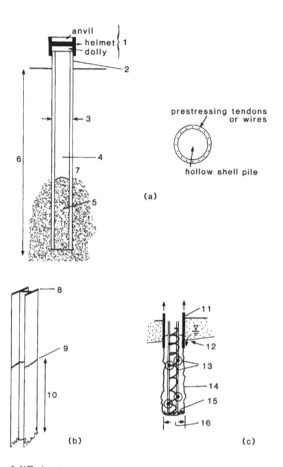

(a)

Figure 3.3 Piles of differing types.
(*a*) Driven, precast, reinforced concrete or steel tubular pile. 1, fabricated steel driving head—designed to distribute impact stresses evenly; 2, cutting pile head to required level must be done carefully or damage and weakness will result; 3, typical diameter 300–450 mm; 4, void (may be filled with concrete); 5, plug of soil forced up inside cylinder; 6, typical length 10–25 m; 7, pile wall thickness must be sufficient to preclude overstressing during driving and handling.
(*b*) Driven steel H-section pile. 8, after driving pile head can be 'flame cut' to level; 9, butt welding is most common method of joining lengths; 10, pile sections are generally approx. 300 × 300 or 350 × 350 mm with weights in the range 75 to 150 kg/m. Lengths up to 26 m are available but 10–15 m lengths are generally easier to handle.
(*c*) Bored, cast-in-place reinforced concrete pile. 11, temporary lining tube lifted upwards and removed after concrete is poured and while it is still plastic; 12, water in sandy layers may cause problems as lining tube is removed; 13, reinforcement 'cage' with 'spacer' wheels to ensure correct cover to reinforcement; 14, surface of hole may be roughened to increase load capacity of pile; 15, some excavation debris may not be removed from base of pile; 16, pile diameter typically 350–550 mm.

  (ii) conducting ground investigations to determine the soil, rock and groundwater properties;
 (iii) conducting investigations into the shape, strength and stability of the structure to be underpinned, and also of any nearby structures that may be affected by the underpinning;
 (iv) selecting a type of pile (or types of piles) suitable for the ground conditions, and able to withstand the groundwater conditions; driven piles must not be used if their installation would adversely affect the environment or the condition of surrounding buildings.
  (v) performing calculations to establish the likely levels at which the bases of the piles should be founded. For many ground conditions calculations may not be relevant, e.g. when the ground comprises alternating layers of hard and soft strata. In such cases judgement based on experience will be necessary; the judgement should be subject to tests during pile installation and load tests on completed piles.

The outcome of the design should be drawings showing in plan the positions of the piles, and if they are to be inclined from the vertical ('raked'), the direction and amount of rake. Sections should be drawn to show how the piles are related to the assumed ground and groundwater conditions, and how they are to be finished at the heads of the piles and then connected to the existing structure.

The design normally also includes a specification, which stipulates the qualities of, and tests on, the materials to be used in making the piles. It also lays down constraints on construction methods, and procedures for testing the load–settlement behaviour of trial piles (special piles made to be tested well above the expected load, to confirm the margin of safety) and procedures for testing piles installed as underpinning. The Institution of Civil Engineers (ICE, 1988) have produced a model *Specification for Piling* and companion *Contract Document and Measurement* which give good guidance in the preparation of contract documents.

3.3.1 Assessment of ground and groundwater conditions

It is preferable that the ground and groundwater conditions across the whole length and breadth of the structure to be underpinned should be known and understood, as only then can it be possible to ascertain the limitations of underpinning by normal methods, and make rational decisions on the types and size of pile that should be installed.

Very often the amount of investigation it is possible to undertake is restricted, either for the practical reasons of lack of spaces to make boreholes and trial pits, because parts of the structure are in occupation and cannot be entered, or because making multiple boreholes or trial pits would be too expensive in proportion to the cost of the works. In such cases judgement is necessary in deciding when enough information has been obtained to make a reasonable design and estimate of likely costs. Such judgement is best entrusted to a person of experience in the field of geotechnics who is especially experienced in piling.

In the most complicated ground conditions it would be true to claim that the ground and groundwater investigations continue until all the underpinning foundations are completed, and that, at best, only a partial knowledge can result from investigation prior to undertaking the works. Nevertheless, embarking upon a potentially difficult, hazardous and expensive undertaking such as underpinning without some understanding of the likely outcome of the methods proposed should be regarded as foolhardy.

3.3.2 Quality and costs of investigations

Ground investigations before anything is built on 'open' sites are usually expected to represent between 0.5% and 5% of the cost of the structural works, although the upper end of this range is not common and is only likely where deep piled foundations are found to be necessary. For the conditions under which investigations are made for underpinning, costs are likely to be within the range of 2% to 15%. A major part of an investigation with piling in mind is likely to be into the shape, reinforcement and condition of the existing structure, as the connections of piles to existing structures may pose difficult problems.

The quality of investigation profoundly influences the reliability of the information obtained. Good guidance on investigation and testing methods and procedures are given in BS 5930 (Site Investigations) and BS 1377 (Methods of Testing Soils) and a comprehensive review of site investigation is given in Clayton *et al.* (1992).

With good guidance on method and procedures, there remains the need to ensure that the results are correctly and economically applied to the project in hand. A good specification of requirements is essential and an engineer or engineering firm specializing in geotechnics (the all-embracing name for soils and rock engineering) is best able to provide this; the geotechnics specialist should also supervise the investigation fieldwork and report on the result. If it is not appropriate to involve a specialist geotechnics engineer, guidance on specifications can be obtained from *Specification for Ground Investigation* (ICE, 1989), which includes a model ground investigation specification. (Note that 'site investigation' is taken to refer to the complete investigation of a site, including the desk study of geology and site history, whereas 'ground investigation' refers specifically to physical investigation of ground materials by borehole, sampling, laboratory test, etc.)

3.3.3 Interpretation of ground and groundwater conditions

Of the conditions which are likely to give rise to foundation failures, unexpectedly weak ground and abundant groundwater released on penetrating into strata of high permeability are the most common.

The 'buildability' of foundations is of as great an importance as bearing capacity. It is therefore essential that each ground investigation should concentrate not only on finding and recording the properties of the strata likely to take a significant part in supporting the possible types of foundations to be used as underpinning, but that the intimate details of the factors which may have greatest influence on the choice of foundation should also be obtained. There is likely to be little scope in most underpinning work for a major redesign of the foundations, without incurring a massive increase in costs.

Weak ground can be easily missed during boring and drilling for samples; particularly in core samples taken by drilling, the weak ground is likely to have been washed away and be part of the 'lost' core. Repeated drillholes with cored samples taken one after the other to produce 'continuous' samples may be necessary to obtain recovery of sufficient samples of weak ground for identification and tests.

In weak (soft) soils the 'continuous' samples may be obtained by taking piston samples taken one after the other, or by the Delft continuous sampler, Details of coring and sampling equipment are given in BS 5930, and the procedures for classifying the soils and rocks, and obtaining their 'index' (basic) properties and strength and compression characteristics are explained by Clayton *et al.* (1992).

Groundwater conditions are identified by observations made during the processes of drilling and boring, and by means of specially constructed standpipes and piezometers in those holes. As the depth to groundwater at each instrument usually takes some time (several hours to several weeks, depending upon the permeability of the ground and of the standpipe or piezometer installation) to reach a constant value, readings of depth to water in all instruments should continue for as long as necessary to be satisfied either that a constant value has been reached, or that the depth to water fluctuates. If the depth to water fluctuates, it is essential to discover why this is, because the highest water level (least depth to water) may be critical either to the underpinning foundation during construction, or affect the foundation performance.

3.3.4 Examples of adverse ground and groundwater conditions

When all that is known of the ground and groundwater conditions at a particular site are placed upon a plan and cross-sections of the site and its surrounding ground, then the process of interpretation can begin. The factual data from two imaginary sites are depicted in Figures 3.4 and 3.5. In both cases piling is the most obvious method of providing underpinning to the existing building.

The problems with the piles underpinning the brick building (Figure 3.4) are likely to occur where they penetrate the silty sand with gravelly layers. Bored

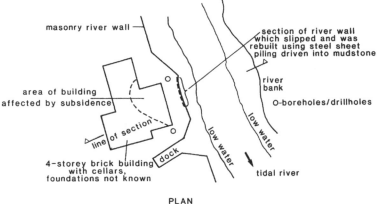

PLAN

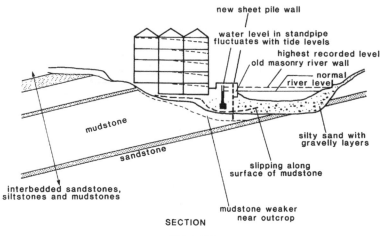

SECTION

Figure 3.4

piles will be especially difficult to construct, as there is a possibility both that an excess of silty sand will be excavated (causing further subsidence) and that water in gravelly layers could displace the wet concrete in the piles (making the piles undersized or 'necked'). Difficulties would also be likely if closely placed displacement piles were adopted, as the weaker mudstone layer would probably cause adjacent piles to uplift or move laterally.

For the case of the proposed factory extension (Figure 3.5) the problems are likely to be that displacement piles cannot easily be driven through the glacial till (unless a separate pre-boring operation is undertaken to provide a 'pilot' hole), and all types of non-displacement piles are likely to run into considerable construction difficulties once they penetrate the water-bearing sand with some gravel. Large-diameter bored piles founding near (but not too near) the base of the glacial till could be the correct solution, but access

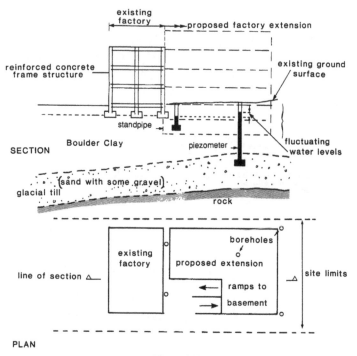

Figure 3.5

difficulties with the large plant required and the high cost would probably make this uneconomical. The most cost-effective solution could be steel H-section piles driven into the underlying rock; it would be necessary to prove that the piles would found within a metre or so of the top of the rock, otherwise the length (and hence the cost) of them could become excessive.

3.3.5 Estimation of pile bearing capacity

3.3.5.1 Pile behaviour. For practically the whole range of possible combinations of pile material and ground material, the pile material is the stronger and stiffer. Therefore, for most practical purposes, it is the ground which fails first, and it is generally only the ground which fails when a pile reaches its maximum load capacity.

Only for timber piles founded in weak rocks and concrete piles founded in strong rocks is the pile likely to be the weaker element, unless there is a defect in the pile or the pile is caused to deform and break in a manner for which it has not been designed (as when a precast pile is lifted with the slings in the wrong places, for example).

Because ground behaviour dominates the process of piles carrying load, pile

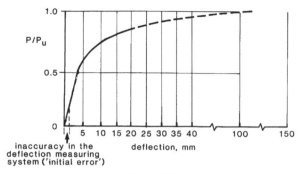

Figure 3.6

movements are also dominated by ground movements. Most ground materials are weak compared with structural materials such as concrete and steel, unless they are cemented or the particles are otherwise tightly interlocked. The ground is therefore likely to have low stiffness with the stress–strain behaviour such that pile movements under 'working' loads are likely to be between 5 and 50 mm.

When a pile is loaded above the allowable load, the movement (also called 'deflection' or 'settlement') increases more and more rapidly with increasing load, until the pile reaches a condition where it will accept no further load. This load is called the pile failure load, (although correctly it should be called the ground failure load) or the pile 'ultimate' load. The methods of carrying out pile loading tests are described in detail in 'Pile Load Testing Procedures' (CIRIA, 1980).

A typical load–deflection curve for a medium-sized pile (450 mm diameter, 15 m long, in stiff clay) is shown in Figure 3.6. It is quite usual to plot pile deflection (measured at the head of the pile) in an upward direction even though the pile in reality moves downward. P is the load applied to the pile and P_u is the failure ('ultimate') load. As shown, it is not unusual for the curve to reveal inaccuracies in the measuring system at low values of P/P_u, and such 'initial error' should be deducted from the measured deflections.

It is apparent that the pile has two modes of behaviour (or rather, the ground supporting the pile is exhibiting two modes of behaviour). Up to a value of P/P_u of about 0.6, the deflection increases more or less linearly with the increasing load. The ground is said to be 'behaving elastically' as if Hooke's Law applied, but in fact this is a 'one-way' elasticity, as unloading the pile from any load reached does not usually result in full recovery, there being some residual or 'non-recoverable' deflection as shown in Figure 3.7. The curves of offloading are usually closely parallel to one another as shown, so that the amount of residual settlement increases with increased load. Usually, when further loading is applied after unloading, the load–deflection curve resumes the same path as if it had not been unloaded. If it does not do this, either the ground or the pile must be suspected of having a defect.

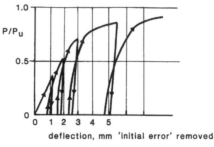

Figure 3.7

When the load is increased above a value of P/P_u of about 0.6, the amount of deflection increases more and more with increasing load, and the ground supporting the pile is said to be behaving 'plastically'. This is particularly marked by large increases in residual (non-recoverable) settlements when the pile is offloaded, as Figure 3.7.

3.3.5.2 Selection of factor of safety. The actual shape of the P/P_u v. deflection curve varies for piles of the same size in different ground conditions, and a selection of curves is shown in Figure 3.8 (deflection being plotted to the same linear scale). These curves are taken from Butler and Morton (1971) in which are considered

(i) a rational approach to the selection of the appropriate load factor (commonly called Factor of Safety, F) relating the ultimate load P_u to the design load P_d, in the equation

$$P_d = \frac{P_u}{F}$$

(ii) criteria for the acceptance of contract piles (piles installed to support a structure) using non-destructive pile loading test procedures.

The relationship of design load P_d to allowable load P_a varies with the circumstances of each structure. If the piles are widely spaced, so that they do not interact upon loading, and there are no down drag or other forces affecting the piles, then $P_a = P_d$. In most cases, however, P_a is somewhat less than P_d, so that the sum of P_a plus interactive forces plus downdrag, etc., equals P_d. The working load P_w applied by the structure must not exceed P_a.

Although Butler and Morton (1971) considered piles in clay, their criteria are equally applicable to piles in all particulate or granular soils. They are not applicable to piles in cemented soils and rocks, although similar criteria may be developed.

Ideally, selection of the appropriate P/P_u v. deflection curve load factor should be made after considering the results of a series of pile tests in which

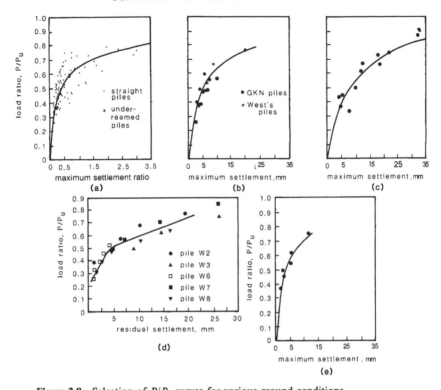

Figure 3.8 Selection of P/P_u curves for various ground conditions.
(a) Load ratio v. maximum settlement ratio; all London clay sites.
(b) Load ratio v. maximum settlement; Hull, silty clays over glacial till.
(c) Load ratio v. maximum settlement; Gorbals, Glasgow, silty fine sand.
(d) Load ratio v. residual settlement; driven precast concrete piles in chalk.
(e) Load ratio v. maximum settlement; Belfast, soft silty clay over Boulder Clay.
(a), (b), (c) and (e) from Butler and Morton (1970); (d) from Lord (1977).

piles of various sizes have been taken to ultimate load in the appropriate ground conditions. The possibility of doing such a series of tests for underpinning at a particular site is remote, and in the majority of cases P_u will have to be obtained by calculation or assessment from case histories. The shape of the P/P_u v. deflection curve can usually be selected with reasonable reliability from the 'library' of curves given in Figure 3.8, or obtained from the results of tests on piles in the soil under consideration, if the soil is not similar to any of those given.

There are two 'factors of safety' relevant to the estimation of the design load P_d of a pile. The first is a factor against failure (against the ultimate load being attained), and this has been found to have a value of about 1.3 to 1.4 when the ultimate load P_u has been obtained; this means in effect that no similar pile is likely to have an ultimate load lower than 1/1.3 or 1/1.4 (0.77 or 0.71) of the expected ultimate load. This factor of safety has little practical value, as pile settlements are likely to be unacceptably large at such high values of P/P_u, as illustrated in Figure 3.8.

The second factor of safety is against a pile suffering excessive deflection (settlement) under the design load P_d. As can be from the P/P_u v. deflection curves in Figure 3.8, the 'point' (or more appropriately 'zone') at which the ground supporting the pile ceases to behave in an 'elastic' manner varies from soil to soil, and lies between about 0.55 and 0.75. At loads above the 'point' value the settlement curves for individual piles tend to vary widely as shown dotted in Figure 3.8(a), and individual piles may undergo quite large settlements.

If it could be guaranteed that every pile made in the given ground conditions would have the same ultimate load P_u, then the design load P_d could be between 0.55 and 0.75 of ultimate, that is the factor of safety could be $1/0.55$ ($= 1.82$) to $1/0.73$ ($= 1.33$). However, as discussed above, the actual values of P_u may be as little as $1/1.4$ ($= 0.71$) of the value of P_u from tests, so it is appropriate to 'scale down' the 'point' values of P/P_u to $0.55 \times 0.71 = 0.39$ and $0.75 \times 0.71 = 0.53$, giving factors of safety in the range $1/0.39 = 2.56$ to $1/0.53 = 1.88$, which have a magnitude more familar to those experienced in designing piles. In practical terms this means that if the ultimate load of a given type and size of pile in given ground conditions is known, then the factor of safety (load factor) should be between 2.0 for a pile in 'stiff' ground to 2.5 for a pile in 'soft' ground. Clay strata at the depths (15 to 30 m) to which piles commonly penetrate are generally 'stiff' if they are overconsolidated, and 'soft' if they are normally consolidated.

In many cases the value of P_u for a given type and size of pile at a given underpinning site will not be known, and an additional factor of safety to allow for the possibility of error in assessing P_u should be made. The magnitude of this additional factor is discussed in the following sections dealing with soils and weak rocks.

3.3.5.3. Estimation of load capacity. Because of the shape of piles they have a 'base' which bears down upon the underlying ground, and a 'shaft' which follows the base down, at the same time shearing the adjacent soil at the interface. In some piles the base is very small; for instance the typical steel H-pile has a 'base' area of only $0.02\,\text{m}^2$, whereas its shaft surface may be 25 to $100\,\text{m}^2$. In large bored piles the base may be enlarged by underreaming to give a base area of $20\,\text{m}^2$, and the shaft area may be also 25 to $100\,\text{m}^2$.

The mechanisms of the resistance offered by the ground are different between shaft and base, and this is recognized in the analysis of pile behaviour and the methods of design. As there is very little interaction between the two mechanisms, the ultimate loads that can be supported by the shaft and the base are assessed separately using the formula

$$P_u = Q_{su} + Q_{bu}$$

where Q_{su} and Q_{bu} are respectively the ultimate resistances of the shaft moving through the ground and the base moving through the ground.

For most piles the factor of safety of 2.0 to 2.5 (or such higher values as are discussed in the following sections) is applied directly to the combined ultimate resistances to give the design load:

$$P_d = \frac{P_u}{F} = \frac{Q_{su} + Q_{bu}}{F}$$

The value of P_a, the allowable load, is then derived from P_d, by taking into account pile interaction and downdrag.

When the pile base area exceeds about one-tenth of the pile shaft area, the movement of the pile base under load dominates the pile behaviour, and a factor of safety of 3.0 (or larger value, as discussed in the following sections) is applied to the base component only, with the magnitude of shaft resistance selected to be appropriate to large movement through the ground:

$$P_d = Q_{su} \text{ (large movement)} + \frac{Q_{bu}}{3}$$

3.3.6 Piles in particular soil and rock types

In the following sections descriptions are given of the problems associated with broad categories of rock and soils.

Rocks are generally intersected by joints and bedding surfaces, the more weathered rocks near to the interface with overlying soil or to the exposed surface of the rock itself, commonly being more frequently intersected than is the case for rocks at depth. The individual pieces of rock between the joints or bedding surfaces may be strong, but the behaviour of the rock mass is dominated by the strength available along the joint and bedding surfaces.

Soils are derived from rocks, and may be the result of weathering of the underlying rock (residual soils) or the result of deposition after having been moved by air, water, glaciers or gravity (transported soils). Some soils are cemented and others contain or are derived from organic materials. From the many ways in which soils are derived and laid down it is to be expected that their properties will be very variable.

Soils may be separated into mainly granular or (non-cohesive) materials such as boulders, cobbles, gravels and sands, and the cohesive materials silt and clay. The engineering distinctions relate mainly to the far faster rate at which water can pass through granular materials than through cohesive materials—hence any excesses in water pressure within the soil dissipate far more rapidly in granular soils. They also relate to the manner in which cohesive materials stick together and to the surfaces of structural materials. Cohesive clay soils are further distinguished into normally consolidated (NC) clays, which have not in the past been subjected to pressures greater than the present overburden pressure, and overconsolidated (OC) clays which have been subjected to pressures greater (sometimes much greater) than the present overburden pressure. The important effect from the point of view of pile design is that OC clays are generally stronger than NC clays of the same composition, but the OC clays may weaken (soften) rapidly after having been strained to generate their peak resistance, as illustrated in Figure 3.9.

3.3.7 Piles in cohesive soils

Cohesive soils include not only the 'fatty' clays such as London, Gault and Oxford clays, but also the silty clays such as the Kimmeridge and Lias, the widely graded stony Boulder Clays of glacial origin, and the silty sandy

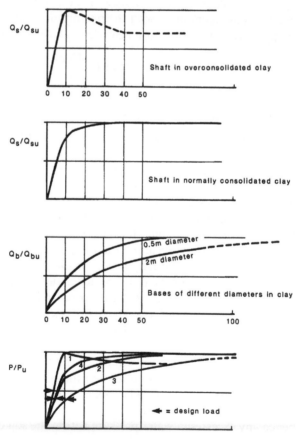

Figure 3.9 Load ratio curves for shafts and bases separately and combined. Clay types, sizes of piles and proportions of loads.
(i) OC clay 1. 0.5 m diameter base, $Q_{su} = 3Q_{bu}$
2. 0.5 m diameter base, $Q_{su} = Q_{bu}/3$
3. 2.0 m diameter base, $Q_{su} = Q_{bu}/9$
(ii) NC clay 4. 0.5 m diameter base, $Q_{su} = 3Q_{bu}$

alluvial clays found in river valleys and estuaries. In temperate climates the clays are nearly always found in a saturated or near saturated condition, except in the upper 2 to 3 metres affected by seasonal weathering and vegetation.

When a pile is loaded, the load *v.* settlement behaviour is different between the base and the shaft, as illustrated in Figure 3.9. Around the cylinder or 'shaft' of a pile only the soil nearest to the pile is dragged downwards, until, as the pile load increases, a shear surface (or shear surfaces) develops within the cohesive soil and close to the surface of the pile, leaving a skin of cohesive soil adhering to the pile, as illustrated in Figure 3.10. The peak adhesion of the shear surface depends upon the pile type, the pile material and the method of pile construction, and guidance is given in CIRIA publication PG5 (DoE/CIRIA, 1978). Pile movement of between 2 and 10 mm from its initial unloaded position is normally sufficient to reach peak adhesion. For over-consolidated soils, the adhesion falls to about 35% of the undrained shear strength if the pile is moved five to ten times the movement for peak adhesion.

Below the base of a pile the soil compresses at first loading; then as the load on the base is increased, it deforms and shear surfaces develop in the soil surrounding the base. With increasing load the pile begins to settle more and more, and either large volume changes have to take place to compress the soil surrounding the base, or the ground surface has to heave upwards, as shown in Figure 3.11 *a* and *b*.

The procedure for the design of piles in cohesive soils is as follows:

(i) Select a cross-section shape and length of pile.
(ii) Select the values of soil strength that will be relevant to the calculation of shaft load capacity and base capacity.
(iii) Calculate separately shaft and base ultimate loads and hence estimate the load *v.* settlement curves for each component.

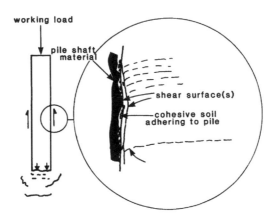

Figure 3.10 Piles in cohesive soil.

(iv) Combine shaft and base loads at various pile settlements to see what design loads (P_d) can be achieved for various pile lengths.

(v) Check that the selected design loads (P_d) have the appropriate factor of safety against exceeding the critical 'point' on the appropriate P/P_u against settlement curve.

(iv) Select the allowable load P_a as a proportion of the design load P_d, if downdrag and pile interaction are likely to affect behaviour. Guidance on the method of calculating downdrag and interaction can be found in Tomlinson (1987) and Whitaker (1976).

It is recommended that the factor of safety used in the calculation in (v) above should not be less than 2.0 for piles in stiff to hard cohesive soils and 2.5 for piles in soft to firm soils, if trial piles are to be made and tested to determine P_u. If no trial piles are to be made and only the calculated values of P_u are to be relied upon, then it is recommended that the factor of safety should be at least 2.5 and 3.0 respectively, with higher values still if the soil information is poor, or the ground conditions are known to vary widely.

Two methods of calculation of pile shaft ultimate load Q_{su} are available. In one the value of adhesion is assumed to be a proportion (α) of the undrained shear strength of the cohesive soil through which the pile passes; in the other method the cohesive soil is assessed as having a friction value β typical for the soil in which it is installed. The α method is explained in Burland et al. (1966), and the β method in Burland (1973). Piles in non-fatty clay such as boulder clay require special consideration, and this is given in CIRIA publication PG5 (1978).

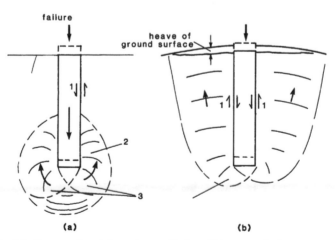

Figure 3.11 (a) Compression takes place entirely within ground, to accommodate pile penetration. (b) Ground heaves in response to pile penetration. 1, shaft adhesion or friction at or beyond peak value; 2, soil compress; 3, soil heavily sheared.

3.3.8 Piles in granular soils

In essence the method of designing piles in granular soils differs only in the way in which the base and shaft components of ultimate load are derived.

Granular soils depend upon the precise details of particle size, orientation, packing and confining stresses for the shear strength they develop. The smallest change in one of the factors caused by disturbance during attempts to take samples can effect a large change in strength. And even if perfect samples could be taken and tested, Terzaghi (1953), in a lecture looking back over the previous 50 years of ground investigation, pointed out:

> The relative density of natural sand deposits commonly varies in an erratic manner in both horizontal and vertical directions. Hence even in the event that one sample has been secured and tested for every foot (0.3 m) of depth along vertical lines spaced 50 feet (15 m) both ways, the density profile constructed on the basis of the laboratory test results may be entirely fictitious. Therefore it is commonly preferable to judge the average density of a natural sand deposit and the variations of density within that deposit on the basis of cone penetration tests in situ.

Very little development has occurred to invalidate that view; the open-shoe sampler and driven cone, together with the jacked cone (cone penetrometer) remain the most effective and economical means of obtaining design information on sands and other granular strata to this day.

Methods of using information from the standard penetration test (SPT) and the cone penetration test (CPT) in the design of piles are described in detail by Tomlinson (1987), Whitaker (1976), and Fleming and Thorburn (1983).

As piles in granular soil are usually long compared to their cross-section dimension, an overall factor of safety to limit settlement to an acceptable amount is normally applied to the combined ultimate base and shaft resistances:

$$P_d = \frac{P_u}{F} = \frac{Q_{su} + Q_{bu}}{F}$$

These values should take into account the manner of construction of the pile, particularly if it is constructed by boring through the granular deposit, in which case the density can be markedly reduced and the strength likewise.

Because of the inherent variability in the behaviour of driven piles a generous factor of safety is appropriate, ranging from 2.5 if several piles are subjected to ultimate load tests, to 4.0 in the absence of good soils information including cone test results.

3.3.9 Piles in weak rocks

In respect of piling, weak rocks are separately identified from strong rocks by the need to consider the interaction of the pile shaft with the surrounding rock, in addition to bearing support derived at the pile base. The rock material is mainly weaker and less stiff than the structural material of the pile. The principal weak rocks are the Chalk, the Keuper Marl (Mercia Mudstone), and the weathered or otherwise weakened superficial few metres of such rocks as the Devonian and the Carboniferous, which are generally shaly mudstones and siltstones. Drilling equipment of sufficient power to drill holes into the rock for bored piles, and hammers big enough to drive tubular and H-piles into the rock, have been available for some 20 years.

As with piles in soil, there needs to be a margin of safety against loading the pile to its ultimate load. Large differences in ultimate load may exist between apparently similar piles, as is evident from a study of the pile tests reported in the ICE symposium, 'Piles in Weak Rock' (ICE, 1977). This indicates that a conservative value of the factor of safety is desirable against the ultimate load, either obtained by testing, or calculated. Values of 2.5 and 4.0 respectively are appropriate.

The wide variations in ultimate capacity are a direct result of the behaviour of the rock being dictated by the strength developed along joint and bedding surfaces; as no two piles will encounter identical arrays of such surfaces, wide variations in bearing capacity at large displacements can be expected.

In one of the contributions to the ICE symposium 'Piles in Weak Rock' (ICE, 1977), Cole and Stroud sought methods of analysing the stiffness of a pile in weathered Coal Measure Series mudstones, siltstones and sandstones. In essence the method for medium to large bored piles assumes that they are formed in 'sockets' in the surface of the weak rock, and that for each pile the applied load is transferred to the rock through the surface of the shaft and base of the socket, there being negligible contribution from the overlying strata. The method proposed recognized that the stiffness of the ground beneath a pile base is commonly some 20 times the stiffness of the rock surrounding the shaft acting in shear. By deducing ground stiffness from ground investigation and pile test results, the load v. settlement performance of each pile was predicted having regard to the details of the strata.

In another paper in the same symposium, Lord examined the behaviour of a selection of common medium-sized proprietary driven piles founded in chalk, and found good agreement with the stiffness moduli of various grades (qualities) of chalk. Other papers give test results on a variety of pile types in various weak rocks, but some of the results have to be carefully considered as they indicate faults in piles and bring into question the appropriateness of certain pile types as weak rock foundations.

The special problems of piles in chalk are described in the publication 'Piling in Chalk' (CIRIA, 1979).

3.3.10 Piles founded on strong rocks

When piles are founded on strong rocks, it must be ensured that the piles cannot move or be moved from the initial area of contact. The surface shape of rock buried beneath soil cannot be known with certainty and thus measures have to be taken to ensure that the piles are well 'keyed' into the rock surface.

If the rock is stronger than the pile material the keying-in needs to be no more than to ensure that the rock materials can sustain the load from the pile, even though they may be locally fractured. Bored piles are for this reason usually given a shallow 'socket' of depth about half the pile diameter.

With steel H-piles it is usual to drive the piles until the number of blows for a given penetration using a rated hammer is more than a specified number, viz. 15 blows for 25 mm. If the rock surface is likely to be irregular, a 'rock point' should be fitted to the base of the pile and this in theory enables the pile to 'wedge' itself into the rock surface, rather than being deflected down dip.

References

1. BS 5930: 1981 Code of Practice for Site Investigations. British Standards Institution, London.
2. BS 1377: 1990 Methods of Testing Soils for Engineering Purposes. BSI, London.
3. Burland J.B., Butler, F.G. and Duncan, P. (1966) The behaviour and design of large diameter bored piles in stiff clay. *Proc. Conf. on Large Bored Piles*. ICE, London.
4. Burland J.B. (1973) Shaft friction of piles in clay. *Ground Eng.* 6 (3).
5. Butler F.G. and Morton K. (1971) Specification and performance of test piles in clay. *Proc. Symp. on Behaviour of Piles*, ICE, London.
6. Clayton, C.R.I., Simons, N.E. and Matthews, M.C. (1992) *Guidance to site and ground investigations. A handbook for engineers.* (2nd edn.) Blackwells, Oxford. In preparation.
7. Department of the Environment (DoE) and Construction Industry Research and Information Association (CIRIA), Piling Development Group.
   ———(1977) A review of bearing pile types. *Report PG1.*
   ———(1977)Review of problems associated with the construction of cast-in-place piles. *Report PG2.*
   ———(1977)The use and influence of bentonite in bored pile construction. *Report PG3.*.
   ———(1977) Integrity testing of piles; a review. *Report PG4.*
   ———(1978) Piling in Boulder Clay and other Glacial Tills. *Report PG5.*
   ———(1979) Piling in Chalk. *Report PG6.*
   ———(1980) Pile load testing procedures. *Report PG7.*
   ———(1980) Survey of problems associated with the installation of displacement piles *Report PG8.*
   ———(1980) Noise and vibration from piling operations. *Report PG9. CIRIA, London.*
8. Fleming, W.G.K. and Thorburn, S. State of the art report on recent piling advances. *Proc. Conf. on Piling and Ground Treatment*, Thomas Telford, London.
9. Institution of Civil Engineers (1977) *Piles in Weak Rock, Proc. ICE Symp.*, ICE, London.
10. Institution of Civil Engineers (1989) *Specification for Ground Investigation with Bill of Quantities.* Thomas Telford, London.
11. Institution of Civil Engineers (1988) *Specification for Piling* and *Contract Documents and Measurement.* Thomas Telford, London.
12. Terzaghi, K. (1953) Fifty years of subsoil exploration. *3rd Int. Conf. on Soil Mechanics and Foundation Engineering*, Zurich.
13. Tomlinson, M.J. (1987) *Pile Design and Construction Practice.* (3rd edn.) Viewpoint Publications, Leatherhead.
14. Whitaker, T. (1976) *The Design of Piled Foundations.* (2nd edn.) Pergamon, Oxford.

4 'Pali radice' structures

F. LIZZI

4.1 Underpinning by means of conventional piles

The introduction of piling techniques for new buildings at the beginning
of this century was one of the great advances in modern construction methods.
Steel piles, precast concrete piles and bored cast-in-situ concrete piles replaced
the old shallow foundation systems. The most important feature of piling
techniques was the opportunity to design foundations on the basis of the loadbearing capacity of a single pile which could be verified, when necessary, by
direct load test. Pile diameters, according to early European Rules, could
not be less than 400 mm.

Considering the success of piling in foundations it was logical to investigate
the possibility of using the same system for underpinning. Driven piles had
to be discounted on account of the vibrations they induced in nearby
structures. Some attempts were made with bored cast-in-place concrete piles
of the type then in use (vertical piles, minimum diameter: 400 mm). The piles
were constructed as close as possible to the walls, to which they were then
connected by reinforced concrete beams. This involved cutting major
openings in walls, before the connection of the piles to the walls. This method
had a short and difficult life, and success was counterbalanced by several
failures.

4.2 Underpinning by means of pushed-into-the-ground piles

This method, described in another section of this book, has been used widely
and is still in use. Nevertheless, it is not applicable where the presence of
boulders, old foundations or other obstructions makes penetration of the
ground with segmental piles difficult or impossible.

4.3 'Pali radice'

The problems and, in some cases, complete failure of earlier systems of
foundation strengthening promoted the introduction of new, more reliable
and safer systems. Such was the situation in 1950–52 when the first patents

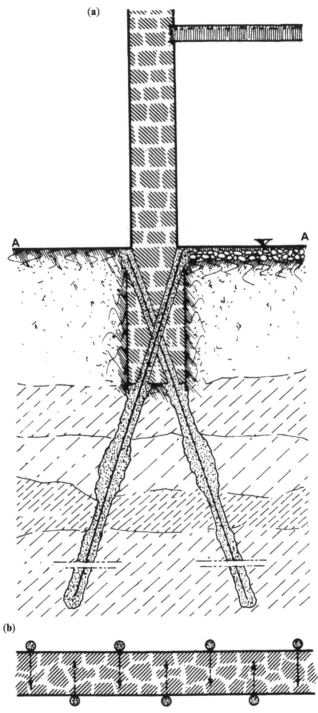

Figure 4.1 Typical scheme of a pali radice underpinning: (*a*) vertical cross-section; (*b*) horizontal cross-section A–A.

for what would afterwards be called 'pali radice' were sought by the author, on behalf of the Italian firm Fondedile of Naples.

A typical underpinning by means of pali radice is shown in Figure 4.1. The foundation reinforcement is formed by a double series of small-diameter piles, rotary drilled through the existing masonry and taken to an adequate depth in the ground below. When concreted, the pile is automatically bonded with the upper structure: there is no need of complementary connecting structures, no risky cuts in the walls, and no disturbance to the building's activities.

The construction of the piles, spaced out along the base of the walls, does not present any risk to the stability of the existing structures. No harmful vibration is involved. The construction of the piles does not introduce any particular stress in the wall or the ground: this is of vital importance in buildings, especially ancient monuments, in which the conservation of the existing, however precarious, equilibrium is of paramount importance. Pali radice can be drilled in any ground, no matter what boulders, old foundations or other obstructions it may contain.

4.3.1 The technology of a palo radice

Figure 4.2 shows the construction sequence of a palo radice.

(i) The drilling is performed by a rotating casing progressively introduced into the ground; the spoil is removed by flushing water or bentonite mud introduced from the top through a rotating swivel.

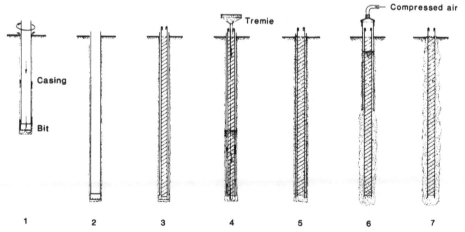

Figure 4.2 Method of construction of a palo radice. (1) Drilling; (2) drilling completed; (3) placing of the reinforcement; (4) placing the grout via tremie pipe; (5) grouting completed; (6) extraction of the casing; (7) palo radice completed.

(ii) Once the drilling has been taken to a convenient depth, the reinforcement is placed: a single bar for smaller diameters (100 mm); a cage or a tube for larger diameters (up to 300 mm). In underpinning, the smaller diameters are generally to be preferred.

(iii) The grout is then placed by tremie pipe, the mix comprising 600 to 800 kg of cement per m³ of sieved sand. It is, therefore, a high-strength grout.

Once the casing has been filled, it is gradually extracted. At the same time, compressed air introduced from the top pushes the mix outside the pipe against the borehole. The pressure of the air is limited to 6 to 8 bars, in order to avoid rupture of the soil ('claquages'), while being sufficient to obtain a very rough outer surface to the pile. This gives the marked adherence to the soil that is an essential characteristic of the palo radice.

4.3.2 The performance of pali radice

The most significant feature of the pali radice used in underpinning work is the response to any movement, however slight, of the structure. This essential feature is due to the technology of its construction in which a palo radice is essentially a friction pile. In Figure 4.3 the diagram shows the load–settlement curves of one of the first load tests for the underpinning work carried out for a school in Naples (1952). The pile, 13 m long, had a diameter of 100 mm and was reinforced by a single bar of 12 mm diameter. The subsoil, characteristic of the volcanic area of Naples, was composed of alternating layers of varying thickness of pozzolana, lapilli and sands. Generally, it may be considered a loose soil formation. It has been noticed since then that even for high loads the settlement of a palo radice is very small, in the order of a

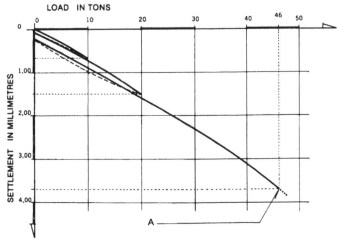

Figure 4.3 The first load test on a palo radice. A, crushing of the toe of the pile ($\sigma = 450 \, \text{kg/cm}^2$).

few millimetres. Being constructed as it is, pali radice underpinning does not supersede the existing foundation. From the start its function is complementary and only when necessary does it contribute to the foundation. The building continues to rest on its old foundation soil: it will call on the piles to assist only if, and to the extent which, it settles. This is the most important aspect which has made the palo radice so popular in underpinning works.

To clarify further, pali radice underpinning can be considered practically inactive at the moment of its construction. If the building has a subsequent, albeit minimal, settlement the piling responds immediately, absorbing part of the load and reducing at the same time the stress on the soil. If, despite this, the building continues to settle, the piles continue to take the load until, finally, the entire building load is supported by them.

In even the most extreme case the settlements would be limited to a few millimetres.

4.3.3 The service load of a palo radice

The service load of a conventional foundation pile for a new building is generally a fraction of the ultimate load. In its assessment many factors have to be considered, including the desired safety factor and, above all, the extent of the acceptable settlement for the new building. For instance, while it may be acceptable to allow high settlements for the piers of a simply supported bridge, the same is not possible for a continuous structure. This consideration, which

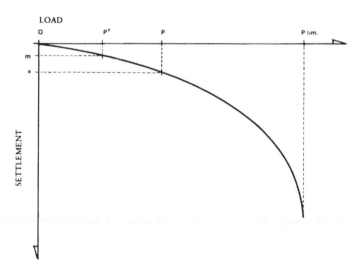

Figure 4.4   The 'working load' for a palo radice in underpinning.

is fundamental for a pile foundation, is even more important for underpinning and so cannot be neglected.

For instance, should x (see Figure. 4.4) be the maximum settlement acceptable for the structure to be underpinned (the value would be in millimetres or fractions of millimetres), the corresponding load to be supplied by the pile would be P. The fact that the ultimate load for the pile is much greater than P is of little importance. It must be stressed that the matter of importance is the acceptable settlement.

It can be seen that although a palo radice used in underpinning is not completely utilized, compared to its full bearing capacity: it could always bear much more than the assigned load. The pile, when its installation has been completed, has no immediate work to do; possibly it will never be completely utilized. As Figure 4.4 makes clear, for each intermediate settlement (m) part of the load (P') is transferred to the pile, while the remaining part ($P - P'$) continues to be supported by the soil.

4.3.4 The 'safety factor' of a pali radice underpinning

It can be seen from the above that the safety factor of the pali radice underpinning is not limited to the safety factor of the piling; it is much greater.

At the commencement of construction of the pali radice underpinning, the safety factor of the existing foundation will be low; but it is obvious that its value will not be less than one. This margin will not be affected by the construction of the pali radice; therefore the new factor of safety after underpinning will be a combination of a proportion of the resistance of the soil and of the resistance of the pali radice. The problem is complex and requires consideration of strain compatibility and group effects. Reference should be made to the principles expressed in Chapter 11 in relation to in-situ reinforcement and fundamental design considerations.

4.3.5 The design of a pali radice underpinning: the ultimate load
of a palo radice

Where case histories, in different soils, do not provide sufficient data about the length and bearing capacity for the various pile diameters, the design problems can be solved by direct load tests.

After load tests have been carried out, some generalisations can be made.

(i) With the exclusion of extremely soft soils, a palo radice develops its maximum bearing capacity (to the compressive strength of the grout in the cross-section) over lengths of not more than 30 m.

(ii) For soils of firm consistency, 20 m is generally sufficient.

Table 4.1 Values of K (in kg/cm^2)

Soil	K
Soft soil	0.5
Loose soil	1.0
Soil of average compactness	1.5
Very stiff soil, gravels, sands	2.0

Table 4.2 Values of I

Diameter of the pile	I
$D = 10$ cm	1.00
$D = 15$ cm	0.90
$D = 20$ cm	0.85
$D = 25$ cm	0.80

(iii) For very stiff soils, lengths of 10–15 m are generally sufficient.

(iv) For compact sands and gravels, the limit varies from 6 to 10 m.

Considering the above, a simple empirical formula has been developed for the ultimate load P_{ult} (kgf) of a palo radice:

$$P_{ult} = \pi D L K I$$

where

D is the nominal diameter (in cm) of the pile, i.e. the drilling diameter;

L is the length of the pile (in cm);

K is a coefficient that represents, in kg/cm^2, the average interaction between the pile and the soil for the whole length. (From the physical point of view, it can represent the shear stress induced at the pile–soil interface or the shear strength of the soil.)

I is a non-dimensional coefficient of form, that depends on the nominal diameter of the pile.

Tables 4.1 and 4.2 give the approximate values of K and I.

4.4 Case histories of underpinning

The versatility of the pali radice system, which can be executed in any ground and site conditions, has encouraged its extensive use. It is by now the almost universally adopted system of underpinning. A small number of case histories, to be considered only as examples, are given below.

A

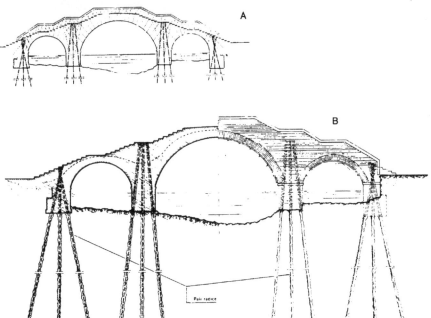

B

Figure 4.5 The 'Tre archi' bridge, Venice, Italy. *A*, scheme of the reinforcement of the upper structure; *B*, scheme of the underpinning.

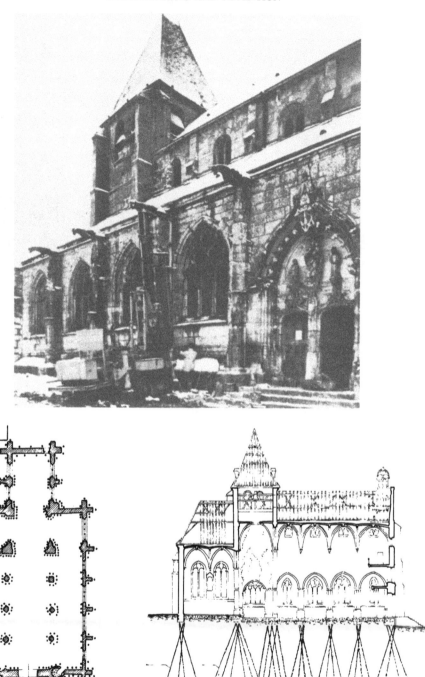

Figure 4.6   Eure, France. Monumental church of Tourny (XV century).

4.4.1 Venice (Italy): Tre Archi Bridge (XVII century)

This is the only bridge in Venice to have three arches (Figure 4.5). It was constructed in the 17th century across one of the most important urban waterways. Following very serious settlement it was originally decided to demolish it. The bridge, however, was fully restored through a complete underpinning with pali radice, a reinforcement of the masonry with the 'reticolo cementato' system (network of grouted steel bars), and strengthening of the vaults by a saddle in reinforced concrete. An appropriate use of resins was made in the works; particularly, they were used in the connection between the existing vault and the saddle to improve the bond between these two elements.

4.4.2 Eure (France): monumental church of Tourny (XV century)

The historic building (Figure 4.6) was in a very bad condition, following marked differential settlements of the foundations. The underpinning with pali radice did not cause any additional harm to the existing structure.

4.4.3 Ghent (Belgium): 'Het Toreken' building (XV century)

The very critical overall static condition of this building (Figure 4.7) demanded a full preliminary restoration of the upper masonry before underpinning with pali radice could begin.

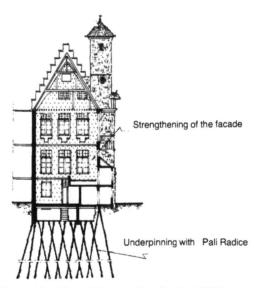

Figure 4.7 Ghent, Belgium. 'Het Toreken' (XV century).

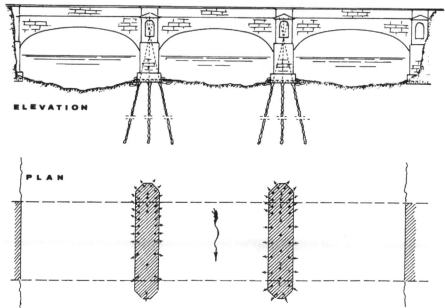

ELEVATION

PLAN

Figure 4.8 Derby, UK. St Mary's Bridge.

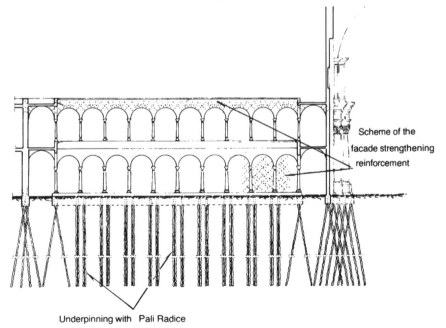

Scheme of the
facade strengthening
reinforcement

Underpinning with Pali Radice

Figure 4.9 Trapani, Italy. Pepoli Museum (XIV century).

4.4.4 Derby (UK): St. Mary's Bridge

St. Mary's Bridge (Figure 4.8) over the River Derwent, in Derby, was built in the years 1778–94, on the model of the bridge of Neuilly in Paris. It was built on the foundations of an old bridge which dated from the time of the Danish invasions in the year 800. Strengthening was required owing to foundation settlement of the two central piers, caused by extensive undermining due to scour. It was therefore carried out by first filling the cavities with bagged concrete and then underpinning with pali radice (diameter $D = 22$ cm) installed from the bridge deck through the masonry and penetrating about 12 m into the clay soil.

4.4.5 Trapani (Italy): Pepoli Museum (XIV century)

The National Pepoli Museum (Figure 4.9), built in calcareous tuff masonry, showed evidence of stress at many points. Those most apparent could be seen in the north-west corner of the cloister porch, along the junction of the Church and the Chapel, and along the West End. Further, it was clearly out-of-plumb, and a remarkable bulging of the four sides of the porch in relation to their central sections could be seen. A preliminary survey indicated that the edifice had been affected by the instability of the subsoil, which because of the type of soils, was particularly affected by movements in the water-table level caused by the exploitation of some nearby wells. Other separations were noticeable in the correspondence of the porch arches with the adjacent structures. As well as restoration of the upper masonry the foundations were underpinned with pali radice.

4.4.6 Florence (Italy): the underpinning of the Ponte Vecchio

This job was carried out in the years 1962–63. The work was necessary owing to the precarious state of the structure, caused by age, the events of the last war (Ponte Vecchio was the only bridge that survived the systematic destruction of all the bridges during the war), and, last but not least, vibration caused by traffic.

As is shown in Figure 4.10, the underpinning is a classic pali radice type, with piles installed through the structures that have to be supported. The upper part of the bridge was not touched, because it was possible to carry out the work from under the arches, installing all the equipment there and using raking pali radice. This was a very important advantage. This underpinning underwent an exceptional and onerous test during the disastrous flood that struck Florence in 1966. The Ponte Vecchio with its small arch spans acted as a sort of dam, which had to bear the very strong impact of the current and the detritus it carried. This catastrophic event made it necessary to enlarge the flow capacity through the arch spans of the Ponte Vecchio. This was done recently by lowering the concrete floor under the arches by 50 centimetres.

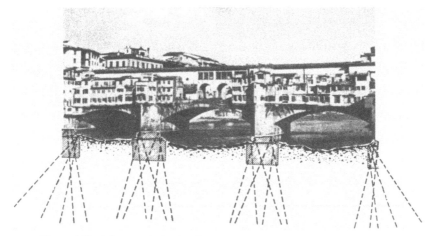

Figure 4.10 Florence, Italy. The Ponte Vecchio. Scheme of the underpinning.

4.5 Reticulated pali radice (RPR)

The support of any man-made construction is the soil. But the natural soil, as it is found in the site selected for the new construction, does not always have sufficient strength for the purpose. To provide this basic requirement, several different systems have been adopted since ancient times. The oldest was based on enlarging the base of the walls of the new construction, in order to reduce the unit stress on the soil to within acceptable limits, consistent with its mechanical characteristics. In very soft soil, overall resistance was increased by means of small wooden piles, driven at close centres at the base of the walls. This system, used since ancient times and with substantial success (consider the case of Venice) was, until recently, the only example of soil reinforcement. The limitations of the available means of driving the piles precluded more extensive use of piling.

The introduction of piling, aimed essentially at supporting indirectly only the vertical loads of the construction, did not solve all the problems which confronted the engineer. There was in some cases a need for a 'direct' reinforcement of the soil. In recent years several very interesting systems of soil reinforcement have been introduced and are, by now, widely used.

On the other hand there are instances in which no state of stress, alteration of the existing equilibrium, variation of the volume, or variation of permeability can be accepted on account of the presence of existing structures because of which even small excavations are not possible. The soil must remain unaltered—where it is and as it is. This is the field of application of reticulated pali radice.

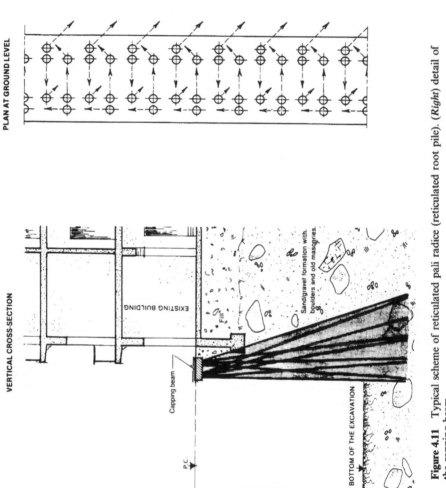

Figure 4.11 Typical scheme of reticulated pali radice (reticulated root pile). (*Right*) detail of the capping beam.

4.5.1 Reticulated pali radice structures

A reticulated pile structure (see Figure 4.11) is a three-dimensional lattice soil/pile structure built directly in the soil according to a pre-determined plan, with many pali radice forming a special resisting network. The piles are all connected at the top by a reinforced concrete capping beam and the piles are the lines of force, while the soil encompassed supplies the weight, rather like a gravity wall. The whole is intended to resist compressive as well as tensile and shear forces.

In the example shown in Figure 4.11, and in the practical applications given in the following pages, the advantages afforded by this original type of structure are seen mainly in the context of the protection of old buildings, when the excavation is to be carried out in close proximity. Other applications also described relate to the prevention of landslips close to existing constructions (buildings, roads, bridges, etc.). On account of the natural rigidity offered by the soil, the settlements and deformations of a reticulated pali radice structure are generally extremely small. The use of these structures is therefore of great advantage for solving problems where the utmost avoidance of soil relaxation is essential.

This structure had its origins in 1950–51, in attempts to find a proper solution to the problem that frequently occurs during the construction of underground structures in towns. The problem indicated in Figure 4.11 is that of constructing a retaining wall in close proximity to existing buildings, without previous excavation and without introducing any relaxation in the buildings' foundation soil. This may have to be done in any type of soil, including those containing boulders, old foundations or other obstructions. The first idea was to strengthen a convenient portion of soil by means of grout injections, in an attempt to consolidate it. However, not all soils can be injected, and even in cases where it is possible, the correct spread of the grout cannot be guaranteed. What is more, a soil, although 'treated' with grout, lacks tensile resistance. It was therefore envisaged that reinforcing elements might be introduced into the soil according to a predetermined plan in order to strengthen it, in a similar way to which the steel bars strengthen the structure in reinforced concrete.

The basic element of the reticulated pali radice structure is therefore the palo radice, whose main characteristic is exceptional adherence to the soil. It is worth noting that the system does not rely on an intergranular improvement of the soil obtained by penetration of the grout used for the construction of the piles, since, as stated above, in impermeable soils, the effect would be non-existent. (If it does exist, so much the better in such cases, as it can be taken into account by reducing the number of piles needed to construct a suitable network.) On the contrary, the system is based on the interaction between pile and soil: in fact, if a pile takes its bearing capacity from the soil, it is logical to presume that the soil, in its turn, is in a position to transfer a stress to a pile.

Provided they are not too far apart, the piles can therefore be relied upon to provide a kind of 'network' effect. The network of piles encompasses and so supports the soil, but at the same time the piles are supported by the soil. The piles' primary purpose is that of introducing reinforcing elements into the soil. The validity of the above assumption, at first intuitive, has been widely confirmed by experience.

In the following sections different applications of reticulated pali radice are described.

4.5.2 The strengthening of the ground in urban areas

As stated above, the problem which first instigated the development of reticulated pali radice structures was that of the protection of existing buildings against alterations induced in the subsoil by natural causes or human action (particularly the construction of underground tunnels in urban areas). This is not a problem of protecting structures already suffering from foundation settlement, but rather one of avoiding damage to currently stable constructions. The protective structure constructed in the soil is intended to separate the foundation soil of the building from the disturbed zone of the excavation. Obviously, it is imperative that the construction of the protective structure must not introduce any alteration in the static equilibrium of the existing buildings. This means that no relaxation or reduction of resistance can be permitted in the soil on which the buildings rest.

This essential requirement suggested the idea of direct reinforcement of the soil, in order to improve its mechanical characteristics in view of the new states of stress set up by the execution of the new works. The basic philosophy of this reinforcement is derived from several general considerations, i.e. a vertical excavation in natural soil presents in general the following difficulties (Figure 4.12):

(i) the loss of soil from the face of the excavation, thereby losing the continuity and compactness possessed, to a greater or lesser extent, by any soil.

(ii) the sliding along possible critical surfaces, a/a, b/b...n/n, so that even if soil loss can be prevented, the failure of an entire section of the soil could occur.

The vertical excavation sketched in Figure 4.12 would be possible, if the soil were appropriately strengthened to form an ideal supporting gravity wall $ABCD$, with the soil itself acting as basic material. The purpose of the pali radice must therefore be twofold, i.e.

(i) the piles must retain the soil and prevent its loss by any 'flow' through the network formed by the piles, causing a reduction of the continuity and unity of the gravimetric mass—it is only under these conditions that the

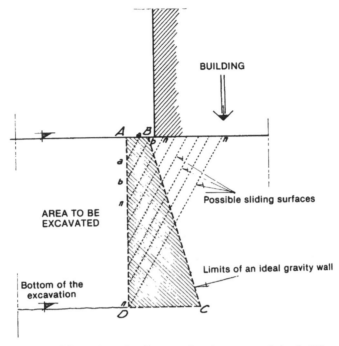

Figure 4.12 Stability scheme for an excavation close to an existing building.

overall behaviour of piles and soil, acting as a uniform whole, can be relied upon;

(ii) the piles must also supply a stitching and nailing of the various layers of the earth, by offering an additional resistance along the possible sliding surfaces. In this way the solid *ABCD* can assume a unitary behaviour as a gravity wall able to support compressive, as well as tensile and shear forces.

4.5.2.1 The protection of buildings close to excavations for underground roads or railways. A typical scheme, used for the Paris underground as well as several other underground roads and railways in urban areas (Milan, Washington and Barcelona underground, Naples Rapid Transit, Salerno subterranean railway, etc.) is shown in Figure 4.13. The reticulated pali radice structure does not impose any restriction on the building activity and, if necessary, can be carried out without disturbing the surface traffic, as shown in Figure 4.13, where the work was carried out by means of a small service tunnel constructed below the sidewalk. The reticulated pali radice wall is intended to supply a cut-off between zone *A* (above the tunnel to be excavated) and zone *B* (below the buildings); consequently the possible relaxation of zone *A* does not affect zone *B*.

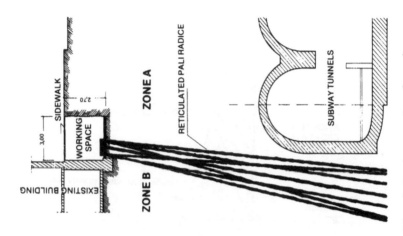

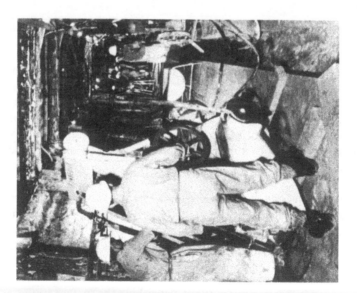

Figure 4.13 Paris Metro—protection of existing buildings during the excavation of tunnels.

4.5.2.2 The protection of buildings directly above construction site of shallow tunnels. In these cases it is impossible to prevent the soil under the building from being affected by the relaxation caused by the tunnel. It is therefore necessary to transfer the load of the structure outside the relaxed area. In Figure 4.14 an example is shown. As a preventive measure, a reinforced concrete structure was constructed in the basement of the building, connecting the building frame to the pali radice piling to make a solid complex. The piling, executed as a multiple reticulated pali radice structure, fulfils the two aims of underpinning the building directly and avoiding the relaxation caused by the tunnel construction, by being extended from zone *A* to zones *B*, outside the protective structure.

Figure 4.14(*a*) Salerno, Italy. Protection of buildings during the construction of the new railway tunnel under the town. Elevation. (1) Reticulated pali radice structures; (2) capping beam in reinforced concrete connecting the reticulated pali radice with the structures above; (3) complementary pali radice for 'stitching' the soil above the tunnel.

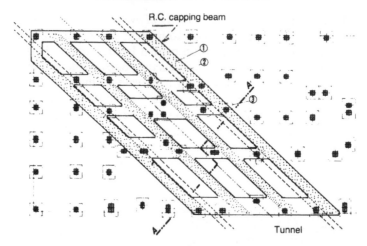

Figure 4.14(*b*) Salerno, Italy. Protection of buildings during construction of the new railway tunnel under the town. Plan. Numbers as in Figure 4.14(*a*).

4.5.3 *The problem of landslides*

Another solution to a problem for which reticulated pali radice structures have proved to be very effective is that of landslides. The direct or indirect protection of a construction (building, road, bridge, etc.) against the risk of landslide should be considered as a problem of underpinning, because it involves the stability of the existing structures.

Any natural surface when inclined has an inherent tendency to slide down. The potential movement is normally prevented by the combined actions of internal friction and cohesion of the soil. When these internal forces decrease in, or lose, their efficiency, the soil slides: this is the case, for instance, in clayey soil, where the increase of moisture content reduces its mechanical resistance. In this case appropriate retaining structures are necessary in order to reach firm strata and keep in place the natural soil above, especially when there are structures resting on the surface. Whenever possible, masonry or concrete retaining walls are constructed; but there are cases where the excavation necessary for placing such retaining walls is not possible without endangering the stability of the whole slope. In such cases a reticulated pali radice retaining wall, buried deeply in the soil to reach the firm strata (Figure 4.15), offers a suitable solution, because

(i) it does not require any excavation;
(ii) it can be constructed in any soil, whatever the permeability and whatever boulders or other obstructions may be present;
(iii) the network of piles does not prevent water circulation in the subsoil, so there is no risk of water accumulating at the back of the wall;
(iv) design can be arranged to counteract any pattern of internal forces.

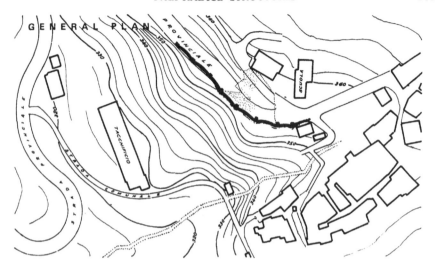

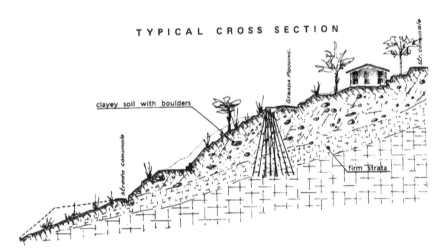

Figure 4.15 Reticulated pali radice structure for stabilizing a landslide in loose soils.

Of course, such a solution is only possible provided some stable formation can be found at a reasonable depth.

4.5.3.1 The problem of landslides in fractured rock or semi-rocky formation. The classic landslide in clayey soil, as discussed in the previous section, is not the only kind of landslide. Recently the construction of highways, dams and other large structures has involved major modifications in the morphology of the existing surface of the natural soil, thus creating a state of instability and risk even in normally stable formations. The increased

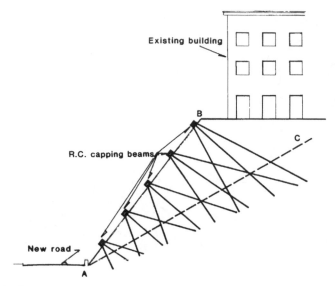

Figure 4.16 Typical example of slope consolidation by means of reticulated pali radice in a semi-rocky formation.

gradient of the slopes—dictated by local exigencies—has created the possibility of landslides. A typical example is sketched in Figure 4.16. The construction of the new road involves a cut *A–B* in the existing slope. Because of the presence of existing buildings, the cut *A–B* must be steeper than the safe inclination *A–C*. Consequently the portion *A–B–C* of the slope needs to be consolidated.

A reticulated pali radice structure, with piles spread in groups connected by a series of small horizontal concrete beams, proves to be a most suitable solution. The three-dimensional pattern of the reinforcement gives an appropriate unity to the unstable rocky foundation, which comes to resemble a form of reinforced masonry resting on the safe inclined surface *A–C*. Additionally, the penetration of the piles for a few metres beyond the critical surface *A–C* acts like 'nails' providing additional resistance to the sliding forces.

The scheme is similar in appearance to the system of prestressed anchoring, but it is substantially different in its philosophy.

(i) First, its purpose is not the connection of the surface with some firm and reliable strata in the subsoil, for which very long anchors are sometimes necessary. The only purpose of the reticulated pali radice structure, from this point of view, is the 'reinforcement' of the wedge *A–B–C*. It is therefore sufficient for the toes of the piles to penetrate only a few metres beyond the critical surface *A–C*.

(ii) A reticulated pali radice structure does not introduce any stress in the soil; this point, as described above, is crucial and fully in accordance with the spirit behind the underpinning of buildings. For the complete preservation of the existing equilibrium is also of paramount importance in the case of a slope in a critical state, while the introduction of forces, as with prestressed anchors, could be dangerous.

(iii) Prestressed anchors depend for their efficiency on the constraints introduced in the soil. Any pile of a reticulated pali radice (RPR) structure is, however, always participating over its full length in the overall resistance of the whole in taking compressive, as well as tensile, stresses, even the case of partial ruptures after a failure. This is because of the high frictional resistance of a palo radice.

(iv) Anchors require a bulky reinforced concrete structure on the surface to unit the very large loads entrusted to them. The RPR structure, based on a denser configuration of piles, requires only small horizontal connections. This is very important for the preservation of the environment that is normally required for slopes in inhabited areas.

Figure 4.17 illustrates example, as described above. The new São Paulo–Santos Highway in Brazil, had to cross a very difficult mountainous region, the Serra do Mar. Cuts in the slopes and the construction of foundation wells (after the traditional system of 'tubuloes') involved in several cases, an alteration of the equilibrium of the soil (weathered gneiss, largely fractured). A three-dimensional reticulated pali radice structure, introduced gradually, without any disturbance in the subsoil, supplied the necessary reinforcement.

4.5.4 The reticulated pali radice structure as a reinforced soil basement

The problem illustrated in Figure 4.18 is another example of the adaptability and versatility of the system of reinforcing the soil with pali radice. Two tunnels on a new line of the Tokyo underground had to be driven near to an existing tall construction, (the 'Panorama Tower'). The problem was the preservation of the stability of the tower.

The problem was solved in a very quick and simple way by a complete underpinning of the existing foundations with a network of piles encaging the soil. The result is a reinforced soil 'underbasement', integrated with the tower to form a single gravity system sunk deep into the soil. Indeed, so deeply rooted was the structure that the construction of the tunnels caused only minor movement. The obvious conclusion is that the integrated soil/pile basement involves a very large volume of the subsoil which cannot, as would have been the case with the existing foundations, be affected by the relaxation resulting from the construction of the tunnels.

TYPICAL CROSS SECTION

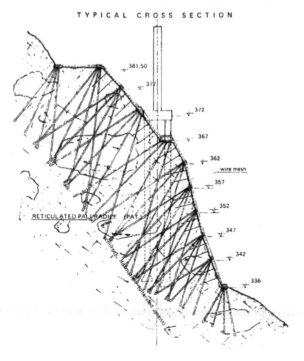

Figure 4.17   São Paulo, Brazil—Santos Highway. Landslide prevention in a semi-rocky formation by means of reticulated pali radice.

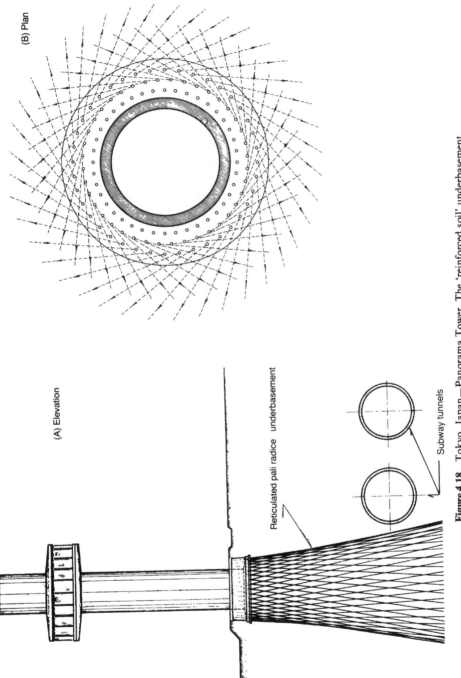

Figure 4.18 Tokyo, Japan—Panorama Tower. The 'reinforced soil' underbasement.

4.5.4.1 Another case of underpinning, to facilitate the construction of a medium depth tunnel beneath an existing building. Figure 4.19 illustrates one of the problems which was solved in connection with the construction of a new highway in the subsoil of the city of Naples. As soon as the excavation of one of the two tunnels approached it, the building showed signs of some very worrying differential settlement. Considering the relative position, in plan and depth, of the building and the tunnels, it was impossible to consider an underpinning of the type indicated in section 4.5.2.2, where the load of the structure is taken in the deep strata 'outside' the tunnel. The solution was the construction of a reticulated pali radice structure through the foundation walls, in order to obtain a reinforced soil sub-foundation 'above' the tunnels, suitable for absorbing the unavoidable soil relaxation introduced by the excavation of the tunnels, without apparent deformations and without disturbing the overall stability of the construction. There was, in fact, general settlement of the area, including the building which did not suffer any damage, as the RPR structure precluded the possibility of differential settlement of the various parts of the structure.

4.5.5 Reticulated pali radice for consolidation of damaged tunnels

Tunnelling in structurally complex formations, such as flysch (a chaotic mix of marl, clay and limestone) is still very difficult. Very frequently, the unbalanced soil thrust causes damage to the lining of the tunnels. Figure 4.20 illustrates the scheme for strengthening the outer shell of a tunnel. The pattern of pali radice forms an arch of reinforced soil reaching far into the surrounding soil. The piles have the function of homogenizing and distributing the stresses exerted by the soil, and of eventually concentrating these stresses within the tunnel lining.

This type of reinforcement should not be confused with 'rock bolting' usually carried out in rock formations to reach sound materials existing just beyond the zone disturbed by the excavation. Here in fact there is no competent material to reach; it is the unstable soil itself which is reinforced and made self-supporting. The design of the reinforced soil arch is based on an evaluation of the average mechanical behaviour of the soil surrounding the lining, which is to become the support of the composite soil–pile structure.

This scheme was successfully adopted in Italy recently to secure a number of highway tunnels which showed signs of imminent failure after their construction. A similar reinforcement was carried out for an old railway tunnel in France, where the invert was progressively suffering uplift.

4.6 The advantages of a reticulated pali radice structure

One of the major advantages of pali radice used for the purpose of underpinning is the simplicity of design, in comparison with the old systems. A

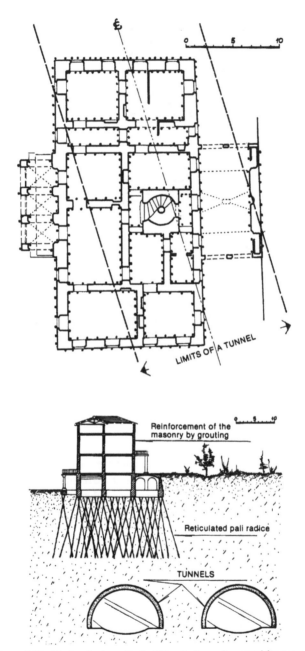

Figure 4.19 Underpinning of a building to be underpassed by a tunnel.

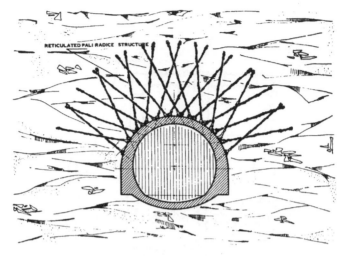

Figure 4.20 Scheme of reinforcement of a tunnel in an unstable clayey formation.

reliable design is not, however, so simple with a reticulated pali radice structure. In this connection, it must first of all be remembered that the purpose of the pile network is twofold:

(i) first, to encompass the soil in order to make a composite soil/pile mass—in general, this depends on the mechanical characteristics of the soil: a good internal friction and, even more, a good cohesion, reduces the number of piles necessary for the purpose;

(ii) second, to supply the necessary reinforcement to the soil/pile structure, in order to resist essentially tensile forces.

The design of a reticulated pali radice structure is therefore based essentially on the 'density' that must be given to the structure itself in order to accomplish these requirements. For this, it is necessary to determine the following:

(a) the number and diameter of the piles necessary to adequately retain the soil in order to enable it to behave as a unit;

(b) the number and diameter of the piles necessary to stitch the different layers between them and to provide the rooting in the soil beneath.

Each of these requirements, though they do not conflict, does envisage a different 'density' for the piles. It is evident that the number of piles will be that required by the prevailing conditions in each case. In general, it may also be said that it is better to use a denser distribution of small-diameter piles, rather than a more thinly scattered one of large-diameter piles.

This general rule may be departed from in specific cases, for example when dealing with reticulated structures required to prevent landslides composed of rocky or semi-rocky fractured materials. This material in its natural condition constitutes a negative element owing to its precarious equilibrium, but if inserted into a reticulated pali radice structure may, on the contrary, become a co-operating element of very great usefulness. In this case, the general encompassing of the soil is always necessary, but even more important is the need to oppose the shear forces, which can, along the critical surfaces, reach very high levels, especially when the landslip is actually moving. In these cases (presence of rocky materials), the possibility of exploiting to best advantage the effect of the shear resistance of the piles suggests the adoption of diameters greater than those normally used, together with an adequate steel reinforcement.

The use of reticulated pali radice structures is by now quite common, and the specific requirements of each case offer the chance of different schemes and sizing. This ability to be adapted to very different situations is a unique element of reticulated pali radice. It must be remembered that the sizing of such a structure—that is, the 'density' to be given to the piles—is primarily a technical problem, in as much as it directly affects the safety factor; but there is also the problem of economics, since a reticulated pali radice structure with too dense a pile layout would be very expensive, and its practical application would therefore become very restricted. The encouraging results of the work already carried out, however, suggest that safety margins adopted may perhaps have been excessive. Since the beginning, therefore, the need has been felt to check the behaviour of the structure in model tests and at full scale, by measurement and instrumentation, to derive suggestions for future applications.

4.6.1 Model tests on reticulated pali radice

The essential purpose of several tests, carried out to full scale and in model reticulated structures, has been to check, at least on a qualitative basis, the interaction between the piles through the soil. The mutual influence between piles located at close spacing (the 'group effect') is well known. In order to check the soil/pile behaviour as a whole, it is necessary first to verify, at different spacings, the mutual influence of parallel piles. It is very difficult to carry out such an investigation on a theoretical basis only, on account of the heterogeneity of natural soils. Some contribution to the understanding of the problem was therefore supplied by tests carried out in artificially made homogeneous soil with reduced-scale model piles.

The scheme of one such test is indicated in Figure 4.21. The soil was coarse sand. The full experiment consisted of four series of tests, with piles respectively 50, 100, 150 and 200 diameters long. Each series was formed of 6 groups of 3 piles each, spaced from 2 to 7 diameters, plus an additional single pile. The results of the tests are summarized in the diagrams showing spacing against efficiency. The term 'efficiency' is indicative of the loadbearing capacity of a pile in a group, compared with the similar loadbearing capacity of

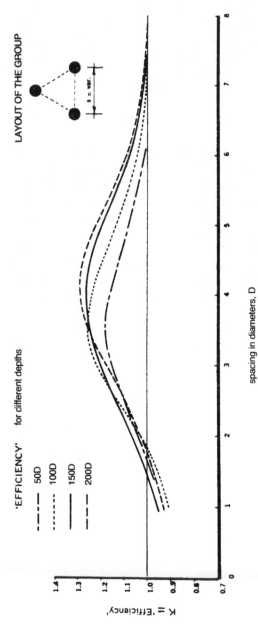

Figure 4.21 'Efficiency' of piles in groups. $K =$ 'efficiency' = (loadbearing capacity of the pile in group)/(loadbearing capacity of the single pile). Soil, sieved coarse sand.

a single pile—practically, the 'group effect'. From the diagrams it can be seen that spacing between 2 and 7 diameters, the loadbearing capacity of the piles in groups is higher than the loadbearing capacity of a single pile. This result is not in agreement with what is currently considered the normal behaviour of piles in groups, where closer positioning of the piles is supposed to reduce their 'efficiency'. But, in the case of the group of pali radice, one must consider the great length of the piles as compared with their diameter. The test demonstrates, therefore, that the load is carried not only by the single piles, but also by the soil they encompass. Obviously the results correspond to the soil and to the particular conditions of the test; nonetheless, full-scale experiments have confirmed that the interaction between parallel piles is effective at spacing well beyond the conventional limit of the three diameters.

This 'group effect' is not the only effect on which a reticulated pali radice structure relies. There is in addition the 'network effect', derived from the three-dimensional pattern in which the piles are arranged, to encompass the soil to make it part of the soil/pile whole. This very important effect

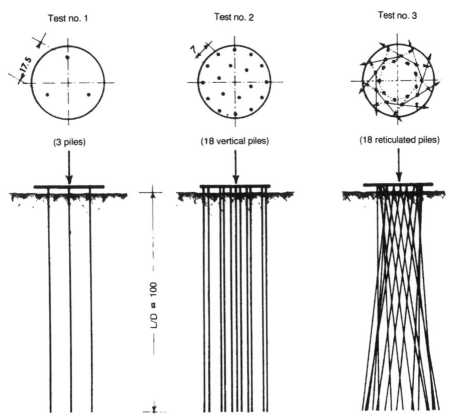

Figure 4.22 'Group effect' and 'net effect'. Layout of the tests. Spacings (s) and length (L) expressed in diameters, D. Soil, sieved coarse sand.

has been checked by field and model tests. One model test is described in Figure 4.22. The soil was coarse sand. Tests were carried out on three groups of piles: one formed by three vertical piles, spaced at 17.5 diameters; the second formed by 18 vertical piles, spaced at 7 diameters; and the third formed by 18 piles again spaced 7 diameters, but arranged in a network, like a basket. The increase in the loadbearing capacity, taking 1 as the value for a pile belonging to group No. 1 (piles widely spaced), was 68% for the piles in group of 18 vertical piles and 122% for the piles in the 18-pile network. Similar results, although not with the same percentages, were found in different soils.

4.6.2 Full-scale tests carried out on reticulated pali radice

The first official test on a full-size structure was carried out in 1957 on behalf of the Milan Underground (Figure 4.23). A full-scale reticulated pali radice retaining wall, 12 m deep, was installed in the Milan subsoil (essentially sand and gravel). The wall consisted of 16 piles (60 mm nominal diameter, each reinforced with a 12 mm steel bar) per running metre of wall. The 60 mm diameter was exceptionally small, as normal minimum diameters are in the range of 100 mm. A reinforced concrete capping beam connected the piles at the top. On the back of the reticulated pali radice structure, a reinforced concrete wall, supporting heavy ballast, was intended to represent the action of a building. The load of the ballast was approximately 40 tonnes per running metre of wall.

Once the installation was completed the front face of the wall was progressively excavated to a maximum depth of 8 m, over a central section of 8 m. The deformation of the reticulated structure was accurately checked during the excavation (first phase), and during a second phase when the load was transferred from the back of the wall directly on to the wall itself.

Despite the size of the applied load, the wall did not show significant deformation, and although this encouraged the Milan Underground Authority to adopt this type of structure for several other difficult jobs (Loggia dei Mercanti, Corso Buenos Aires, Via Boccaccio, Molinello and Turro Viaducts, etc.) it did not contribute significantly to the evaluation of the ultimate resistance of the reticulated structure. The doubt remained, therefore, that the structure had, perhaps, been overdesigned. On other jobs strain gauges were installed on some of the piles to check the presence of stresses, and in these cases as well, although the structure fully satisfied the purposes for which it was adopted, no significant stresses have been detected.

This particular aspect (rigidity of reticulated pali radice structures) will be dealt with in the next section.

4.6.3 The reticulated pali radice structure from the physical point of view

The experience gathered from both the many works carried out and the direct loading tests explains the true physical meaning of the behaviour of a reticulated structure.

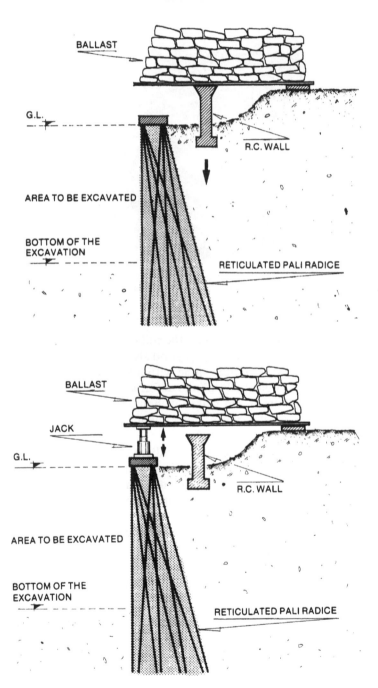

Figure 4.23(*a*)   Scheme of the load test on a reticulated pali radice retaining wall. The scheme: top, first stage; bottom, second stage.

Figure 4.23(*b***)**   Scheme of the load test on a reticulated pali radice retaining wall. The site.

The central factor in this structure is the soil with its mechanical properties (density, internal friction, cohesion, etc.) and also its fragility and very limited deformability. A network of piles alone without the soil has, in practice, no resistance at all and a potentially infinite deformation. If inserted through the soil, the network retains it and makes it 'homogeneous' by connecting the various parts and so allowing a full collaboration by supplying, where required, a resistance—even if a small one—to tension and shear forces. This resistance is available only within very small limits of deformation, since the limiting factor is the failure of the soil. At this stage it may appear that the contribution of the piles is not very significant, since their full resistance is not developed. This is true, but the small contribution they make is nonetheless essential.

One cannot expect significant deformation to precede the final failure of a reticulated structure built in a natural soil. A soil reaching a deformation near the maximum rate allowed by its mechanical characteristics becomes fissured, loses its continuity and fails without being supported by the reticulated structure: similarly in reinforced concrete, if the concrete is crushed the stability of the structure can no longer be supplied by the reinforcement alone. In looking for experimental confirmation of the way a reticulated pali radice structure works, therefore, it would be necessary to use exceptionally sensitive measuring instruments, capable of checking even the minimal movement of the encompassed soil, as well as the smallest deformations and stresses arising in the piles.

4.6.4 *The design of a reticulated pali radice structure*

Considering the above, the design of a reticulated pali radice structure is not an easy task. In the very complex soil–pile interaction, there are many factors whose influence on the final behaviour of the structure cannot be conveniently assessed. There is the soil, with its heterogeneity at different levels, each of which is characterized by different geometrical and geotechnical parameters (cohesion, internal friction, specific gravity, modulus of elasticity, subgrade reaction, etc.). There are the piles with their geometrical patterns and their characteristics of resistance, modulus of elasticity, skin friction etc. Apart from the fact that the nature of the relationship between so many different parameters is practically unknown, there is also the difficulty of collecting correct values for the parameters themselves.

On the other hand a design may be considered reliable only when different designers come to the same conclusions for the same problem. It is well known that this cannot be expected in some geotechnical problems. It is therefore preferable to base the design of a reticulated pali radice structure on some simple assumptions based on the concept of 'reinforced soil', and so similar to those currently used for reinforced concrete. Two typical cases will be considered: the reticulated pali radice gravity retaining wall, and reticulated pali radice for general soil reinforcement.

4.6.4.1 The pile–soil stiffness factor. The pile–soil stiffness factor (m) is a key parameter whose value has a great influence on the results of numerical

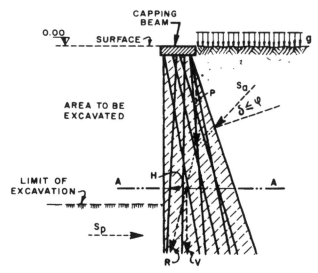

Figure 4.24 Reticulated pali radice gravity retaining wall.

Table 4.3(a)

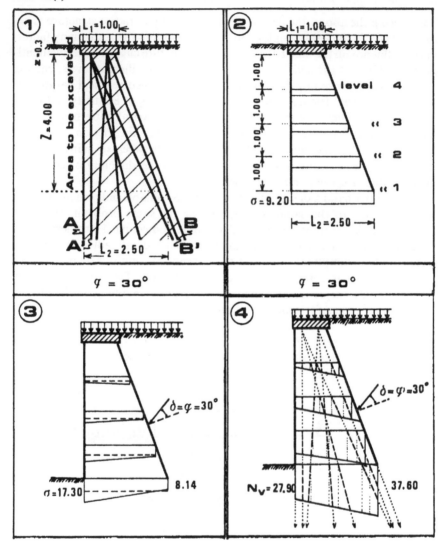

calculations. It can be assumed as:

$$m = \frac{E_p}{E_s}$$

where
　　E_p = modulus of elasticity of the pile
　　E_s = modulus of elasticity of the soil.

Table 4.3(b)

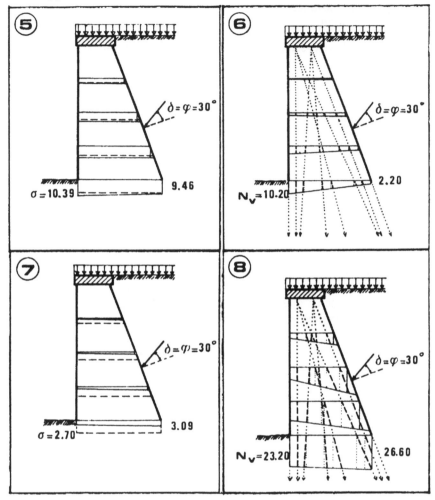

4.6.4.2 *Design of a reticulated pali radice gravity retaining wall.* This scheme is most suitably applied in cases where conventional systems are impractical or impossible, such as in 'cut and cover' constructions for subways where the stability of nearby buildings has to be fully protected, and where, owing to difficult subsoil conditions, other solutions—such as diaphragm walls—cannot be used; or for landslide prevention in loose soil, where it is absolutely impossible to excavate for the construction of conventional retaining walls.

Figure 4.24 illustrates the typical arrangement of a reticulated pali radice gravity retaining wall. The forces acting on the structure are

S_a = active load
S_p = passive resistance
P = vertical load

On a horizontal cross-section $A-A$, the forces can be reduced to

V = vertical force (load)
H = horizontal force (shear).

For the design, different assumptions can be made depending upon the particular soil characteristics in order to collect several ultimate values for the soil and pile stresses. Table 4.3 shows a possible calculation, with reference to a reticulated pali radice retaining wall.

Geotechnical data

angle of internal friction:	$\phi = 30°$
unit weight of the soil:	$\gamma = 18\,kN/m^3$
cohesion:	$c = 0$
surcharge:	$g = 20\,kN/m^2$

Pali radice data

diameter:	$b = 150\,mm$
steel reinforcement: one bar, diameter:	$d = 18\,mm$
number of piles (linear metre of wall):	$n = 10$
Pile–soil stiffness factor:	$m = 40$
Vertical unit stress on the soil:	$\sigma(N/cm^2)$
Load on piles (vertical component):	N_v (kN)

(1) shows the scheme of the reticulated structure; (2) shows the stresses on the soil before the excavation; in (3), the vertical load as well as the forces introduced by the active pressure, after the excavation, are assumed to be supported by the soil only (no piles). The maximum values for the unit stress on the soil, σ, are obtained. In (4), the same forces as for (3) are assumed to be supported by the piles only. The maximum values for the load on the piles are obtained. In (5) and (6) the forces introduced by the excavation are assumed to be supported by the combined action of soil and piles according to the pile–soil stiffness factor, m. Some intermediate values are obtained, both for the soil and the piles. In (7) and (8) it is assumed that, ultimately, all forces, including the dead weight before the excavation, are supported by the combined soil–pile action. Some reference values are obtained.

It is up to the geotechnical engineer to decide the most acceptable values, considering that a very efficient soil–pile interaction can be more easily obtained in stiff soil. As for the resistance to sliding of the structure due to the load produced by the active earth pressure, it is resisted, normally, by the soil only, through the frictional resistance and the cohesion, if any. Therefore the condition of stability is:

$$P \tan \phi + Ac \geq S_0$$

where

P = total vertical load acting on the horizontal section;

A = area of the section

S_0 = horizontal component of the active load.

Only in the case of a semi-rocky fractured formation is the shear resistance entrusted mainly to the pile reinforcement.

4.6.4.3 Design of a reticulated pali radice structure for general soil reinforcement in stiff or semi-rocky formations. This is the scheme generally adopted for soil reinforcement to stabilize a landslide in stiff or semi-rocky formations (Figure 4.25). In addition to the normal geotechnical parameters of the soil (at least as an average), and the geometrical, structural, and mechanical data of the elementary palo radice, it is also necessary to collect data on the possible external forces acting on the proposed reticulated structure as a whole. Once the above elements are collected, the design of the reticulated structure may proceed. The problems to be solved are generally related to the behaviour of the composite soil–pile structure as subjected to compressive (and tensile), as well as shear, stresses. In all cases, the problem is to calculate the contribution made by the piles to the resistance of the natural soil.

The first step is to determine the 'critical' sliding surface, i.e. the surface which has the minimum factor of safety. Several methods are in use, but analysis by an experienced geotechnical engineer is absolutely necessary. The purpose of the network of pali radice is twofold: first, to encompass the soil portion above the 'critical' surface, and second, to 'nail' this surface, so supplying additional shear forces. The design of this 'nailing' can be carried

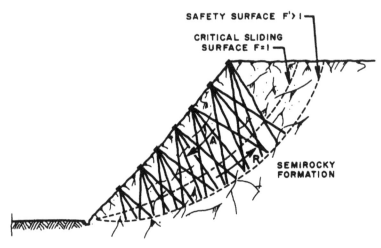

Figure 4.25 Reticulated pali radice structure for landslide prevention in stiff or semi-rocky formation.

out as follows: once the factor of safety F_x has been assessed, the design of the reticulated structure (number and diameter of piles, reinforcement etc.) is obtained by the following formula:

$$F_x = (R + R')/A \geq 1$$

where

R = total resistive forces on the 'critical' surface

A = motive forces on the same surface

R' = the additional shear resistance provided by the piles.

This additional resistance depends on either the shear resistance of the piles or the resistance of the soil to be 'cut' by the piles, whichever is the smaller. In semi-rocky or very stiff soils the shear resistance of the piles is more important than in loose soils, where the resistance of the soil is the limiting value.

4.6.4.4 Reticulated pali radice structures in rocky formations. This case occurs more frequently than might be expected, because the instability of such natural formations involves serious problems of safety. The main principle in the use of reticulated pali radice structures in these situations is very simple: the object is to form a strengthening three-dimensional network of sufficient thickness to produce a kind of cyclopean masonry wall, in which the connecting element is supplied by the piles (Figure 4.26). No surface capping structures are necessary, except for some wire mesh, reinforced by cables to guard against the fall of small pieces of rock.

4.6.5 The design of a reticulated pali radice structure—conclusions

In the absence of a more reliable theoretical approach, calculations roughly based on the concept of 'reinforced soil', and so similar to those currently used for reinforced concrete, may be used. A better guide, however, can now be obtained by an extensive examination of works carried out. Reflection upon the facts, and the collection and selection of results to obtain some data of general validity is, at the moment, the best way to determine future designs. In addition, the advantages and opportunities offered by the modern systems of calculation must be fully encouraged in order

(a) to explore the influence on the final result of the several different parameters of the problem;
(b) to investigate the possible forces, as well as the possible deformation of the structures, according to the different hypotheses.

A computer could be of great help in a survey of this kind, since the problem could be posed as a logical matter of analysis and synthesis, supported, if possible, by more and more appropriate instrumentation.

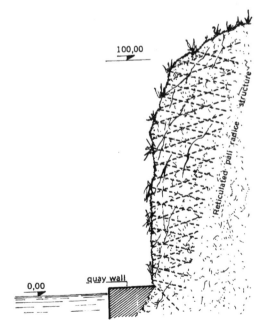

Figure 4.26 Capri, Italy. Reticulated pali radice for consolidation of a rock face. Lower part of diagram shows a typical cross-section.

It must be concluded that for this, as for many other geotechnical problems, it is not possible to have at one's disposal an exhaustive system of calculation ready to be applied with complete safety. The design of a reticulated pali radice structure, therefore, requires the attention of a qualified expert. It is to be hoped that the interest raised in this particular structure will lead to some improvement in future design approaches.

4.7 The problem of stability of a tower

The reinforcement of the foundations of a tower is a unique problem in the field of the consolidation of ancient monuments. The stability of a tower is always extremely delicate and the risk of a disastrous failure arising from its uncertain stability during the reinforcement works must always be borne in mind, besides which, the erection of preventive shoring is not very simple, for two fundamental reasons. Firstly, with masonry shoring (buttresses) there is the risk of increasing the unit stresses on the soil adjacent to the tower, resulting in a possible consequent failure. Secondly, a metal shoring, on the other hand, is subject to variable stresses induced by variations in temperature. In the following sections some further case histories are illustrated, in addition to the reinforcement of the foundations of the Panorama Tower in Tokyo, already described in section 4.5.4.

4.7.1 The Burano Bell Tower, Venice, Italy (XVI century)

The consolidation of the foundations of this famous tower (Figure 4.27), a well-known feature in the landscape of the Venetian lagoon, was a very difficult problem which involved the following (Figure 4.28).

(a) Preliminary preparatory work. Owing to the very poor quality of the soil and, even more, the extreme urgency, it was not possible to study and carry out adequate preparatory work. The tower was in danger of imminent collapse. The problem of an emergency safeguard was solved by means of two opposing groups of a few provisional pali radice, prestressed respectively in compression and in tension, in such a way as to create a stabilizing moment. For this purpose the piles were drilled, as usual, through the existing foundation masonry, but were not connected to it. Instead, they could slide in their holes under the action of two series of jacks, one for introducing compressive forces on the leaning side, the other for introducing tensile forces on the opposite side. The jacks acted against a sturdy steel frame constructed in the base of the tower.

Figure 4.27 The leaning Burano Bell Tower, Venice, Italy.

(b) Underpinning. Once a reasonable stability of the tower had been obtained the underpinning, by means of pali radice, was carried out. The aforementioned provisional piles were not taken into account. The pali radice underpinning, on account of the close spacing of the piles, may be considered also as a reinforced soil basement, gravimetrically associated with the tower. The system of pali radice can be easily compared to the root structure of a tree. It is connected to the structure in such a way as to react both to tensile and compressive stresses. This is extremely important. The centre of gravity of this gravimetric complex is very near to the level of the soil, the advantages of which are obvious.

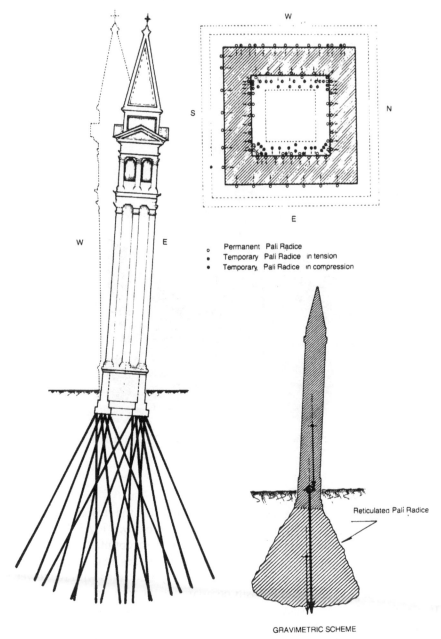

Permanent Pali Radice
Temporary Pali Radice in tension
Temporary Pali Radice in compression

Reticulated Pali Radice

GRAVIMETRIC SCHEME

Figure 4.28   Burano Bell Tower—scheme of the underpinning by means of pali radice.

Figure 4.29   Mosul, Iraq. The Al-Hadba leaning minaret.

4.7.2 The strengthening of the Al-Hadba Minaret, Mosul, Iraq (XII century)

Dating back to the 12th century, this minaret (Figure 4.29) has always, for both historical and religious reasons, been a famous monument beloved of the Iraqi people, who have often taken it as one of the very symbols of the cultural heritage of their nation. Like the majority of other minarets of the area, exposed as they are to the northwest winds, the Al-Hadba has shown over the years a marked tendency to lean towards the south-east. This phenomenon had resulted in an ever-increasing and decidedly dangerous swaying in recent years, introducing extra compressive stresses on the leaning side and tensile stresses on the opposite side. The precious architectural brickwork ornamentations were crumbling and becoming detached from the main structure. The original masonry could not resist such phenomena, made as it was of poor-quality, chalky-based bricks and mortars, due to which it had undergone a slow and progressive deterioration.

As may be seen in Figures 4.30 and 4.31, a complex and systematic intervention was made, both above and below ground level, yet this was completed in only nine months of effective work.

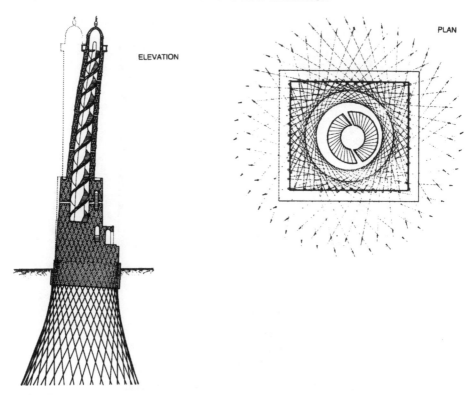

ELEVATION

PLAN

Figure 4.30 The Al-Hadba leaning minaret. Scheme of the underpinning by means of pali radice.

4.7.3 The Tower of Pisa

The fact that the problem of the consolidation of the Leaning Tower of Pisa is still waiting for a solution, after more than eight centuries, is just as amazing as the problem itself. Thousands of proposals collected during this time cram the archives of the authority in charge of the preservation of this world-renowned monument. Nevertheless, up to now no decision has been taken, although, in recent years, an international invitation resulted in the submission of a substantial number of proposals from all over the world, from Mexico to Japan, to the special International Committee.

It must be recalled that, in that past, some minor restoration works carried out in the immediate subsoil of the tower led to an abrupt and unexpected increase in the inclination of the monument. Perhaps this alarmed the authority in charge and suggested the need for the maximum of prudence. Yet, apart from the unique historic and artistic renown of the Tower of Pisa, its

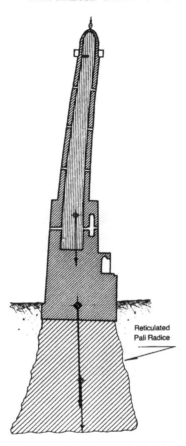

Figure 4.31 The Al-Hadba leaning minaret. Gravimetric scheme.

problems of stability are probably no more complex than the two analogous problems illustrated above (Burano Bell Tower and Mosul Minaret).

Figure 4.32 shows the three structures side by side to the same scale. The reader will note, in the ratio between height H and minimum base dimension B, the greater slenderness and greater instability of the Burano Tower and the Mosul Minaret, when compared with the Tower of Pisa. But what is more important is that the movement of the first two towers at the commencement of the strengthening contracts were extremely active, while the movement of the Tower of Pisa is almost negligible, and can be detected only by very sophisticated instruments.

Figure 4.33 illustrates the design proposal submitted by Fondedile S.p.A. It was designed by a group formed by the author (F. Lizzi), and by E. Giangreco,

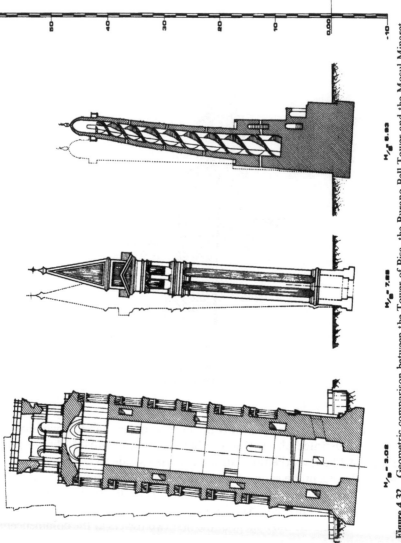

Figure 4.32   Geometric comparison between the Tower of Pisa, the Burano Bell Tower and the Mosul Minaret.

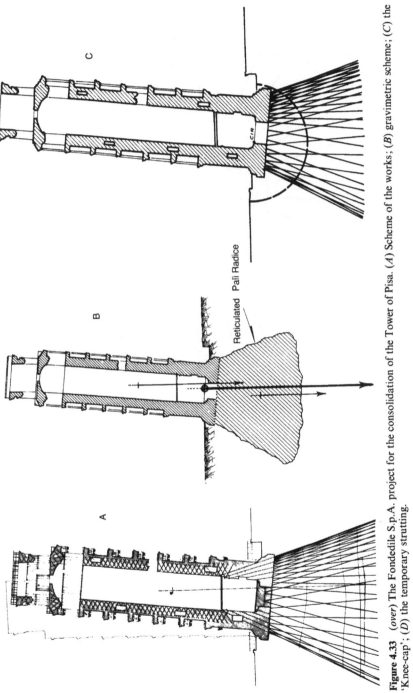

Figure 4.33 (over) The Fondedile S.p.A. project for the consolidation of the Tower of Pisa. (A) Scheme of the works; (B) gravimetric scheme; (C) the 'Knee-cap'; (D) the temporary strutting.

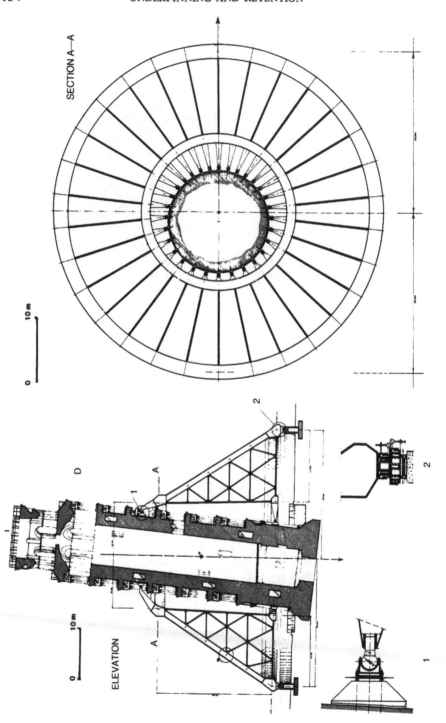

SECTION A–A

10 m

0

ELEVATION

10 m

0

J. Kerisel, and R. Morandi, with the collaboration of G. Carnevale and F. Grasso. The most significant aspect of the proposal is the fact that it is based on techniques and technologies (such as pali radice) which have been widely tested over thirty years of experience. Here the concept emphasized above must be recalled, in which a dense pali radice underpinning, in addition to being a true foundation, may also be considered as a soil–pile 'reinforced soil' basement, closely bonded to, and associated with, the upper structure, with the resultant overall centre of gravity very close to the ground surface.

There is, however, another way of looking at it, namely that proposed by Professor J. Kerisel: the 'root pile' underpinning makes a positive crossing of the critical sliding surface (a kind of 'knee-cap'), which, on the basis of the available data, may be supposed to be under the Tower.

4.7.3.1 Temporary strutting for the Tower of Pisa. An important part of the project concerns the temporary strutting. In fact, in addition to the project for consolidation, the International Invitation asked for a preliminary structure to be installed before the execution of any work, to obviate the risk of any

Figure 4.34   Layout of the temporary struts.

movement, however minimal, of the tower during the phase of underpinning and strengthening.

The structure proposed by Fondedile is a reticular steel 'bell', which surrounds the tower (Figures 4.33, 4.34). The reaction between the building and the propping structure is induced automatically, in case of emergency, by the reaction between the tower and two solid metal rings, belonging to the 'bell' and situated at different levels. Under normal conditions, the two rings have simple contact with the tower, by means of bearings, which slide vertically along the surface of the tower; there is therefore, no transmission of stresses. In the case of emergency (a sudden increase in inclination), the tower reacts against the two rings, developing a stabilizing couple (only a couple), that prevents the tower from leaning further. Besides the advantages of not increasing the load on the soil, therefore the support offers immediate relief to the instability of the monument.

As well as the above-mentioned vertical bearings, the 'bell' is supplied with sliding horizontal bearings set on the outer base ring. They ensure that any variation of temperature gives rise only to an expansion (or contraction) of the

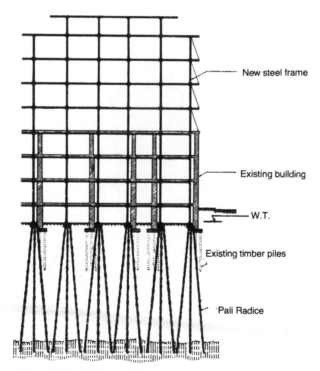

New steel frame

Existing building

W.T.

Existing timber piles

Pali Radice

Figure 4.35 Adding additional floors to an existing building.

steel structure as a whole; so, as the vertical as well as the horizontal bearings allow for such deformation, no action occurs against the tower. This insensitivity to temperature would exist also if the 'bell' were actually supporting the tower, because the reaction between the building and the 'bell' would not hamper the free expansion (or contraction) of both the vertical and horizontal bearings.

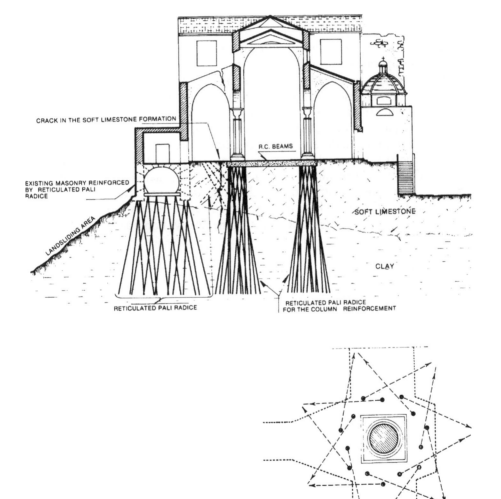

Figure 4.36 The Cathedral, Agrigento, Italy. (*Top*) consolidation works (*Bottom*) scheme, in plan, of the column underpinning.

4.8 Special case histories

Some further case histories, described below, may illustrate better the adaptability of the pali radice systems.

4.8.1 Adding new storeys to an existing building in Naples

As shown in Figure 4.35, the existing building was originally built with only three storeys, besides the basement floor. It was necessary to increase the height of the building by five additional storeys; but the existing offices had to be kept in operation throughout the work.

   The problem was solved by constructing foundation beams supported by pali radice in the basement floor, adjacent to, but neatly separated from, the existing foundations. Steel columns were then erected, starting from these beams and passing through the existing floors up to the roof. This work was quickly carried out in the evenings after office hours. The working site for the construction of the five additional storeys in steel frames was then installed on the roof of the old building, without hampering the activities of the offices below. Once the construction of the five additional storeys had been completed, it was possible to transfer the offices to the new floors and then demolish the old structures, incorporating the relative floors as part of the new structure.

4.8.2 The consolidation of an old monument in Sicily, in a landslide zone

The ancient Cathedral of Agrigento is founded on a soft limestone layer of varying thickness, resting on a clay formation. The limestone was fractured in all directions and a large crack had appeared in recent years along the entire length of the floor of the left aisle, and also outside the church. The clay, being subject to slow plastic deformation in the area of contact with limestone, and exposed to a marked landslide movement outside the monument, tended to slide down, leaving the upper limestone almost suspended in a cantilever state: hence the large crack and the differential settlement of the various parts of the church.

   Owing to the extreme delicacy of the situation, the work was divided into two contracts separated by some years (Figures 4.36 and 4.37).

4.8.2.1 Prevention of landslide. The first phase of the work had the aim of stopping the landslide movement. The novelty of this first part of the work lies in the fact that the retaining structure (reticulated pali radice) was executed from a previously strengthened underground chamber. These reinforced structures became, therefore, the capping beam of the system. The underlying

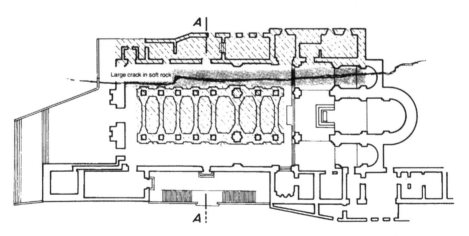

Figure 4.37 The Cathedral, Agrigento, Consolidation works. (*Top*) the entrance; (*bottom*) zone of intervention, in plan.

reticulated pali radice structure behaves like a retaining wall, the weight of which is supplied by the soil and the overlying masonry, while the resisting frame is formed by the pali radice, founded in the deep clay well beyond the unstable area. This reticulated pali radice wall also serves as underpinning for the external columns of the left aisle. Also included in this first phase of the

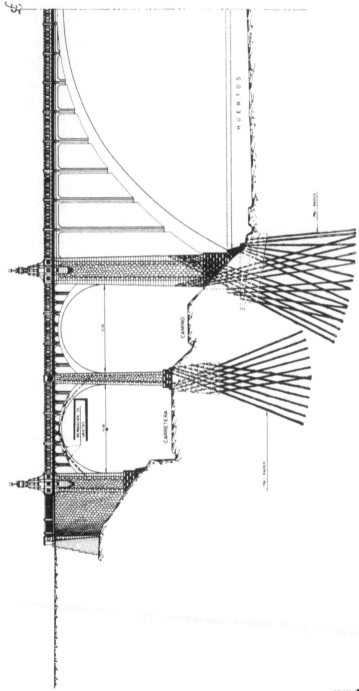

Figure 4.38

work were the stitching and strengthening of the large fracture in the church floor, by means of cementitious pressure grouting.

4.8.2.2 Underpinning of the central nave. After some years during which the Cathedral was kept under observation, the effectiveness of the retaining wall was confirmed, so the second phase of the work, involving the underpinning of the central nave, was carried out. Figures 4.36 and 4.37 show the details of this underpinning. The characteristic layout of the piles in *doppia rigata* (double helical rows) must be mentioned. This arrangement is normally adopted for pali radice in the underpinning of towers and slender structures having limited cross-sections (see the case of Panorama Tower in Tokyo, section 4.5.4.) To improve the connection between the reinforcing structures the piles were linked below the floor level, by horizontal reinforced concrete beams. It was an essential feature of these works that they allowed complete protection of the environment, as well as of the church and the slope on which it is built.

4.8.3 The underpinning of an old bridge in Spain

In the late sixties, the Teruel bridge (Figure 4.38), subjected over a long period to foundation settlement and to some slow landslide phenomena, suddenly became an emergency that demanded prompt remedial measures. The underpinning by means of pali radice has proved very suitable for the gradual improvement of the stability of the abutments. At the same time, the dense

Figure 4.38 Teruel, Spain. The old bridge.

Figure 4.39   Milan, Italy. Consolidation of an old building close to the subway cutting to be excavated.

complex of piles required for the underpinning formed a reticulated structure well rooted in the soil. The scheme adopted, therefore, satisfies at the same time the exigencies of both underpinning and landslide prevention.

4.8.4 Another case of underpinning and RPR retaining wall— the Milan Underground

The construction of a cutting so close to an existing, and very important, historical building raised concern about the stability of the edifice. It was

Figure 4.40 Stabilization and partial straightening of a building in South Italy subject to differential settlement during its construction.

feared that the digging of the cutting, although carried out with much care, might produce relaxation of the subsoil, with dangerous consequences for the overall stability of the building. The edifice was already in a precarious condition, and therefore not in a position to withstand even a minimal alteration of equilibrium. This condition necessitated a precautionary reinforcing of the foundations, and their protection against possible relaxation caused by the excavation works. To these ends, the foundation masonry was locked into a reticulated pali radice structure, executed through the masonry and extended to well beneath the lower excavation level. In this way, the reticulated structure performs the twin functions of underpinning the edifice (to give it a stable new foundation), and acting as a gravity retaining wall.

4.8.5 Stabilization and partial straightening of leaning edifices by means of pali radice

There have been several cases in which the use of pali radice in two phases proved to be successful, first straightening and then stabilizing buildings leaning in consequence of differential settlement. This is the case illustrated in Figures 4.40–4.42 where the foundation structure of a huge building in South Italy is illustrated. A minor building is located adjacent to the main construction.

The soil, of very poor strength, was not suitable for the building, which started to settle during its construction, and when approaching the fifth floor reached the rate of about 20 mm per week on the north side, and a smaller amount on the south side. Consequently there was a marked progressive inclination towards the north end. Fortunately the upper structure was

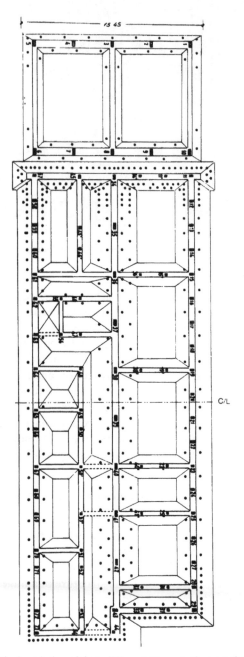

Figure 4.41   Plan of the foundation of the building of Figure 4.40, strengthened with pali radice.

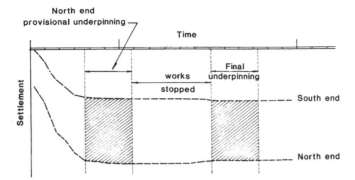

Figure 4.42   The leaning building of Figures 4.40–4.41. The settlement.

Figure 4.43   Santos, Brazil. Leaning buildings.

formed by a sturdy reinforced concrete frame, which supported the serious deformation of the foundation level without any damage.

The contract, which had to be carried out quickly, was divided into two phases.

(i) In the first provisional phase, the north end only was underpinned with some 30 pali radice. This first phase was sufficient to stop this end moving. In the meantime the south end continued its slow settlement, with consequent partial straightening of the building.

(ii) After a period of six months the underpinning was resumed at both ends

Figure 4.44 York, UK. Bootham Bar—the underpinning.

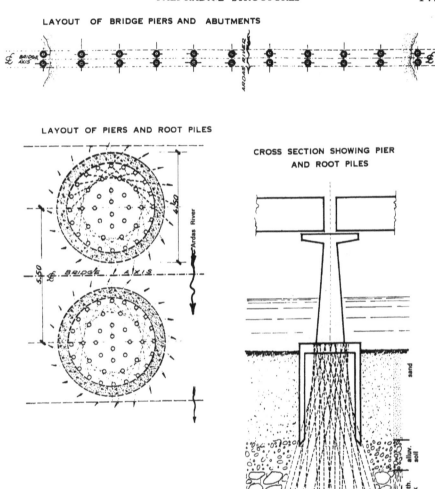

LAYOUT OF BRIDGE PIERS AND ABUTMENTS

LAYOUT OF PIERS AND ROOT PILES

CROSS SECTION SHOWING PIER
AND ROOT PILES

Figure 4.45 Bridge on the River Ardas, Greece. Caisson foundations in difficult soil strengthened with pali radice.

(including additional piles at the north end to replace the 30 first-phase piles, which were no longer considered reliable) and carried out towards the centre of the building.

A similar job was carried out in Santos (Brazil) where, on account of the poor-quality subsoil some tall buildings were subject, after construction, to differential settlement (Figure 4.43).

4.8.6 A case of foundation settlement owing to traffic — Bootham Bar, York, U.K.

This is an ancient monument of great historical importance. It is one of the four old gateways to the city of York, which was founded by the Romans. It was built by the Normans in the late Norman style in the twelfth century, on the foundations of the old Roman gate. The subsoil consists of silty sands. The work was necessary owing to serious foundation settlement, caused by vibrations induced by traffic and variations in the water table. The underpinning was carried out by 114 mm nominal diameter pali radice having a working load which, owing to the delicacy of the situation, was limited to only 10 tonnes, even though the test loads gave results above 50 tonnes without failure.

4.8.7 Integration of an inadequate foundation in difficult soil — the bridge on the River Ardas (Greece)

This is a long bridge of 10 spans. In the original project each pier was intended to be supported by a pair of reinforced concrete caissons (4.50 m diameter). But, on account of a very difficult layer of weathered rock and boulders, the caissons could not be sunk to the depth required. The foundation was completed by a network of pali radice, drilled through the caissons.

4.9 The development of micropiles

When pali radice were first applied, early in the fifties, the regulations then in force for piling (in particular the German DIN), prescribed a diameter of not less than 400 mm for cast-in-place piles. This created problems when the palo radice was introduced as it was difficult to believe that a pile having a diameter of only 100 mm could bear loads of up to 30–40 tonnes or more. Fortunately a load test for values in the order of fifty tonnes involved only very modest expense, but supplied a convincing, definitive proof of the effectiveness of the system.

Pali radice made a name for itself in Italy and abroad, although in the USA there were some difficulties, as cast-in-place piles are not very popular in that country. After about twenty years (in the 70s), with the expiry of the first patents of pali radice, other similar piles were proposed. They were generically called 'micropiles'. Some of them in practice conform to the characteristics of pali radice, while others have been characterized as steel piles, cemented into the soil.

4.10 Steel pipe micropiles, cemented into the soil

It has been said before that the high bearing capacity of pali radice, compared with their small diameter, is their most favourable characteristic. But in a palo radice, as in any concrete pile, the bearing capacity has its limit in the crushing resistance of the cross-section of the shaft.

The tendency therefore arose to increase the steel reinforcement to obtain more resistant sections. Finally micropiling, consisting substantially of very heavy metal pipes (or structural beams), which could bear considerably higher loads were proposed. They are securely cemented into the subsoil.

One such micropile ('Tubfix') is illustrated in Figure 4.46, where the phases of construction are depicted.

(1) The drilling is carried out with any system suitable for the particular soil, including flushing of the spoil with bentonite mud.
(2) Once the fixed depth is reached, a steel pipe (reinforcement) is introduced. At its lower end, the pipe is provided with a series of holes protected by rubber sleeves (valves).
(3) Through these valves a cement mix is injected to fill the annular space between the hole and the pipe.

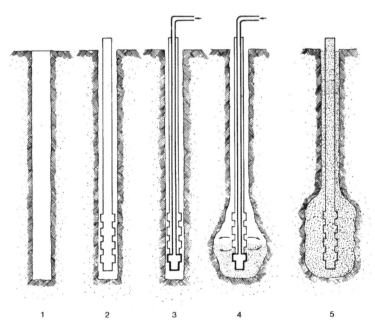

Figure 4.46 Micropile 'Tubfix'—stages of execution.

(4) Before the complete hardening of the mix, the lower valves are pressure grouted in order to form a kind of bulb. These special valves allow for some regrouting, if necessary.

The high strength of the steel offers an increased loadbearing capacity. On the other hand some other points must be considered.

(a) The adhesion between the pipe and the soil, in metal piles, is obtained by means of fluid cement grout, injected through the pipe. Obviously, the thin cement grout crust so obtained cannot ensure a very effective adhesion between steel and grout or in turn, between the grout and the soil. It is necessary to inject in the lowest part of the pipe, where it is possible to obtain higher pressures. The bearing capacity is obtained in practice by a kind of point bearing supplied by the deeper layers, instead of by skin friction as is the case with pali radice. Figure 4.47 shows two typical curves of load transfer from the pile to soil, for a palo radice and a metal micropile respectively. For this reason, the settlement of a metal micropile is in the order of some centimetres, and not millimetres as in the case with a palo radice. Figure 4.48 shows the load–settlement graphs for a palo radice and a steel micropile. Both were constructed on the same site for a building that needed underpinning and a partial reconstruction with additional new structures. The same figure shows the characteristics of the subsoil. The diagrams demonstrate that metal micropiles, even if (owing to

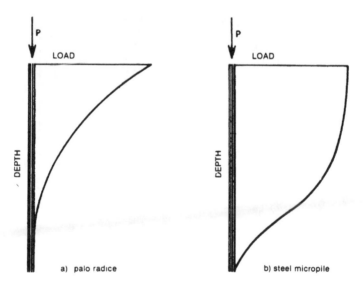

Figure 4.47   Typical load transfer curves from the pile to the soil, for a palo radice and for a steel micropile.

the resistance of the steel) they are potentially fit for supporting greater loads, cannot be used for underpinning, because of their high settlement. In fact, for the underpinning of the building, pali radice were used, while the steel micropiles could be used as foundations for the new structures.

(b) The risk of corrosion in steel pipe micropiles cannot be ignored.
(c) Buckling, which is not normally taken into consideration in the case of very slender piles, cannot be ignored in the case of steel micropiles which are heavily loaded and not adequately supported by the soil in the upper layers. For this reason, such piles are only installed vertically.

4.11 The preloading on micropiles

As shown above, the palo radice, if used carefully within well-defined, and not excessive, load limits, is able, with its prompt response, to give definitive support to foundations experiencing problems. But, as has also been shown, the bearing capacity of a palo radice can be very high, much higher than required: it cannot be utilized, because the corresponding settlements would not be acceptable for a building in distress. This has sometimes suggested the idea of preloading the micropiles, in order to make better use of their bearing capacity without exposing the building foundation to the corresponding greater settlement. This idea must be rejected (except, of course, for temporary use in particular cases) for the following reasons.

(a) The pre-loading introduces stresses into both soil and building which can disturb the state of balance, the consequences of which can be very severe.
(b) The factor of safety of such an underpinning has to rely only on the bearing capacity of the piles, without the essential contribution of the soil (see section 4.3.4).
(c) The connection between the piles and the structure has to be postponed until the time when the piling is complete, at least in part, and in a state of preloading. Instead of a progressive improvement, as in the case with normal pali radice underpinning, a long period of distress would occur, terminating only with the completion of the structures above ground level.

4.12 Micropiles for new foundations in difficult soils

Conventional foundation piles, especially large-diameter piles cannot be drilled easily in any soil. Sometimes the presence of boulders, old foundations or other obstructions in the ground makes the drilling of such piles difficult. It would usually be more convenient to substitute for the large-diameter

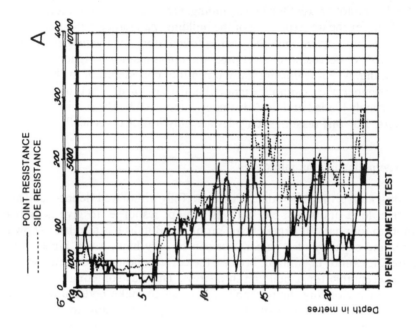

b) PENETROMETER TEST

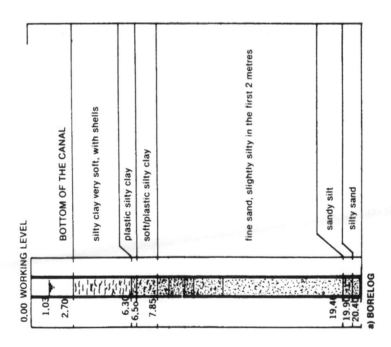

a) BORELOG

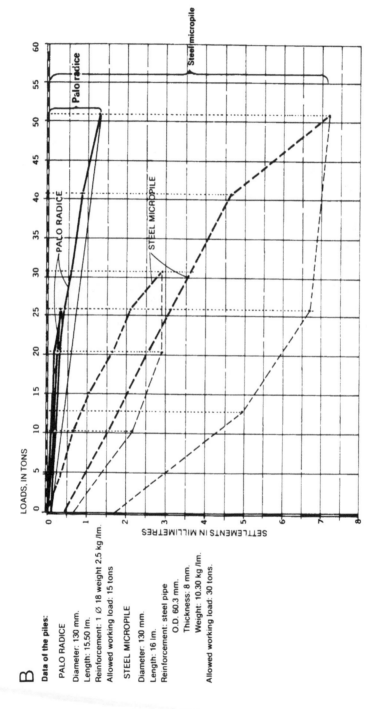

Figure 4.48 Load tests carried out on the same site on a palo radice and on a steel micropile. (*A*) the site investigation; (*B*) load/settlement charts.

Figure 4.49   Reticulated root pile foundation, in difficult soil, for a tall viaduct in Naples, Italy.

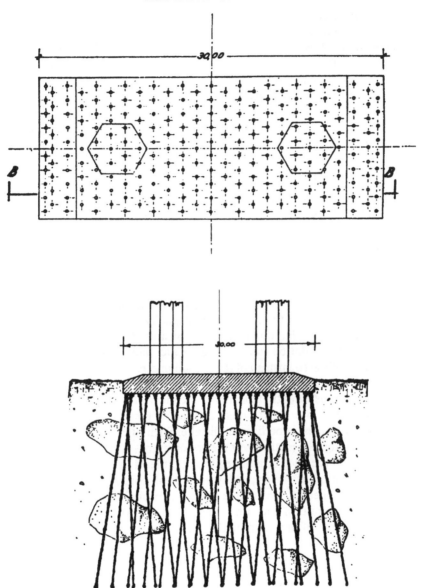

Figure 4.50 Scheme of the foundation for the viaduct of Figure 4.49.

piles an adequate number of minipiles, which on account of their more flexible technology, can be more easily drilled through any obstruction.

Figures 4.49 and 4.50 illustrate the foundation of a very high modern high-way viaduct. The ground comprised old quarry fill, including large boulders, so the large diameter piles adopted along the other sections of the same highway could not be constructed in this particular section. Instead, a minipile foundation proved to be very efficient and easily executed. The boulders were penetrated without any difficulty, so becoming part of the foundation. As in the case of landslides, once reinforced with root piles, the boulders became a positive element of support.

It is worth noting the pattern adopted for the piling: it is a reticulated pali radice scheme, in which the loadbearing capacity of the single piles is increased by a 'group effect'—owing to the density of the piles (one per square metre approximately)—and by a 'network effect', owing to their mutual inclinations.

The foundation can therefore be considered from two aspects:

(a) As a normal pile foundation relying on the loadbearing capacity of the single pile. This was the basis of the design, derived from load tests.

(b) As a 'reinforced soil foundation', where all the structure (piles, soil, boulders, etc.) is contributing to the loadbearing capacity of the foundation. This is a result of the reinforcement introduced by the piles, which connect and, as it were, 'integrate' the different ground conditions.

The loadbearing capacity of the 'reinforced soil' according to this second viewpoint is greater than the loadbearing capacity of the total piles acting individually as in (a). This was clearly demonstrated in practice, when the settlement of the foundation (once of the viaduct had been completed) gave results of practically zero, instead of the few millimetres that could be expected from the load tests carried out on single piles.

This result demonstrates that the modern tendency to increase the diameter of foundation piles, in order for them to support greater loads in the range of thousands of tonnes, is perhaps a mistake. Such piles, spaced widely, are generally very long—sometimes 50 metres or more—because they have to rely essentially on the deep strata, neglecting (at least in terms of safety) the possible contribution of the upper layers of weak soil. A denser pattern of smaller-diameter piles, of reasonable length, is probably more suitable for involving the contribution and enhancing of the soil, and enhancing its role as principal support of any man-made construction.

Bibliography

1. Lizzi, F. (1964). Root pattern piles underpinning. In *Symposium on Bearing Capacity of Piles*, Roorkee, India.
2. Zanetto, L. (1975). Aspetti tecnologici della costruzione dei micropali. In *6° Ciclo Annuale di Conferenze Dedicato ai Problemi di Meccanica dei Terreni*, Politecnico di Torino.
3. Lizzi, F. (1976). Pieu de fondation à 'cellule de précharge'. In *Revue 'Construction'*, June 1976, Paris.
4. Lizzi, F. (1977). Practical engineering in structurally complex formations. In *International Symposium on the 'Geotechnics of Structurally Complex Formations'*, Capri.
5. Lizzi, F. (1978). 'Reticulated root piles' to correct landslides. In *ASCE Convention*, October 1978, Chicago.
6. Lizzi, F. (1981). *The Static Restoration of Monuments*. S.A.G.E.P. Publisher, Genoa.

5 The Pynford underpinning method

J.F.S PRYKE

5.1 Introduction

As the years pass, an ever widening range of civil engineering techniques are invented and refined, and then gradually become routine. This applies as much to underpinning, strengthening and shoring techniques as to other branches of civil engineering. Thus often more than one effective solution to a problem is available (references 1, 2). The choice is then determined by cost. It is important to compare the total cost to a client of different methods. Thus, lintel beams may be inserted in brick or stone masonry walls by a number of methods (reference 3) and the amount and cost of temporary shoring and of repairing incidental damage to the building will vary depending upon the method chosen, the particular location of the beam in the building, the quality of the finishes, and whether or not it is necessary to replace them. The cost at any particular time will also depend upon the workload and location of particular specialists in relation to the general contractor using older and more widely understood and practised methods. The total cost to the client is the sum of the cost of the prime construction, that is the lintel with associated shoring, and of the secondary work, that is the necessary incidental repairs and reinstatement.

The scene is continually changing as methods evolve and new ones are introduced. For example, Pynford reduced the manhours required to construct identical beam and pier underpinning schemes by a factor of three between 1947 and 1970. The process continues and the cost to the client has continued to fall in real terms since then, due partly to continuing innovation and partly to the growth in number of specialist contractors.

It should also be noted that this chapter is primarily about the use of the Pynford method to construct cast-in-situ reinforced concrete Pynford beams in existing loadbearing masonry walls, of either stone or brick. A wide range of problems, arising from the failure of foundations of masonry structures or from the need to alter them, and that were solved safely and effectively using Pynford beams, are described to explain the method and encourage innovative applications by the reader. Thus the chapter discusses underpinning at high level as well as at foundation level. If a reinforced concrete beam or slab in a masonry wall or pier can assist in the solution of a problem, then it can be constructed safely using the method. In the years from 1940, no wall has been encountered by Pynford in which beams could not be safely constructed using the techniques which will be described in this chapter.

The reader is advised to read Chapter 1, which emphasizes the advantage of acquiring a basic knowledge of civil, structural and geotechnical engineering theory and practice before seeking to develop the specialist skills required to understand older structures and to design and supervise underpinning. No book can be a substitute for experience, but perhaps careful reading can help the newcomer to underpinning problems to approach them with more confidence, and the experienced engineer to extend and develop his or her thinking.

Understanding the structure to be underpinned is crucial. The loadbearing structures of all Ancient building, to use Thorburn's categories, and of many Recent and Modern ones, are constructed with the traditional 'craft' materials, that is stone, brick and mortar to carry compressive loads, and timber and small amounts of iron used as ties and connectors, where tensile and bending strength are required. Materials were laboriously hand-crafted before steam power spread in the mid-19th century, and were thus expensive, and often re-used. Buildings were also continuously changed and adapted as one generation of owners succeeded another. In some cases too many changes, each small and safe in itself, accumulate to bring the old buildings near collapse and make them difficult to alter. Where walls or piers are seriously overstressed, the building must be strengthened temporarily before it can be altered. The engineer must be especially wary when evidence of alteration exists. Chapter 1 emphasizes the causes of stress concentration, especially in stone masonry buildings.

The engineer must understand how a building and the supporting soil 'work' to carry the imposed loads. He should have a three-dimensional stress model in his mind and consider at all stages of the work the effects upon this stress model. Amongst others, Heyman has done valuable work in understanding the engineering of masonry, and his book *The Masonry Arch* (1982) (reference 4) is strongly recommended to engineers unfamiliar with his work.

Force (stress × area) causes deformation (strain), and the engineer should also seek to imagine the deformations under load of his stress model and particularly the effect of widely differing material moduli. For example, a flexible beam under a stiff brick wall will not carry a uniform load. Work on composite action such as that by Wood (references 5, 6) highlights the enormous differences between theory and reality that can occur. There is a major discontinuity in both material strength and potential deformation when building loads pass into the supporting subsoil. Foundations are designed to reduce the high stresses in the building materials to the lower stresses that the ground can safely support, to reduce ground movement and consequent deformation of the structure to acceptable levels, and to reduce cracks, if any, to acceptable sizes.

It is now practicable and economical to introduce forces into a building of the same order of magnitude as the building weight using hydraulic jacks. Thus, major modification of the stress patterns in a building can be made, or the anticipated deformation of the new permanent underpinning structures or temporary shoring can be cancelled out to maintain existing stress patterns. A detailed discussion of shoring and jacking falls outside the scope of this chapter and book.

Underpinning cannot be sensibly conceived or safely achieved without considering in detail how it is to be executed. A problem is created where the contractual arrangements tend to separate design from construction and where the system can impose a cost penalty on designers who seek to control the method of working after a contract has been let. At the commencement of a project the designer must clearly define the objective of the work and select in broad principle the preferred and most economical solution, perhaps after negotiation with a number of specialists. Then, once the team is chosen, the detailing should take account of the techniques to be used and of the sequence of working.

Finally, it should be remembered at all times that buildings have always been assembled by people, often working in cold and uncomfortable circumstances, subject in all ages to the same need to survive and earn a living that drives us today, tempted to skimp and take 'short cuts' just as we are, getting the same thrill of achievement which rewards us when the task is completed. Those who have little interest in and respect for what our forebears achieved should not seek to underpin and adapt their work to our needs.

5.2 Design

The sizes of completed permanent underpinning work will be checked for stress and deformation using the design aids and Codes of Practice, familiar to structural and geotechnical engineers, and normal safety factors. However, the designer should also check stresses and, where there is doubt, deformations, at all stages of the work. It should be assumed that at any stage the work may stop, and no sequence of operations should commence that it is not safe to delay for an extended period at some intermediate stage. Work can be delayed by many factors outside the control of the designer, such as weather, accidents to personnel, materials failing to meet specifications or legal injunctions applied for by neighbours. At all stages it should be possible to make the work safe with a minimum of further work. For example, a small hole cut for a needle can quickly be filled in or wedged up if the hole reveals an unexpectedly weak wall that cannot easily be supported without further work, and the necessary materials for such temporary packing should be available on site before the hole is cut.

The safety factors used when designing components should take account of their importance and contribution to overall safety. There is ample precedent for such an approach. For example, the ratio of working loads to failure loads of chains and cables varies depending upon the use, with the smallest percentage, and thus highest safety factors, being set when people are being carried. Thus if the removal of a member could lead to a major collapse of part of a building, then that member must either be made extra strong or, if possible, the sequence of operations should be revised to avoid the need for the critical member or operation. Design calculations are based on sometimes highly idealized models of structural behaviour. Codes of Practice which incorporate safety factors give satisfactory results for new work, but the stress distribution assumptions should be used with caution when designing underpinning and shoring. There are, of course, limits to the load that individual members may carry that can easily be identified. A hard spot in a stone masonry column will not be called upon to carry more than the total load on that column. A part of a foundation will never need to carry more than the whole building weight. It is helpful in this context to think of all the structure beneath a particular level as the 'foundation' for all that is above that level. Such exaggerations would overestimate the maximum, but they help to create a realistic attitude to temporary works.

In very rigid structures, such as brick buildings, constructed with strong bricks and mortar, small deflections can dramatically alter the load pattern and increase stresses by factors of two or more. When the 2500-tonne Sjommanshjemmet in Norway was moved and raised in 1983 (reference 7) it was carried on 42 jacks linked in ten operating groups. Increasing the pressure from the average working pressure of 365 bars to about 540 bars raised the building about 6 mm above the jack group being operated without visible sign of distress or cracking in the building and, in some cases, halved the pressure in adjoining jack groups up to 5 m distant. The only way to be certain of the load in support members in a building is to measure it. If the forces in permanent or temporary support members are critical they should be applied by jacks and the pressure measured. This is particularly important if there is a critical maximum force that must not be exceeded.

Conventional calculations for the settlement of supports under continuous beams reveal dramatic changes in stress when small deformations occur. It is usually assumed that these changes do not occur, without detriment to the performance of most structures. However, large load changes do occur and must be designed for, particularly in the intermediate stages of projects for underpinning rigid buildings. Underpinning is potentially dangerous and labour intensive. Material costs are usually small in relation to total costs. The cost of increasing the strength of critical members is small in relation to overall costs. It dramatically reduces risk, and may even contribute to overall economy by increasing confidence on site, and thus the speed at which the work can be completed.

Some simple rules and checks are also helpful when devising schemes for underpinning and alterations. The designer should first consider what would have been designed to fulfil the need if the building were being newly constructed. For example, it might be decided that a house on a shallow strip foundation that is subsiding due to a patch of poor ground beneath part of the foundation would have remained stable if the original foundation had been a frame of ground beams supported by mass concrete piers of different depths to suit the changes in level of the stable stratum. One should treat with considerable suspicion a remedial underpinning proposal very different from this in concept and anticipated performance. Widening the shallow foundation would be much less likely to succeed than deepening it. Increasing the foundation depth beneath those walls where the poor ground was deeper than the shallow foundations could be achieved by traditional continuous underpinning, beam and pier underpinning, beam and pile underpinning, or micropiling without a ground beam (reference 8) and the choice will be governed by overall cost. Another general rule is that the most fragile or doubtful areas of the structure are best tackled early in the project. A building usually has much strength added by 'non-loadbearing' partitions and other 'non-structural' members such as door and window frames or infill panels. This is well known but the magnitude of the effect is often underestimated. Strain gauges fixed in the steel frame of the Air Ministry building in Whitehall in the early 1960s revealed stress levels between 10% and 20% of those designed for (reference 9), a finding which triggered off much of the work on composite action since that date. By starting with the weakest areas one maintains these secondary load paths and increases safety factors, although designers should not take account of the benefit. A corollary of this is that demolishing 'non-loadbearing' elements in an old building can shed load back on to deteriorated or modified main loadbearing walls and make them unsafe. Emergency work was required on a major central London site recently for just this reason.

Any alteration or underpinning project will require parts of the structure to be undermined or cut away. Most brick or masonry walls are designed either to avoid excessive slenderness, or to keep out the weather, or both, and the compressive strength at foundation level may exceed the weight of a building by a factor of ten or more. Thus holes can be cut through the foundations to remove a significant portion of the wall without any risk of crushing failure. This is not always the case, particularly where brick- or ashlar-faced walls have weak cores. Thorburn suggests grouting as one method of strengthening a wall before underpinning. Another solution is to clamp the wall with plates or timbers on either side bolted together, thus changing the structural condition from unconfined to triaxial compression, and raising the strength. A third solution is to bond in ties using epoxy resin or cement grout; epoxies set faster and bond better to clean stone or brick surfaces, but they are more expensive. A fourth solution for highly stressed walls or piers is to relieve the load before cutting away, usually by propping floors or beams carrying load on to the pier or wall. It is most important to appreciate that if the whole load imposed upon a pier is to be relieved, then all the beams carrying load to the pier above working level must be independently propped, unless very careful calculations are made of the upward support that the beam can give acting as a cantilever tailed down by the floor dead load only. Such calculations may indicate that some of the beams at highest level may not need to be propped.

A final and most important rule is that great care must be taken when excavating around or beside existing foundations to ensure that removal of lateral restraint does not weaken the soil beneath the foundations and initiate subsidence or collapse.

5.3 The Pynford underpinning method

To construct a beam in a loadbearing wall a horizontal slot must be formed in the wall into which the beam is either placed in one piece or assembled in segments which are then joined together. 'Beam slots' cannot be simply cut without supporting the wall above as it will collapse. If the beam is to be placed in one piece, which would be the case if a steel 'I' section were chosen, the wall above is usually supported by needles, that is short beams projecting through holes cut in the wall above the beam slot and propped on either side (Figure 5.1). Thus the whole wall load is carried on narrow sections of packing or wedging on top of the needles which would typically be placed at 1 m to 1.5 m centres, raising the compressive stresses locally by a factor of between 5 and 20 if the load is equally divided by careful wedging, and by even more if it is not. The needles may bend under the load and the props on either side of the walls will rest on independent pads or a bearer beam, in turn supported on unloaded ground or even on paving or floors propped from level to level. Small wonder that some cracking and 'settling down' is expected when such work is carried out. Supplying, fixing and removing needles will be fairly expensive even if the beam is near ground level, and the wall will be damaged by holes cut above beam level which ultimately have to be rebuilt. However, the beam will be a cheap length of rolled steel joist (RSJ).

Alternative solutions, such as forming the beam in sections subsequently fixed together, are more expensive but avoid the needling and repair work above beam level (reference 3). Holes can be cut at intervals and segments placed in position along the line of the beam. The segments will need to be accurately bedded down and then packed or pinned up above. The segments may be of precast concrete, placed with small gaps between them which are packed prior to threading tendons and post-tensioning the blocks to form a prestressed concrete beam, or they may be steel sections assembled and pinned in place in sequence and bolted together. Segmental beams have not proved very effective commercially, perhaps mainly because there is not a large market for a standard product, and thus each beam needs to be individually designed and produced.

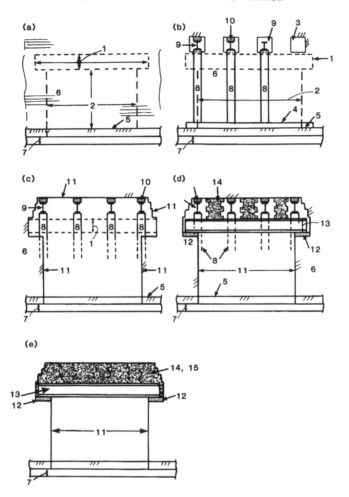

Figure 5.1 Needling. 1, dotted outlines of beam to be inserted: 2, proposed new opening; 3, holes cut through wall *above* new supporting beam; 4, temporary bearer plate on existing solid floor; 5, existing solid floor; 6, wall in which opening is required; 7, foundation; 8, temporary props supported by 4; 9, temporary needle beams; 10, temporary wedging on needles; 11, wall cut away to provide space for the new beam and to form the required opening; 12, padstones to support new beam; 13, new beam; 14, new brickwork built in and pinned up between needles; 15 needles and props removed and brickwork and pinning up completed.

The Pynford beam system is similar to the segmental beam in many respects. The wall loads are supported along the line of the beam slot by stools which are assembled from a small number of standard parts which can be manufactured in large numbers. The original Pynford stool (Figure 5.2) was assembled from a pair of standard precast reinforced concrete U-blocks and two or more precast reinforced concrete plates, bedded together with mortar joints. By adding plates in the middle, the depth of the stool can be varied

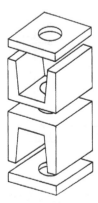

Figure 5.2 The Pynford Stool—inventor Fordham Pryke.

to match the required depths of beam. A variety of different stool designs are now used, the main difference being that the U-blocks and centre plate(s) are replaced by some form of steel strut. These stools are built into holes about 0.3 m wide cut in the wall, generally at 0.9 m centre to centre. The gap over the stool is then pinned up with a strong, low water:cement ratio mortar, 1:1 portland cement and sharp sand, well rammed in. The stool replaces the support given by the brickwork to be cut away. When the mortar and pinning has hardened, the brickwork between the stools is cut away leaving the wall load to be transferred across the beam slot by the stools (Figure 5.3).

The beam is completed by threading reinforcement through and alongside the stools, attaching shear reinforcement, which is generally assembled from pairs of U-shaped bars, placing formwork on either side of the wall, concreting to within about 50 mm of the underside of the wall above, and finally pinning

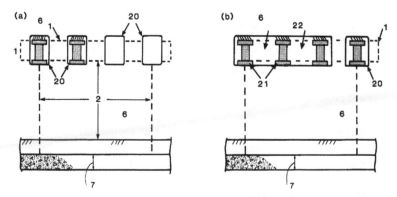

Figure 5.3 Key as for Figure 5.1. 20, holes cut *along line* of required beam for Pynford stools; 21, Pynford stools built into holes and pinned up; 22, brickwork between stools cut away to form *beam slot*.

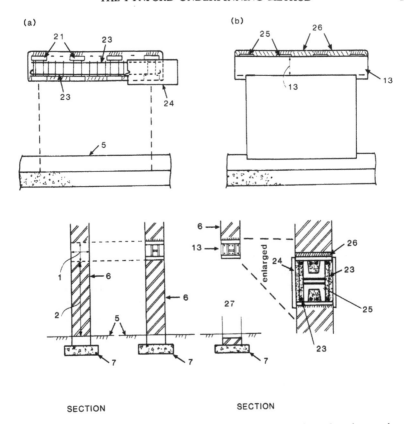

Figure 5.4 Key as for Figures 5.1, 5.3. 23, reinforcement threaded through and around stools and shear reinforcement placed; 24, formwork fixed on either side of the wall; 25, concrete poured around stools and reinforcement to form reinforced concrete beam; 26, gap above beam and between stools pinned up; 27, wall cut away to form opening after concrete has hardened.

up between the stools (Figure 5.4). The stools are cast into the beam, which thus directly carries the load on them without a second load transfer sequence of operations when the supporting wall is cut away beneath the beam to form the opening which the beam will have been designed to bridge.

Stresses are raised by a factor of about 4 immediately above and below the stools but will be virtually unaltered more than 0.6 m above and below the beam slots. Settlement caused by transferring the building load to the stools is negligible. It is primarily that due to direct compression of the stool assembly, as the method prevents the brickwork from loosening before the stools are pinned up in position.

The method is successful because it is easily adapted to the endless variety of beam depths and widths required in practice. Nibs and piers can be underpinned by placing extra stools beneath them and widening the beam and intersections and corners can easily be arranged thus allowing frameworks of

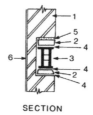

SECTION

Figure 5.5 The 'beam slot' does not occupy the full width of the wall. 1, wall; 2, concrete plates; 3, steel Pynford stool; 4, mortar joints; 5, pinning up; 6, surface to remain undisturbed.

beams to be constructed at no more cost per metre run than simple straight lintels. In thicker walls top and bottom plates can be enlarged, and in very thick walls two or more stools can be placed side by side to support the full width of the wall. If the wall is weak or crumbling, stools can be placed at closer centres and/or the top and bottom plates can be made wider to reduce the stress raising factor to two or less. In very thick walls the first stools can be pinned in position before the full width of the wall is cut away. The work can be carried out from one side of a wall, thus reducing disturbance to a minimum, and in masonry of a reasonable quality the beams may be narrower than the wall, leaving a protective skin of brickwork or stonework undisturbed in order to reduce disturbance and reinstatement costs (Figure 5.5). When used for underpinning foundations, Pynford beams are usually positioned just below dpc level (Figure 5.6). If the foundations are very

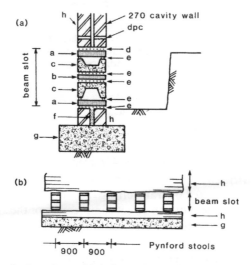

Figure 5.6 (*a*) Pynford stool assembly in beam slot. (*b*) Pynford stools at 900 c/c. *a*, reinforced concrete top and bottom plates; *b*, optional centre plates to increase beam depth; *c*, reinforced concrete U-blocks; *d*, pinning up; *e*, mortar beds; *f*, cavity filled; *g*, concrete foundation; *h*, brickwork above and below beam.

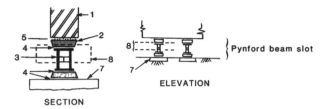

SECTION   ELEVATION

Figure 5.7 Beam slot extends below the existing foundation level. Key as for Figure 5.5; 7, temporary foundation pad; 8, original foundation.

shallow there will be insufficient sound construction beneath the stools to distribute the load on them, and precast or cast-in-situ reinforced concrete pads are then used bearing directly on the ground, which will have already been consolidated by the building load (Figure 5.7).

The fact that the system maintains the existing load paths and stress patterns until the new structure is complete is a very important feature and enables the engineer to guarantee that the parts being supported will be virtually undisturbed by the work.

A number of illustrated case studies selected from Pynford archives follow to illustrate the use of the method in practice. They cover a range of problems chosen to illustrate the versatility of the method.

5.4 Case studies

Case Study no. 1: Underpinning at Imperial College, London

A good example of the use of stools to construct reinforced concrete beams as part of a major underpinning project was the work at Imperial College, London. Imperial College was constructed, with heavy loadbearing brick walls, in the 19th century as a major centre for teaching and research in science and engineering. The College was completely reconstructed in the late 1950s and the first stage was to excavate a very deep basement at the rear of the existing buildings (Figure 5.8). Away from the building the basement excavation was sheeted with contiguous bored pile walls to minimize vibration and noise, but the very heavy surcharge loads generated by the building persuaded the engineers to underpin the main building walls directly. The walls comprised a series of heavily loaded piers each approximately 1 m thick × 2 m wide. The staircase tower wall was over 35 m high, with brick work at the base 560 mm thick. The most heavily loaded pier carried approximately 4000 kN and the staircase wall loading was approximately 250 kN/m. The depth of the basement excavation was required to be 9 m below existing basement floor level and the walls were underpinned to a depth of 11 m using hand excavation methods. The ground was 6 m of river terrace gravel

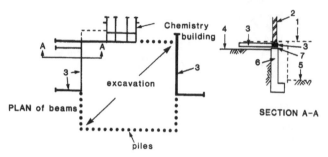

Figure 5.8 Imperial College. 1, basement level; 2, wall to be underpinned; 3, Pynford beam; 4, existing foundation; 5, bottom of excavation; 6, underpinning; 7, pinning up beneath beam.

overlying London Clay, and the water table stood at approximately 2 m below basement floor level.

The underpinning wall required to retain the site beneath the building was formed in pits approximately $3 m^2$ on plan, and the Pynford method was used to construct a strong horizontal beam just below existing basement level to distribute the wall loads uniformly over approximately 3 m lengths of foundation on either side of the primary underpinning piers, which were positioned at about 8 m centres, that is beneath the main loadbearing piers of the lecture room block wall, and at a similar spacing along the staircase wall and the low rise link block between these sections. Return beams at corners adjoining the work and inside the building beneath crosswalls carried the loads of these walls partly away from and partly on the top of the underpinning, thus both relieving heavy surcharge loads close to, and tying back the top of, the underpinning.

The beam was reinforced with a roughly equal area of steel rods top and bottom as the moments that are generated when the main piers are constructed reverse when the panels between them are excavated to complete the underpinning. This is typical of beam and pier underpinning schemes, that is, the beams act as temporary shoring to carry the building loads over the underpinning excavations using the existing foundations on either side as temporary support, and the moments then reverse when the excavations are concreted and pinned up and the ground between either continues to sink or is excavated, as in this case, to complete the underpinning.

When the work was complete the beam was continuously supported in this case and thus the function of the beam is entirely one of temporary support during the excavations. A simple approach to the design was made. A point of contraflexure was assumed at mid-point of the support zone, thus allowing the moments to be easily calculated for the initial condition. Simple continuous beam theory was used to check the bending moments after the main piers had been constructed. The excavation sequence was controlled so that no two piers closer than 16 m apart were excavated at one time.

Figure 5.9

Figure 5.10

The main underpinning piers were reinforced and designed to be strutted from a central dumpling except in the corners where diagonal strutting was used between the penultimate piers.

It is clearly vital that the excavation technique used should eliminate the loss of ground during excavation, and that the excavations should be closely and securely timbered.

Figure 5.9 shows an early stage in the beam construction. A series of holes have been cut beneath the staircase walls, and Pynford stools built in. They bear upon the existing foundation and are assembled with precast concrete plates above and below a steel strut comprising six legs welded into top and bottom steel plates with bracing at intervals. Clearly the brickwork remaining between the stools could not carry the full wall load. The sequence in which the stools are fixed is arranged so that there is always ample support for the wall. In this case every fourth stool was fixed in the first phase of the sequence followed by intermediate stools. At this stage every other stool would have been inserted and pinned up to carry a full load. The stool fixing sequence is then repeated following a similar sequence. Thus at no time is more than approximately 12% of the wall support cut away. Figure 5.10 shows reinforcement fixed between the stools and the formwork partially constructed. Note that the beam is slightly wider than the wall and that the shear reinforcement is closely spaced on either side of the stool, but that there are no stirrups in the stool position. The stool construction provides adequate shear reinforcement as the vertical legs are welded into top and bottom steel plates to form continuous shear loops.

It will be noted that there is ample space for pouring concrete between the stools and beside them, and that the arrangement of the top bars is designed to allow for adequate concrete compaction using poker vibrators. The stools project above the upper surface of the concrete. To complete the beam construction the gaps between the stools will be pinned up. Figure 5.11 shows the base of a main pier bearing upon a main concrete foundation 1.5 m deep. The pier was underpinned with a group of 8 stools, and the stool strength and fixing pattern was designed to ensure that the pier was adequately supported at all times (Figure 5.12). Figure 5.13 shows the stooling-up sequence well advanced. Four stools have been fixed and the workman is cutting away for the fifth and sixth stools. It will be evident that the beam depth must be chosen to allow adequate working space. The nominal beam depth was 675 mm and the overall beam slot was 825 mm, that is the equivalent of 11 courses of brickwork. Note the two-pin stool beside the shovel, which has yet to be pinned up to carry the light infill panel. These figures emphasize the dangers inherent in underpinning operations and the importance of properly designing the stools and the stooling sequence. Figure 5.14 shows the stooling sequence complete and the working area tidied in preparation for fixing the reinforcement. The simplicity of the method and the absence of needles and shoring enables tidy working and

Figure 5.11

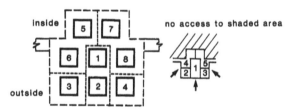

Figure 5.12 (Left) Plan of brick pier showing fixing sequence for Pynford stools. (Right) Plan of pier showing five stages in which beam slot was cut out. Arrows show access positions.

assists the supervisor to ensure a high standard of workmanship and safety. In Figure 5.15 the reinforcement has been partially fixed. Reinforcing bars and shear reinforcement have been placed and tied at the rear, and stirrups are being placed in position for the front reinforcement. Clearly, the main bars must be threaded into position by sliding in and working backwards and forwards in some cases. Thus, careful detailing is essential. It is important to ensure that bars are not bent into position as there is a danger that the workman may use the steel stools for this purpose thus applying lateral leverage and possibly dislodging them. Figure 5.16 shows the completed beam with the gaps between the stools being pinned up. The importance of

Figure 5.13

Figure 5.14

Figure 5.15

Figure 5.16

Figure 5.17

good detailing if the work on site is to proceed with minimum difficulty will be readily appreciated.

The underpinning excavations beneath the Pynford beam were close-boarded and well-strutted. As the depth increased pumps were installed. Ample room was provided for the men working in the hole to stand clear of the bucket being used to remove the spoil. The excavation was carried out through waterlogged sandy gravel in which normal planking and strutting methods are impossible. Pynford developed an effective excavation shield for this project which enabled excavation without loss of fines, checked by pumping through settling tanks. Gaps are left at intervals in the timbering and filled with a filter medium. This produced a large drainage surface as the piers were excavated, and reduced the velocity of groundwater flow into the excavations and thus the tendency to move fines.

The completed underpinning is shown in Figure 5.17. The heavily strutted contiguous bored pile wall is to the right. The Pynford beam is seen in proportion to the building being carried.

Approximately 90% of foundation problems affecting domestic dwellings occur on the shrinkable clay beds of the south-east of England (reference 9). One of the driest spells recorded in England ended in September 1976. Figure 5.18 compares the number of problems reported to Pynford during 1976 with the average of the numbers reported in preceding years. It will

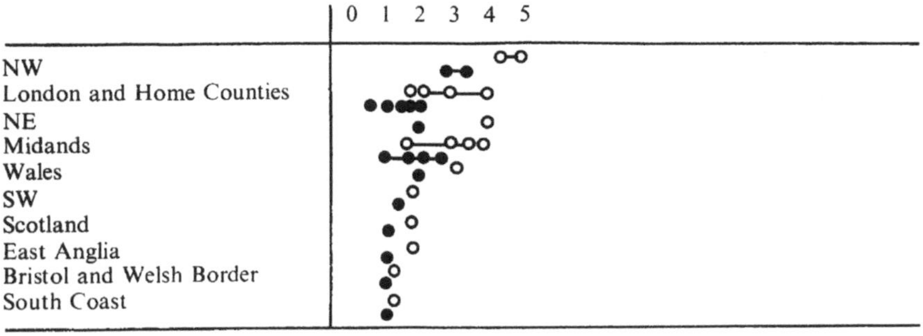

0 1 2 3 4 5
NW
London and Home Counties
NE
Midands
Wales
SW
Scotland
East Anglia
Bristol and Welsh Border
South Coast

Figure 5.18 The extent of the subsidence problem—1976 compared with previous years. ○ 1966–76; ● 1972–76.

be seen that 1976 was between two and four times worse than the average of the previous ten years in south-east England. Stated differently, this means that the exceptional year of 1976 was no more than the accumulation of about three 'ordinary' years. 1976 subsidence damage claims are reported to have been in excess of £100m. The average annual cost of unacceptable differential movement of house foundations is very large. Most domestic

beam
subsidence
FL
GL
Pynford beam
pin up
TYPICAL SECTION

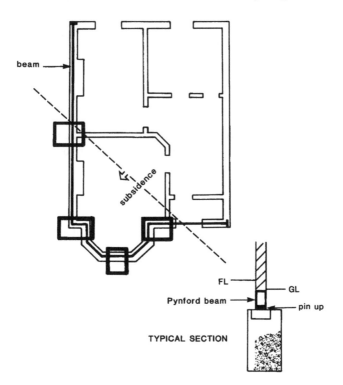

Figure 5.19 (*continued*)

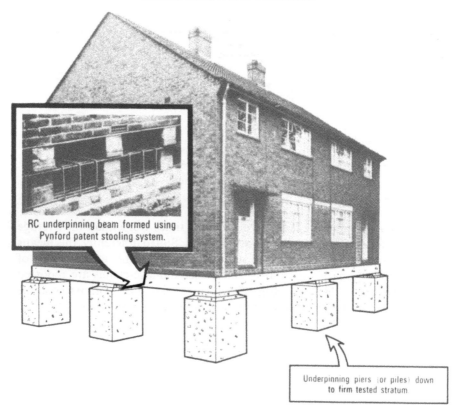

RC underpinning beam formed using
Pynford patent stooling system.

Underpinning piers (or piles) down
to firm tested stratum

Figure 5.19

foundation failures are caused by ground movement in the upper two metres or so of the ground and can be solved by underpinning to a depth of between 2m and 3m.

Hand excavation is relatively simple in clay soils and the most economical method of underpinning to this depth is generally to construct continuous beams just below ground level, supported on mass concrete piers spaced at intervals. A typical underpinning scheme for the corner of a house is shown in Figure 5.19. Subsidence of the foundations of external walls of the front left-hand room has been identified. A Pynford beam extends from the rear left-hand corner to the front right-hand corner and 4 piers support the unstable walls. The Pynford underpinning beam is extended on to stable foundations to support the wall during excavation and to form transition zones where the deep underpinning changes level to the original foundations.

In Chapter 1 criteria for assessing damage are discussed. It is important that cracks are not judged in isolation but that the total crack pattern is assessed. Figure 5.20 shows a distorted arch in Carlisle Cathedral in the north of England. Severe subsidence of the right-hand pier has caused

Figure 5.20

the damage, but the adaptable nature of the masonry and repairs carried
out over the centuries have enabled the movement to continue until it
stabilized without structural collapse. Clearly, such excessive movement is
undesirable and could have been prevented by underpinning at an earlier
stage. Equally clearly, underpinning was not necessary to prevent collapse,
and it should be recognized that much underpinning is carried out to maintain
acceptable serviceability limits of distortion in the superstructure and to
protect the value of the owners' investment. This is particularly relevant for
privately owned houses which are difficult to sell if foundation movement is
evident and which cannot then be fully insured.

Case Study no. 2: Underpinning a house in Wimbledon, London

Figure 5.21 shows, in schematic form, the property, which stood on sloping ground near a road junction. The ground level had been lowered three metres for the construction of a block of flats without adequate strutting. The ground began to slip and at the rear of the house the garden collapsed into the excavation. Cracks over 5 mm wide appeared in the left-hand walls of the house and smaller cracks appeared towards the right. The foundation was stabilized by forming a frame of Pynford beams just below foundation level beneath all the external walls and the internal walls of the garage. The plan lent itself to the design as the walls could all be approached from outside or from the garage, thus eliminating disturbance to the ground floor of the house. The loadbearing crosswall to the left of the front door was considered to be sufficiently far from the adjoining excavation to be buttressed at front and rear by the external underpinning. The piers supporting the underpinning beam were excavated to a depth of 4 m at the left-hand end of the property and those further to the right were progressively reduced in depth. Figure 5.22 shows the stage of beam construction. The reinforcement is fixed in position prior to fixing formwork and concreting. Beam sizes are generally chosen to minimize the need for shear reinforcement. Construction of support piers can proceed soon after the Pynford underpinning beams have been cast and before the pinning up has been completed between the stools. Pinning up between the stools is rarely necessary for structural reasons because the stools are cast into the beam.

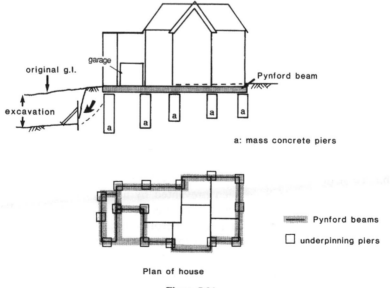

Plan of house

Figure 5.21

Figure 5.22

Case Study no. 3: St Nicholas' Church, Charlwood, Surrey

The Church of St Nicholas was constructed in four stages between 1080 and 1480. The external walls of the church are about 900 mm thick, and mainly constructed of irregular flat stones in weak mortar with larger tie stones at random to strengthen the construction, and dressed stonework surrounding doors and windows. The internal walls are of similar construction as they were first built as external walls.

The Church has a double pitched roof, partly of stone and partly tiled, with a central valley gutter. Rainwater is collected in gutters and discharges via downpipes into an open brick channel around the church walls which drains into a nearby brook. The south side of the church was suffering long-term continuing subsidence of the foundations, and severe lateral movement of the south wall had occurred at eaves level.

Site investigations identified clay shrinkage due to natural atmospheric conditions and the extraction of moisture by tree roots as the principal causes of foundation movement, compounded by outward thrust from the roof and by the shallow depth and poor quality of the foundation construction. The foundations consisted of a 1 m-deep layer of random stones set in a matrix of grey/brown clay, bearing upon a 3 m deep firm brown clay over dense dry shale. The foundations and the clay just beneath them were heavily colonized to a depth of 2 m by the roots of a large conifer tree growing near the south-west corner of the church.

There was an ancient and very valuable wall painting on the inner face of part of the south wall. The Parish Church Council (PCC) naturally required that there should be no damage or cracking whatsoever to this wall painting as a result of the underpinning works. To ensure this, the sequence of stool fixing was very carefully planned and added protection was given by temporary raking shores set against the part of the wall that was inclined most severely.

The PCC also required that there should be no disturbance inside the church. The Pynford stooling system of underpinning enabled all the work to be executed from an external access trench, and normal church services and functions were able to continue throughout the period of underpinning.

Pynford beams extended alongside the whole of the south wall, around the porch and returned 8 m along the west wall and 10 m along the east wall, and were supported by four mass concrete piers positioned along the south wall and two more at the corners of the porch bearing upon the dense shale at a depth of 3.2 m below ground level. As has already been stated, when only part of a building is being underpinned using a beam and pier system, it is good practice to extend the beams well beyond the piers at each end of the section that is being refounded. These extensions are termed 'transition beams' as they reinforce the foundation at the junction between deep and shallow

Figure 5.23

Figure 5.24

foundations. They are designed as if they were supported by 'virtual' piers at their ends. The underpinning is illustrated by Figures 5.23 and 5.24.

The sequence of work was to erect the shoring against the most tilted section of the south wall, to cut away for and fix the stools, to cut away between the stools, to fix the reinforcement and formwork, concrete and pin up between the stools and to excavate, concrete and pin up the piers and finally to remove the shoring. The beams were stooled up and concreted in four sections as an additional precaution in view of the importance of the wall painting.

After the underpinning was completed, cracks in the walls were repaired by epoxy resin injection and tie bars were fixed at high level to provide additional lateral restraint for the tilted south wall.

Case Study no. 4: Beam and pile underpinning for a factory near Nottingham

The factory had been constructed on a raft on a site that had been backfilled with a high proportion of domestic refuse. Before construction the ground had been improved by the installation of a pattern of stone columns using the vibro-replacement method of ground treatment. After a few years differential settlement began to develop. The cause was identified as decomposition of organic material in the filling resulting in a loss of confinement of the stone columns and progressive collapse. One consequence of this particular case was that, as a general rule, columns are no longer installed as a means of improving sites containing a high organic content. However, the ground improvement method offers very high cost savings, and on another Pynford project a raft has been installed with provision for jacking should differential subsidence develop. This project has proved very successful and to date no jacking has been required.

Figure 5.25

The failure of the Nottingham foundation resulted in a claim being made against the ground improvement contractor, and the building was underpinned using Pynford beams supported by bored piles. Figure 5.25 shows the caisson piling rig in position in preparation for boring a cast-in-situ reinforced concrete pile. Figure 5.26 shows an internal corner. Framework is fixed on the left-hand side of the picture and is being fixed at rear right. An internal cross-beam is shown projecting beyond the wall on to a pile.

Piles may be used in a number of ways to support underpinning. These include piles placed in pairs, one in compression and one in tension, to support cantilever caps. Piles may be used in pairs on either side of a wall to carry a simple pile cap or a cross-member formed as part of the underpinning beam frame. This requires internal piling and all the disturbance that this causes. For light structures piles may also be installed singly with cantilever caps, provided that they can be adequately reinforced for bending. However, a minimum number of piles is usually required when the arrangements are similar to those shown in this case. If piles can be installed in doorways or openings, one then approaches most nearly to the ideal case, that is, the 'open site' beam and pile foundation with the piles installed directly beneath the walls.

Figure 5.26

Case Study no. 5: Redeveloping major retail premises,
Oxford Street, London

Another example of the use of the Pynford stooling method for shoring is illustrated by Figure 5.27. Very large numbers of flues are found in the cross-walls of 19th-century buildings which were heated by numerous open fires. A typical flue size is 350×225 mm, and adjoining flues are generally separated by half brick walls, 112 mm wide. At high level the void ratio in thick walls may be 30% or more.

The project required the removal of a length of wall approximately 10 m high at the bottom of a crosswall, formed entirely of chimney flues, over 7 m long and over 25 m high. A conventional prop and needle scheme had been designed and was being installed when it was realized that the brickwork would be heavily overstressed above the needles.

To enable the work to proceed, Pynford designed and constructed a beam just above the needles which could safely transfer the wall loads to the needle positions. To construct the beam slot it was clearly necessary to ensure that the network of slender walls could be fully supported by the stooling process, as a significant increase in stress levels could not be allowed. It should also be noted that it is inherently more difficult to maintain the integrity of the brickwork above a beam slot than below it. This is of particular importance when dealing with Recent buildings, and particularly those constructed with relatively soft lime mortar. The brickwork is strong and tight when under compression, but when the compression forces are relieved the brickwork can easily be loosened, thus stool holes must be kept narrow if the brickwork above the beam slot is to be supported. It is clearly much less of a problem to avoid loosening the brickwork below the beam slot as it is held in place by gravity.

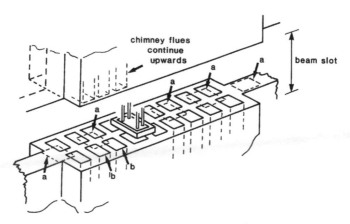

Figure 5.27 *a*, stool positions shown dotted; *b*, chimney flues; *c*, bag pushed down flue to block it; *d*, concrete filling flues; *e*, special Pynford stool; *f*, channel as top plate; *g*, concrete lintels; *h*, pinning up; *j*, mortar beds.

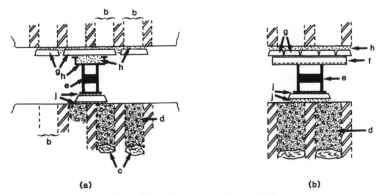

Figure 5.28 Key as for Figure 5.27.

In this particular case it was decided to provide total support above the beam slot and to use conventional stool and pad detailing below the beam slot. This was possible and practicable because the chimney construction below the beam slot could be strengthened by filling the flues with concrete, whereas this could not be done above the beam slot. A stool was designed with a steel channel replacing the upper steel plate (Figure 5.28a) and a series of small precast concrete planks replacing the upper concrete plate (Figure 5.28b). The stools were positioned at the intersection of alternate crosswalls between the flues. The upper concrete plate was changed to a set of lintels to make it possible to ensure that all the narrow walls separating the flues were completely supported. Finally, the gap beneath the lintels in the channel was pinned to relieve the load on the edges. The wall was successfully stooled up in this way, reinforcement was fixed and the beam was concreted. It was then possible to needle and prop safely to carry the load of the wall down to basement level.

Case Study no. 6: Pynford beams used for shoring

When altering larger buildings of recent origin it is often the case that beams or columns are found passing through the position in which it is desired to construct the Pynford beam.

(a) Case 6A. The base of a substantial steel column was found to be just above the top of the proposed underpinning beam. In such a case the base of the column is too small for a significant proportion to be cut away for a stooling sequence. The column load must first be relieved, and this was achieved by bolting a steel beam to the column shaft just above beam level. This was propped by Pynford stools on either side of the column at a safe distance from the column base so that the holes cut for the stools would not disturb the column support. Figures 5.29 and 5.30 show this in detail. The stooling operation is complete and reinforcement is fixed prior to concreting the beam.

Figure 5.29

One supporting stool can be seen on the left-hand side of the picture, and the gap in which shoring jacks are installed can be clearly identified. To the right of the picture, beside the stabilizing props, the column can be seen. This has two stools fixed and pinned up between the base.

(b) Case 6B: a column beside a wall. In this example the basement in a shop unit in a terrace of properties was being deepened. The adjoining property had been reconstructed with new floors supported by columns ranged alongside the party walls and closely abutting them (Figure 5.31). The columns adjoining the party wall that was to be deepened were founded on bases above the level of the underpinning. There was a restaurant at ground-floor level with a serving counter immediately beside the wall and it would have been

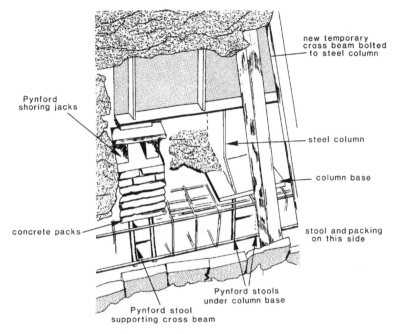

Figure 5.30

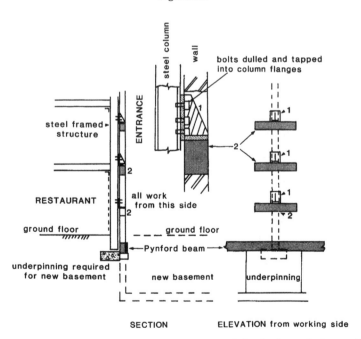

Figure 5.31 Case *6B*. 1, new steel bracket fixed to column; 2, Pynford spreader beams distribute column load on brickwork.

impossible to screen off a working area around the columns without closing the restaurant and paying heavy compensation. The columns were screened with vitrolite panelling and the problem was solved by transferring the column load to the party wall, spreading it at three levels on Pynford beams formed on the working side. The load was transferred to the beams by cutting carefully positioned holes immediately behind the columns, drilling and tapping the flange closest to the wall and fixing a bracket. With the load relieved it was then possible to underpin the column from the opposite side of the wall.

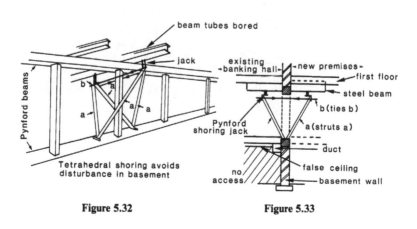

Figure 5.32   **Figure 5.33**

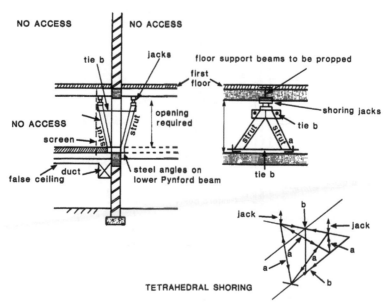

Figure 5.34   (See also Figures 5.32, 5.33). Tetrahedral shoring: 4 No timber struts (*a*); 4 No steel tines (*b*).

(c)   Case 6C: the treatment of steel beams running into the 'beam slot'. In this typical case a frame of beams and columns was being constructed to remove a wall at ground-floor level to extend a banking premises. The beams could not be propped down to basement floor level as the disturbance in the basement could not be tolerated. Tetrahedral shoring was designed to prop the first-floor level beams from the spreader beam forming the bottom member of the frame. The general arrangement is shown in Figure 5.32. Figures 5.32, 5.33 and 5.34 show the tetrahedral shoring. Pynford shoring jacks were used to preload the shoring before the padstone supporting the beam was removed. The jacks are tightened until the beam is seen to just lift off the seating, thus ensuring that the actual load is exactly matched. Where the web would obstruct the reinforcement, holes are cut with an oxyacetylene torch to allow reinforcing bars to pass through. In some cases it may be necessary to fix additional brackets to the web to reduce bearing pressures on the concrete to acceptable levels.

Case Study no. 7: Strengthening medieval stonework at Winchester Cathedral

A number of columns developed vertical splits between the springing of the arches in the main nave and the base of the column at floor level. These splits

Figure 5.35

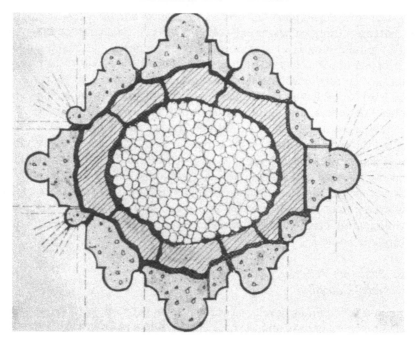

Figure 5.36

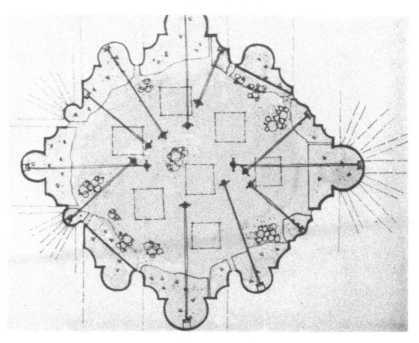

Figure 5.37

occurred in each face of the column which was roughly square in plan with the diagonals following the axes of the cathedral. In the worst case the split had widened to about 80 mm wide at mid-height tapering upwards and downwards. Figure 5.35 shows a general view of the column with scaffolding erected around it. Plumbing showed a distinct barrelling outwards of the column at mid-height. Figure 5.38 shows a view of another column where a crack can be clearly seen.

To strengthen the column, reinforced concrete discs were placed at approximately 3 m centres within the height of the column. In the worst column, three discs were inserted, and in other less severely damaged columns the number was reduced. Figures 5.36 and 5.37 show plans of a typical column supporting the nave walls. The cathedral was refaced in Perpendicular style in

Figure 5.38

the 14th century. Figure 5.36 shows clearly the new stone facings, the 14th-century mortar behind the facings, and the Norman core. The quality of the cores was variable and in some cases the load of the columns had clearly been transferred to the stone facings. Remedial work had been carried out by Nash in the 19th century on nearby columns, and this work was being extended.

Figure 5.37 shows a plan of the concrete discs. The disc was formed by cutting a hole in the side of the column and then cutting away the core in stages, replacing it with stools as the work proceeded. The position of the stools is marked. The stonework was then bonded into this core with phosphorbronze bolts anchored into the stonework and embedded in the concrete disc. This work was carried out in the early 1960s, before modern epoxy resin bonding and grouting techniques had been developed. It is a clear example of the way in which stooling can be used to insert reinforced concrete into loadbearing masonry.

Clearly, holes could not be safely cut in a column in this way, even to remove part of the plan area, unless the column were first strengthened. Nash constructed fairly elaborate shoring for this purpose, a model of which is still to be seen at the cathedral. For the purpose of the Pynford operation, the columns were strengthened by steel bands secured tightly around the columns at distances varying between 0.3 and 1.5 m. Pairs of bands were arranged immediately above and below each beam slot. This can be clearly seen in Figure 5.38. It should be appreciated that cutting out and refixing stones in columns and walls is a common masonry practice. In this case the columns were strengthened by banding and the cutting away was extended into the core using the stooling technique.

It is hoped that the foregoing text and illustrations will have clearly explained the Pynford method. The method has been used on many occasions to form reinforced concrete frames or lintels in buildings. An example of such work is now briefly described.

Case Study no. 8: Altering a Tudor structure—Trinity College, Cambridge

When the senior combination room at ground-floor level beneath the Master's Lodge at Trinity College, Cambridge, was extended in the early

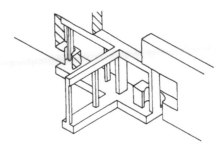

Figure 5.39

Figure 5.40

Figure 5.41

Figure 5.42

Figure 5.43

1960s, a reinforced concrete framework was inserted using the Pynford stooling system. Figure 5.39 shows the framework in isometric view. Figure 5.40 shows the stooling and reinforcing complete for the foundation level beams just prior to concreting. The work was being carried out to remove the walls at the rear and on the right between ground- and first-floor level. The top of the frame was 75 mm below first-floor level. The alternative method of constructing the frame in structural steelwork would have required needling of the rear wall above first-floor level. For this purpose the unique Tudor panelling in the Master's Lodge at first-floor level would have had to be removed and the very heavy cost of doing this work was saved by using the stooling method. Figure 5.41 shows construction of the beam slot at first-floor level nearly completed. It is interesting to compare the very small area of steel

Figure 5.44

carrying the rear wall, that is 28 no. 25 mm diameter circular steel legs, with the much larger area of Tudor stone masonry that has been cut away. This dramatically illustrates the discontinuity in construction methods and materials that occurred when steel was introduced as a new building material. In Figure 5.42 the reinforced concrete frame is complete. Some of the walls have been cut away to form the colum slots. In Figure 5.43, the masonry between the columns has been removed and the building work is in progress.

Moving brick and stonemasonry structures can be readily accomplished using modern cranes or rolling and sliding gear, and Pynford have completed many operations of this type. It is important to have a rigid and stiff chassis beneath the structures that are being moved, and the final case included in this chapter illustates an example of such work.

Figure 5.45

Case Study no. 9: Moving a historic school house at Warrington, Lancashire

The historic old school house at Warrington is another example of a brick structure that was moved on a Pynford reinforced concrete chassis. This 800-tonne Georgian brick building, which was of considerable historic importance to the city of Warrington in Lancashire, was moved over 15 m to make room for a road widening improvement scheme. The building was strengthened using epoxy resin techniques, and some tie bars attached to timber plates. A reinforced concrete frame was then constructed just below foundation level. This is clearly seen in Figures 5.44 and 5.45. A second reinforced concrete frame was then constructed beneath the first, leaving a gap within which the sliding gear was installed. This lower frame was extended to the new site, and the beams were infilled to form a raft foundation. Figure 5.44 shows the building just prior to the move, with the River Mersey in the background. Figure 5.45 shows the move at halfway stage, with building sliding on grease skates and towed by winches (see reference 10).

References

1. International Association for Bridge and Structural Engineering, *Proc. Symp. September 1983*. IABSE, London.
2. Institution of Civil Engineers (1982), *Proc. ICE Conf. Repair and Renewal of Buildings*, Thomas Telford, London.
3. Pryke, J.F.S. (1982) Underpinning, framing, jacking-up and moving brick and stone masonry structures. In Ref. 2.
4. Heyman, J. (1982) *The Masonry Arch*. Ellis Horwood, Chichester.
5. Building Research Establishment, Research Paper No. 13.
6. Wood, R.H. Composite action of brickwork supported by beams. BRE Publication.
7. Norwegian hotel moved in biggest ever lift. *New Civil Engineer*, 2nd June 1983.
8. Pryke, J.F.S. (1983) Relevelling, raising and re-siting historic buildings. In Ref. 1.
9. Pryke, J.F.S. (1979) Differential foundation movement of domestic buildings in South-East England. Distribution, investigation, causes and remedies. Adapted from *Proc. Conf. on Settlement of Structures*, Cambridge, April 1974. British Geotechnical Society, London.
10. Stables, A.A.J. A Moving Experience. *Industrial Nottingham*, March 1984.

6 The Bullivant systems

H. BRADBURY

6.1 Introduction

Underpinning as a branch of foundation engineering has evolved and continues to develop by the application of several disciplines—particularly the material sciences, geotechnology, civil and structural engineering, and concrete technology—coupled with individual flair and experience. Underpinning is a continuously improving operation as new materials and ideas emerge in the light of experience. Direct involvement in underpinning work is the best way to gain actual knowledge and practical skills. Theoretical knowledge is necessary in order to optimize sound design and techniques. The Bullivant underpinning systems display innovation and specialization, and quality assurance is a central theme in the methods of construction.

6.2 Historical background

Established in 1971 by Roger Bullivant, the Bullivant Company has grown largely through innovative approaches to problems, coupled with extensive research and development to establish the reliability of new underpinning techniques. The Bullivant approach to underpinning is governed by the firm belief that structures respond to load, temperature, moisture, and ground movements in a manner related specifically to their nature of construction. Earlier house construction used lime mortars and, when a building moved, the movement was not always detrimental. Later, when Portland cement mortar came to be used, any movement usually led to cracking in the brickwork. Bullivant introduced minipiling methods to solve the problems in domestic properties. This technique prospered, so much so that it encouraged new ideas and developments, such as the Bullivant Pre-Cast Concrete Pile using a single reinforcing bar and a method of hydraulic jointing of the piles protected by patents. This chapter describes the various Bullivant underpinning systems for conversion, protection and remedial works.

6.3 Definition of problems

Underpinning solutions for a specific project require careful consideration to be given to the following matters.

198

1. A adequate investigation of the ground and groundwater regime.
2. An investigation of the construction and condition of existing structures, including—where possible—sight of the original plans in local authority archives and any structural changes carried out since the original construction.
3. Problems arising from any deterioration due to inaccurate or poor workmanship.
4. The possible effect of underpinning work on adjacent structures, together with any need to shore or prop these and the existing building.
5. The need to repair brickwork in buildings in advance of underpinning.
6. The decision as to the cause of failure, heave, settlement, etc.
7. The variations in ground conditions over the curtilage of the building.
8. The development of elastic and permanent strain caused by subsequent ground removal in the vicinity.
9. Any restrictions on noise and the effects of vibration or percussion, together with any nuisance caused.
10. Space availability, including restricted access and headroom.
11. Legal rights of adjacent owners regarding the withdrawal of support.
12. The fact that the designer/specifier is often different to the contractor.
13. Selecting the right type of pile suitable for the overall conditions commensurate with the designer's requirements and the environment.

Firms specializing in underpinning would normally consider all of the above when devising economic solutions. Expertise in piled underpinning systems is concentrated within a few specialist contractors, and work is carried out under varying contractual arrangements, which are often inadequate for the technical nature of the work. This situation reflects earlier underpinning methods using mass concrete, which was—and still is—often carried out by builders.

Since piles are structural components forming important parts of a foundation system, it is prudent for a qualified and experienced engineer to be responsible for their design, and for ensuring that the materials and workmanship are suitable, adequate and to specification. In piled underpinning it is often the specialist piling contractor who designs, specifies and supervises the work, and accepts responsibility for it. To understand the stability of a building, particularly domestic housing, where the majority of underpinning occurs and its structural behaviour when a piled underpinning solution is used, it is necessary to understand how the building would behave under the normal design criteria of wind, and superimposed dead and live loading. Wind forces tend to cause rotation at foundation level. This is resisted effectively by the walls (both external and internal) or by floors. This lateral stability is covered and clearly defined by the building regulations and the various applicable Codes of Practice. A domestic structure can be likened to a stiff box that is restrained from rotation about its base and constrained from inward movement by internal support walls, floors and roof. The potential hazard area is the ground support. Failure of the ground to support a structure is demonstrated by the external walls moving downwards and outwards, with evidence of cracking to brickwork. Traditional mass concrete underpinning has been generally used in the past, but is costly, disruptive and time consuming. Since 1977, the advent of the minipile has given the engineer a more speedy, less-disruptive, more versatile and cost-effective method—almost without limitation in use.

A wall of a house does not generally move inwards, and when vertical displacement occurs it causes a horizontal stress failure in the wall, together with an outward and downwards movement. Jointed piles situated in a vertical line directly below the outside edge of the wall, but under the foundation, will resist loading and prevent further movement. This tends to rotate the wall back towards the inside of the house where adequate restraint already exists. Consequently, provided that the shear capability of the combined connected footing and pile is properly calculated and has adequate interface connection, the property is in equilibrium.

Loss of equilibrium or instability, therefore, is often due to either settlement or heave. In many buildings, particularly the older ones, a prime factor is ground movement due to a variety of causes—either natural or man-made. The former includes heave and shrinkage associated with clays. Heave and shrinkage of this type are associated with drought conditions or the effect of tree growth causing zones of desiccation, and require isolation of the structure from the ground affected by the moisture changes, or the use of deep traditional foundations or underpinning. Man-made causes can be due to the effect of mining subsidence, nearby basement construction, tunnelling and de-watering by abstraction—for construction or other purposes. Additional causes include the effects of original or new defective workmanship and methods, and chemical attack on concrete foundations. Diagnosis, therefore, consists of a thorough investigation to determine the real cause(s) of building movements and the approach needed to cure the problem.

6.4 Underpinning systems

The main difficulty in piling in buildings is usually spatial (e.g. restriction on headroom and access). This has been overcome in the Bullivant systems by the development and use of special compact plant and equipment. These have evolved to cater both for causation and spatial constraints, and can be grouped into three main categories:

   (i) Foundation stabilization requiring installation from one side of a wall only, e.g. angle piles; jack-down piles; pile and reinforced concrete raft;

cantilever pile and beam, cantilever ring beam; conventional pad and beam underpinning; pile and needle beam.

(ii) Foundation stabilization requiring installation on both sides of a wall, e.g. dual angle piles; pile and beam systems.

(iii) Specialist systems, e.g. micro- and minipiles; pressure grouting.

6.4.1 Angle piles (Figure 6.1)

This system of underpinning can be used on all types of strip foundation. The existing foundation is pre-drilled using air-flushed rotary pressure equipment and cased steel-driven, or solid or hollow-stem augered piles installed through the pre-drilled hole with the casing terminating at the underside of the foundation. The pile is then concreted, and reinforcement included up through the existing foundation. Depending on the conditions, angle piles can be inserted from one or both sides of a wall. Loadings related to pile sizes from 40 to 250 kN can be catered for.

Advantages:

(1) Large excavations are not required, therefore disruption is kept to a minimum.

(2) Economical at depths greater than 1.0 m.

(3) Conservative design when pile centres do not exceed 1.0 m.

(4) High load capability.

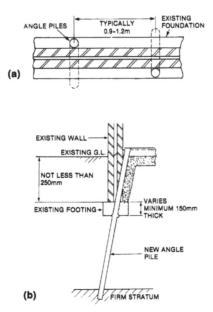

Figure 6.1 Angle piles: (a) plan of foundation; (b) section through foundation.

(5) Suitable for restricted access.
(6) Can be installed from one side only, if required and certain lateral restraint conditions exist.

6.4.2 Jack-down piles (Figure 6.2)

The silent and vibrationless installation of piles can be achieved by jacking techniques. The dead load of the structure, ground anchors, and/or kentledge are used to mobilize sufficient resistance in excess of the required working load of the pile, to provide a factor of safety. Piles can be installed to accept loads in excess of 1000 kN within low headrooms of less than 2.5 m. After the provision of suitable reaction points (e.g. reinforced concrete beams and rafts, steel grillage, kentledge, ground anchorages or existing foundations), heavy steel or precast concrete casings are jacked in sections to the required load. The ensuing piles are then concreted and bonded to the structure being supported.

Advantages:

(1) Every pile is tested as it is installed.
(2) Silent and vibration-free.
(3) The jack rigs are operated by hydraulic power packs, which can be sited away from the area of works.

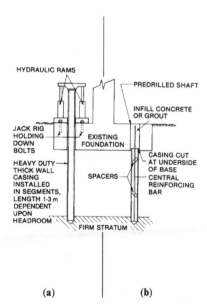

Figure 6.2 Jack-down piles: (*a*) section during installation; (*b*) section through completed foundation.

(4) Clean, dust-free—piles have been installed in operational food factories.
(5) Piles can be installed within 260 mm of a wall to the centre line of the pile, and 500 mm into an internal corner.
(6) Because of their unique size and adaptability, the jack rigs can be manhandled into the most difficult areas, with a maximum operating headroom of 1.8 m but an access size requirement of only 1.2 m–0.7 m.

6.4.3 Pile and reinforced concrete raft (Figure 6.3)

The system comprises the installation of minipiles capped with a reinforced concrete raft incorporating needle beams and reinforced concrete ring beams

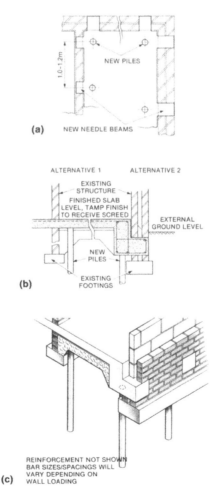

Figure 6.3 Pile and reinforced concrete raft: (*a*) plan of foundation; (*b*) section through foundation; (*c*) view of piles and raft.

as necessary, to support walls and provide lateral restraint and a new floor. It is used where complete rooms are to be underpinned, and includes the provision of new suspended internal floor slabs. This system can also be used where exposure or shrinking soils exist. The piles are either driven, drilled or jacked inside the premises at centres determined by the loadings, and needle beams are provided at centres up to 1.2 m together with reinforced concrete ring beams and floor slabs.

Advantages:

(1) Provides lateral and transverse ties throughout the structure, with piled foundations for internal walls and floors.

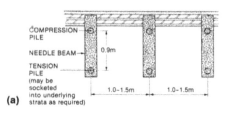

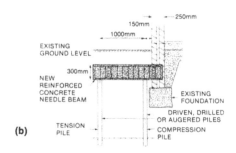

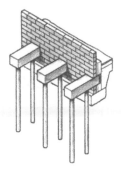

Figure 6.4 Cantilever pile and beam: (*a*) plan of foundation; (*b*) section through foundation; (*c*) view of completed foundation.

(2) Provides fully suspended replacement ground-floor slab.
(3) Especially suitable where access externally is restricted, and lateral and longitudinal restraint is required.
(4) Minimizes disruption to existing external services (i.e. drains and service pipes) and consequential reinstatement.
(5) Recommended for total structure isolation from underlying strata, particularly in clay heave/shrinkage situations.

6.4.4 Cantilever pile and beam (Figure 6.4)

The cantilever pile and beam system involves the installation of minipiles in pairs—one as a tension pile and the other acting in compression—connected by a reinforced concrete beam or concrete-encased steel needle beam supporting the wall. The piles are either driven, drilled or augered at pre-determined centres, usually 1 to 1.5 m laterally, and carry a reinforced concrete or concrete-encased steel needle supporting the walls.

Advantages:

(1) Large excavations are not required, therefore disruption is kept to a minimum.
(2) Access from one side of a building. Occupants do not necessarily need to be relocated.

6.4.5 Cantilever ring beam (Figure 6.5)

The cantilever ring beam is similar to the cantilever pile and beam, but in this case, the tension pile is connected by a ring beam external to the wall being supported. Cantilever ring beams also incorporate a needle beam below the wall but above the existing foundation.

Advantages:

(1) Large excavations are not required, therefore disruption is kept to a minimum.
(2) Access from one side of a building. Occupants do not have to be relocated.
(3) Provides longitudinal stability to distressed walls.

6.4.6 Conventional pad and beam (Figure 6.6)

This involves the insertion of intermittent concrete pads down to firm strata, with reinforced concrete beams spanning between the pads and under the walls to provide support. The system is applicable to all types of shallow foundations where depths of excavation to firm strata are in excess of 1.5 m. Where depth to suitable strata is excessive, the pads can be supported by minipiling.

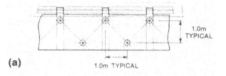

(a)

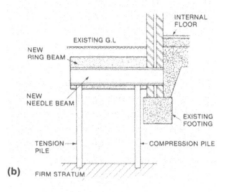

(b)

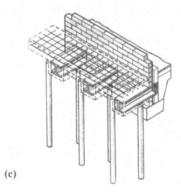

(c)

Figure 6.5 Cantilever ring beam: (a) plan of foundation; (b) section through foundation; (c) view of ring beam under construction.

Advantages:

(1) Cost effective at depths of 1.5 m–3.5 m below ground level, with heavy-foundation loads.
(2) Can be engineered from one side of a wall. Occupants do not necessarily need to be relocated.
(3) Work can be carried out in areas of difficult and restricted access.
(4) System suited to support stone-wall construction.
(5) Should be considered when dealing with heave-susceptible clay soils.
(6) For situations where the bearing strata are at greater depth, the pads can be supported by mini piles.

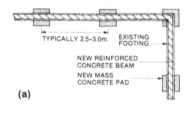

TYPICALLY 2.5–3.0m   EXISTING FOOTING

NEW REINFORCED CONCRETE BEAM

NEW MASS CONCRETE PAD

(a)

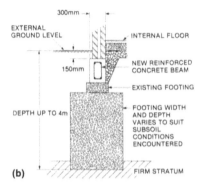

300mm

EXTERNAL GROUND LEVEL

INTERNAL FLOOR

150mm

NEW REINFORCED CONCRETE BEAM

EXISTING FOOTING

DEPTH UP TO 4m

FOOTING WIDTH AND DEPTH VARIES TO SUIT SUBSOIL CONDITIONS ENCOUNTERED

(b)

FIRM STRATUM

Figure 6.6 Conventional pad and beam: (*a*) plan of foundation; (*b*) section through foundation.

6.4.7 Pile and needle beam (Figure 6.7)

Pile and needle beams are essentially minipiles—one vertical and one raked—connected to the wall by a reinforced concrete needle beam. The existing foundation is loaded, and the initial raking pile is placed beyond the base central line. The vertical pile is then offset in plan, and the two piles are connected to the existing wall by the drilled connection through the existing foundation, and by a reinforced concrete needle beam. Loadings can be from 40 to 90 kN.

Advantages:

(1) Rapid construction, compared with traditional underpinning.
(2) Economical.
(3) No internal work necessary, therefore occupants do not have to be relocated.

6.4.8 Dual angle piles (Figure 6.8)

These are used for all types of foundations or bases founded at depths in excess of 400 mm below ground level, in order to stabilize walls by the use of pairs of piles crossing each other at an angle. Existing foundation is pre-drilled, and permanently cased steel-driven, or solid or hollow-stem augered piles are then installed through the pre-drilled holes with the casings

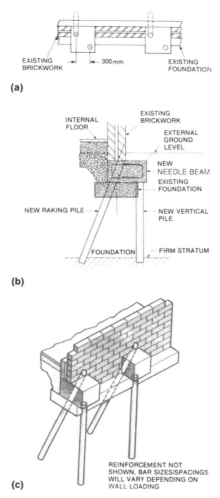

Figure 6.7 Pile and needle beam: (a) plan of foundation; (b) section through foundation; (c) view of foundation.

terminating at the underside of the existing foundation. Piles are then constructed by concreting and reinforcing to extend up through the foundation to ground level.

Advantages:

(1) High load capacity.
(2) Suitable for restricted access.
(3) Can be used in situations where existing foundation quality is poor, obviating the use of the staggered single angle pile.

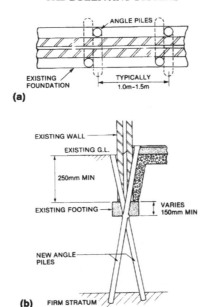

Figure 6.8 Dual angle pile: (*a*) plan of foundation; (*b*) section through foundation.

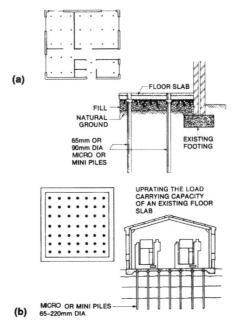

Figure 6.9 Micro- or minipiles: (*a*) plan of floor and section through floor; (*b*) plan of foundation and section through foundation.

6.4.9 Micro- or minipiles (Figure 6.9)

This type of underpinning is used to stabilize or uprate existing domestic or light industrial floor slabs by installing micro- or minipiles through pre-drilled holes in the floor slab at centres determined by the loading characteristics and the existing slab construction. This obviates large breakouts and excavation. It is a very clean method of repair.

Advantages:

(1) Suitable for restricted access.
(2) Capable of being installed in factory environments without the need to shut down.
(3) Recommended for cost-effective provision of machine foundations.

6.4.10 Pressure grouting (Figure 6.10)

The pressure grouting system is used under floor slabs on poorly compacted fills, or where voids occur under subsided floors. Holes are drilled to a pre-determined grid pattern based on trial hole investigations, and a suitable cement: pulverized fuel ash grout injected under controlled pressures to fill the voids and/or lift the settled slabs back to their original position. Pin piles may be incorporated if required.

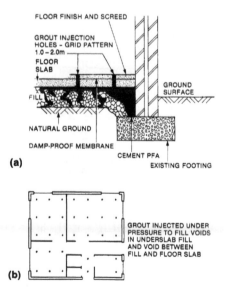

Figure 6.10 Pressure grouting: (*a*) section through foundation; (*b*) plan of grout injection holes.

6.5 Case studies

6.5.1 Tower 4, York City Walls

Tower 4 stands proud of York City Walls and is built on an embankment of medieval fill. The tower had been subsiding for a number of years, and previous efforts had been made to reduce the load on the embankment by replacing part of the core of the tower with polystyrene blocks. Settlement continued, and it was decided to stabilize the tower by introducing a series of minipiles. Ten piles of nominal 150 mm diameter and having safe working loads of 100 kN were required to be installed through the existing masonry and rubble core, then penetrating the medieval fill before founding in the underlying dense sands and gravels.

The Bullivant Angle Pile system was used because of the following difficulties:

(1) the random rubble core of the walls collapsing into the bored hole;
(2) obstructions within the medieval fill. Driven piles could be introduced once pre-boring of the walls had taken place; and
(3) collapsing and scouring of the medieval fill during pre-boring.

A purpose-built miniature rig was used to pre-bore both the walls and the medieval fill for the installation of Odex casings. A rotary percussive airflush drilling system prevented drill foam or flushing water discolouring the ancient masonry or scouring the embankment fill. The Odex casings, once through the medieval fill, were to be drypacked off at their base and then driven on to achieve an adequate set using an internal steel mandrel hoisted by a purpose-built hydraulic winch. On commencement of the drilling, it was discovered that the rubble fill (which had been pre-grouted) and the embankment fill did not require the use of Odex casing. The piles were therefore drilled and cased immediately and, after being driven to the required set, were reinforced with a single T16 bar, and grouted up into the walls using a high strength cement grout placed through an air-operated pump and mixer.

6.5.2 West Princes Street, Glasgow

The property was underpinned as part of a general rehabilitation scheme and, due to the large amount of movement and distress the buildings had suffered, a method offering low vibration and minimal disturbance to the existing foundations was required.

After removal of the floors, trenches were excavated on both sides of the external walls, to permit the construction of reinforced concrete tie-beams. The tie-beams had been designed to distribute the load to the structure and to enable the piles to be jacked against the tie-beams to a load of 400 kN.

Pockets were formed through the walls at 1.4 m centres, steel H-section needles were threaded through. The beams were encased in concrete, and linked internally and externally using reinforced concrete beams; the reinforcement consisted of 20 mm and 25 mm high-yield steel bars top and bottom, with twin links at 200 mm centres. Depending on the condition of the existing wall, a method of concrete for stone replacement at the needle position was used to allow all of the needle ends to be tied together within one concrete pour. At each proposed pile position, pockets were formed through the beams using cone-shaped polystyrene tapering from 175 mm diameter to 225 mm diameter. The reinforcement detail was adapted at the pile position to cater for the additional forces induced during the jacking procedure, and four 24 mm-diameter holding down bolts 500 mm long were cast into the beam at the pile positions.

After curing of the concrete and removal of the polystyrene formers, twin hydraulic rams (operated from a remote power pack) were bolted to the concrete beams, and 140 mm-diameter, closed-end, heavy-walled steel casings were jacked through the beams in 1.4 m lengths. The pile was extended by adding sections jointed using spigot/socket joints; this operation continued until a load of 400 kN was reached. The steel casing terminated at the underside of the concrete tie-beam, and the core was filled using 35 N/mm² concrete to form the tie-in to the beams.

6.5.3 Castle Bytham

A detached dwelling at Castle Bytham had suffered distress as a result of being founded partly on limestone bedrock and partly on what was thought to be a backfilled quarry. The pad and beam method of construction was proposed, as it would give the best resistance to any possible lateral movement of the fill materials.

The scheme involved construction of nine pad bases founded at a depth of 2.0–3.0 m below ground level. The span between the bases depended on wall loading but averaged 3.0 m with pads being positioned to maximize the use of a standard beam size. The reinforced concrete beam was constructed on top of the existing concrete foundation within the depth of the substructure building. The schedule of proposed works included an initial investigation to assess the extent of the works. This was undertaken within the excavation for the three pad bases at the limits of the underpinning works. From this investigation, the remaining pad bases were then designed. Normally the reinforced concrete beam would be installed at the outset in this type of operation, as this would reduce possible further movement during the execution of the works.

The pad bases were excavated by hand through clay fill material with limestone inclusions, down to a maximum of 3.0 m below ground level, and were 'tied' into weathered limestone. Four of the bases were cast with rebates, to allow for the installation of hydraulic jacks. These jacks would subsequently be used to jack the superstructure back to level, thus minimizing the amount of rebuilding required at the major fractures. Construction of the reinforced concrete beams was completed in four stages, with no two adjacent spans being removed at the same time. The existing sub-structure was broken-out, and 'cast in' adjustable stools were introduced at 600–1000 mm centres depending on the quality of the superstructure brickwork and the load involved. Beam reinforcement was introduced to pass either side of these stools, and the beams were concreted using ready-mixed concrete placed by hand. Following a suitable period of curing of the concrete works—assessed by testing of concrete cubes—the superstructure was jacked back to level, with hydraulic jacks placed between the mass concrete pads and reinforced concrete beams. The resulting void between the pads and beams was temporarily shimmed using steel plate, and the void was then pressure grouted on removal of the jacks to ensure a permanent and adequate load transfer.

6.5.4 The Town Mill, Guildford

There was a need to strengthen the existing foundations of the Town Mill, to allow for increased and re-distributed loadings arising from a major rehabilitation programme. In addition, there was a need to provide new foundations for new loadbearing internal walls. A further problem was that the building is a listed structure, and thus foundations have to be strengthened and all alterations made without detriment to the external walls. Moreover, additional problems were posed by the type and depth of the existing foundations, access limitations, working hours restrictions, poor ground conditions, and by the fact that the building was surrounded by water on two sides.

An extensive investigation revealed foundations consisting of mass brickwork to varying depths, and incorporating various culverts, intake works and other subterranean mill features, overlying one or sometimes two horizontal timber grillages—spiked together with old metal tie pins. These grillages were then spiked onto timber piles that had been driven into the chalk substrata. Generally, the brickwork depth below floor slab was approximately 3.0 m, with anything from 150 to 300 mm of timber grillage below the brickwork. The timber piles penetrated the sands and gravels for approximately 2.0 m into a substrata of chalk with flints.

The solution involved the drilling of the existing foundations at an angle both internally and externally, and the installation through the drilled holes of 165 mm nominal diameter minipiles, rated to safe working loads of 100 to 150 kN, and installed at approximately 750 mm centres along the perimeter and internal load bearing walls. The piles were designed with a centrally positioned T16 bar, with a permanent steel casing installed to the underside of the brickwork. The load transfer was then made by the bond from the new in-situ concrete of the pile, through the drilled hole in the brickwork.

A specially designed 180 mm-diameter drilling bit was developed to tackle the combination of brickwork and sodden timber of the grillages. This particularly difficult drilling task was tackled using standard mini-drilling rigs, and 'down-the-hole' hammers equipped with the specially designed 18 mm-diameter bits. Plenty of compressed air was on hand to exhaust the arisings and, in some areas, removal of the drilled brickwork was assisted by the use of drilling foams. Once the holes were drilled, 150 m casings were driven to depths varying from 8.0–12.0 m. On the two sides surrounded by water, a scaffold platform was erected to allow access for the mini-drill rigs. To maintain the original appearance, brickwork was removed at the hole positions and, after piling, was replaced to provide an exact match. Finally, the various ancillary works including ground-beam construction, brickwall construction, void filling, and reinstatement were completed.

6.5.5 Courthouse in Eire

The inadvertent removal of some timber piles beneath two stone columns by a thrust-boring contractor, led to some distress in the building. Thrust boring ceased as soon as the problem manifested itself, and the stone arches above were quickly shored-up. Further problems included the presence of a 1050 mm-diameter thrust-bore pipe under the stone columns, and the fact that the main column was located immediately adjacent to an existing egg-shaped 1000 mm-diameter brick culvert.

The solution involved emplacing a number of interconnecting, concrete-surrounded, universal column section needle beams, which would sit on spreader pads or beams, and which, in turn, would be supported on 165 mm-diameter, segmentally hollow-stem augered minipiles. The total load of the main L-shaped pier was 1200 kN, which was distributed onto 12 piles, with each pile carrying 100 kN.

The operation of installing the piles so close to the brick culvert and the connecting thrust-bore pipe was delicate and had to be executed without vibration to either the structure or the culvert. It was decided to use hollow-stem augering methods to minimize vibration, and to facilitate the pile installation through approximately 2.0 m of soft clays overlying dense sands and gravels. Restricted access and height meant having to use mini-drilling rigs with 1.0 m-length segmental augers. Having installed the piles to the close tolerances, reinforced concrete pads and spreader beams were constructed on either sides of the piers. A phased operation to install the universal column section then took place, with load transfer being achieved by the use of Armourex high-strength grout. The beams were then encased in concrete and reinstated.

6.5.6 Alvaston, Derby

Twenty-six houses required major structural rehabilitation, from under-pinning of foundations through to complete modernization. Most of the properties were three-bedroom semi-detached houses with approximately one metre of space between the buildings. One of the main requirements was that all underpinning works would be carried out externally.

The main areas of distress were centred around the side gable walls, with failures of the front and rear returns from the gable walls. Compounding the problems of limited access conditions and external-only working was the time factor, as the work had to be carried out in under 12 weeks; in addition, over 500 lineal metres of underpinning up to a depth of 3.0 m had to be completed in this time—a rate of over 8 lineal metres per day.

The Pier System was chosen for its speed of installation, structural integrity, and its ability to be installed in very restricted access conditions. Approximately 550 piers were installed up to 3.0 m in depth. This included a test on one of the piers, which was loaded to 60 kN — 1.5 times working load—and produced a maximum deflection of 0.16 mm with 66% recovery. In conclusion, the Pier System proved to be a viable and rapid alternative to traditional mass concrete underpinning.

6.5.7 Minster, Sheppey

An investigation determined that damage was occurring chiefly to the flank wall of the property. This was the result of shrinkage of the underlying Claygate Beds strata, caused by the root activity of an adjacent hedge. A solution had to be devised to limit the works to the external confines of the property at the flank wall area, together with returns to the front and rear of approximately 6 m at each location. As the distance to the site boundary local to the flank wall was limited to 500 mm, it was necessary to remove the hedge at this locality to enable the work to proceed.

Due to the fact that the original construction was a narrow strip foundation founded at 1.2 m below ground level within the Claygate Beds (a clayey, silty, fine sand), the Pier System was used to provide (i) piers down the face of the original footings; and (ii) reinforced concrete knuckles above the footing level to support the structure. To introduce further stability, diagonal needles were emplaced at the corners of the flank wall, and all the piers and needles were connected via a tie-beam constructed along the face of the existing wall. Either a driven 220 mm steel-cased pile or a 250 mm augered cast in-situ pile would be satisfactory. Furthermore, as the Claygate Beds had only a small clayey content (although affected by roots), no form of anti-heave precaution was deemed to be required.

However, as a precautionary measure it was decided to provide a 2 m length of slip sleeve to the upper level of the pile, either by pre-augering and driving the steel-cased pile through the 250 mm-diameter sleeve, or by providing the sleeve to the augered pile construction. Initially a driven pile was attempted, but this resulted in a shallow pile with severe vibration problems. This system was therefore aborted, and 250 mm-diameter augered piles were installed to a depth of 6 m. The Claygate Beds contained more sand with depth but, with a firm base, an open-bore situation occurred and the piles were constructed with the 2 m sleeve and a single reinforcing bar installed to the design locations. The reinforced concrete works were then constructed at a suitable depth below the existing ground level to support the structure as required.

6.5.8 Merridale, Wolverhampton

Formerly a sandstone quarry, the site is now occupied by a detached dwelling, which had suffered due to settlement. The underpinning system used was a cantilever reinforced concrete beam cut into the sub-structure brickwork and supported on two 100 mm-diameter cased piles. The compression pile was a normal driven pile but the tension pile had to be anchored into the sandstone bedrock at the base of the quarry. This was achieved using rotary percussive drilling techniques to form a socket at the base of the tension pile casing.

The existing service pipes and cables were located and marked prior to the execution of the following sequence of operations.

(1) The locations of the needles were marked on walls and ground.
(2) Needle beam trenches 1.2 m long × 300 m wide were excavated to 400 mm minimum below ground level.
(3) 100 mm-diameter steel casing was driven using a 95 Grundomat for compression piles.
(4) 100 mm-diameter steel casings, with open ends, were driven using a 95 Grundomat. This was achieved by fixing polythene over the end of the open tube and filling an approximately 500 mm length of tube with dry sand/gravel/cement (a drypack). The tube was then bottom-driven to refusal.
(5) A standard drilling rig with air flush and a 75 mm-diameter 'down-the-hole' hammer (on sectional 1000 m-long stems) was used to drill a socket into the sandstone, through the bottom of the casing, to the design length of 1500 mm.
(6) After placing one 12 mm-diameter central high-yield reinforcing bar with spacers, the piles were concreted using the mixed 35 N/mm^2 concrete.
(7) The pockets for the needle beams were cut out of the sub-structure brickwork on a 'hit one, miss one' basis to minimize the temporary weakening of the existing foundations.

(8) The high-yield reinforcement cage was installed in the trench with suitable spacing blocks. The projecting bar from the piles was bent over and tied into the cage and, after inspection, the site-mixed concrete was placed using a vibrating poker.

6.5.9 Factory ground-floor slab failure

Unacceptable differential settlements of concrete slab-on-solid floors occurred. This was the result of the failure of ground treatment, using depth vibrators to form stone columns. Five industrial factory buildings were affected by the floor-slab movements. Ground investigation works revealed that the site consisted of made ground varying in depth from 2–5 m, overlying a weak band of friable sandy clay of some 1–2 m thick. This, in turn, was underlain by a band of peat 40–1200 mm thick resting on clay. Floors had sunk 160 mm in the worst case, but, on average, had settled 75 mm. Internal offices built off the slabs were seriously distressed and doors had either been jammed completely or had been adjusted by cutting them to fit the misaligned door frames.

The brief was to provide new slabs throughout, supported on mini precast concrete piles. Work was to be carried out on a progressive basis so that business could continue with the least possible disruption. The solution proposed was to grout-jack the slabs back to line and level; this, in turn, would straighten and line the office walls and mezzanine floors, and lift the brickwork to the original position.

In order to establish the practicality and economy of the proposals, trials were undertaken in one of the factories, and the office slab was grouted and lifted over two days. The floor was tiled with plastic industrial floor covering and, in the worst area, had sunk 100 mm. Where the slab adjoining the outside wall was supported on a beam foundation (which was, in turn, supported at the piled column footings), there was no floor settlement. Prior to commencement of the trial, the positions of the stone columns were determined. A mobile grouting unit loaded with 7 tonnes of pulverized fuel ash was used. The grout mix was 10:1 pulverized fuel ash:cement and was applied over an area of 90 square metres; it was anticipated that 6–8 tonnes of pulverized fuel ash would be required. It was necessary to form a curtain of grout to limit the extent of slab to be lifted. Grouting to the perimeter of the slab commenced first, at a section of slab that had subsided most.

After pumping for 20 min, cracking and movements were detected and the slab could be seen to be rising. The injection nozzle was continually moved to new positions to effect the lift profile required. Cracks started to appear in the factory slab away from the offices, as the top of the slab was now being subjected to stresses. Even though the slab was reinforced with mesh, cracking was now becoming visible. The office walls built on the floor slab were also making stress noises, and some of the original cracks in the wall were now closing. Injection of the grout continued through strategic holes drilled through the slab inside the offices, until the slab was within 80% of its original position. After a further 1.5 h when 9 tonnes of grout had been injected—the slab was back to within 10 mm of the original level. Restoring floor slabs to original line and level was not necessary, as such slabs are rarely constructed to that degree of accuracy. Final levelling of the surface finish was achieved using a levelling screed. An interesting aspect of straightening walls and floors in this way was that after completion the doors were misaligned, as they had been gradually altered to allow them to fit into increasingly warped door frames.

A scheme was prepared based on all slabs being grout-packed and supported on new piles. In one unit there was a mezzanine floor and a printing press; this unit had not suffered extensive subsidence, and the borehole report indicated that the peat was thinnest under the unit. It was decided, however, that since long-term settlement could arise, remedial work should also be undertaken on this unit. For proper operation of the printing press, the line and level of the machine had to conform to close working tolerances. The solution to this problem was to pile this unit throughout, whilst isolating—by saw-cutting—the floor on which the printing press stood. This formed a free-standing base slab, which was piled in exactly the positions required for the holding-down bolts. In addition, a 65 mm-diameter pile was emplaced, both to form an integral pile support, and to provide a fastening support for the printing press.

The mezzanine floor column loads were determined, and two piles per column were provided at or close to each column base. The installation procedure was again to drill through the floor slab and column base so as to allow the pile to be driven to a load set. The pile was then connected to the slab. Precast concrete piles were generally used but, where stone columns obstructed, heavy-wall steel pile casing was used to overcome the difficulty of driving.

6.5.10 Fylde Coast

Severe subsidence problems resulted from extreme thicknesses of soft clays, loose sands, silt, and peat. The high costs of suitable remedial measures resulted in numerous insurance claims which were dealt with on the basis of diminution of value. Several properties on a new estate began to exhibit signs of excessive settlement. These were four-bedroom detached houses which were constructed on substantial raft foundations, extending up to 2.0 m beyond the perimeter of the walls. These had settled unevenly, producing tilts of up to 200 mm.

The solution devised was to install small-diameter piles up to 12.0 m in depth through pre-drilled holes in the raft from within the property. Twin rolled steel joist beams were then placed across the heads of pairs of piles, resin bolts were installed into the raft, hydraulic jacks were placed on the beams, and a further beam was placed across the head of the jacks. The top beams were attached to the resin bolts such that the jack force would react against the piles via the lower beams and be transferred to the raft via the upper beam and resin bolts. Approximately 40 jacks per property were employed. Prior to commencing jacking, the system was 'locked off' on the beams and bolts, and the perimeter projection of the raft was detached by saw-cutting. By careful and controlled application of load to the jacks, it was then possible effectively to pull the raft back to its original position within a matter of hours. Due to the nature of the system employed, it was possible to safely suspend the operation at any time, and to hold the raft in any desired position secure in the knowledge that, should jack failure occur, there would be no risk to the structure. On completion of the lifting, the piles were resin-bonded to the raft, and the void created beneath the foundation was grouted. After this, the jacks, beams and resin bolts were removed, and the property was restored to habitable condition.

7 Ground freezing

J.S. HARRIS

7.1 Introduction

Artificially frozen ground is used by civil and mining engineers alike as a means of ground stabilization to provide support and to exclude ground-water. Since first used in South Wales in 1862, the process has become recognized worldwide as a reliable method of dealing with adverse ground conditions for both temporary and permanent applications.

As a temporary expedient, freezing has been used for excavations (shafts, tunnels, foundations and storage), stabilizing ground slides, underpinning buildings, sampling weak soils, constructing temporary access roads over boggy ground, and the exploitation of ore reserves in lake bottoms.

Permanent freezing systems are rare outside the arctic regions of the USSR and North America, but include control of Young's Modulus in gravels beneath radar sites, maintenance of frozen soil beneath heated buildings and overhead pipelines founded on permafrost, and in-ground storage containers for cryogenic liquids.

Whereas the principal uses of the method are associated with shaft sinking or tunnel headings, the dramatic increase in strength that is achieved as the temperature is lowered is of particular advantage for temporary load support purposes.

As it is transient in nature and does not affect the water table or groundwater quality, the process finds favour when environmental con-siderations are significant.

7.2 The method

The artificial ground freezing (AGF) process is one of the more versatile geotechnical methods available to engineers, there being few limitations on account of scale, site conditions, soil type or the presence of groundwater. Figure 7.1 indicates the versatility of AGF over a full range of soil types so long as they contain (or can be made to contain) water. Unless there is a heat source (e.g. fast-moving water) in the immediate neighbourhood of the ground being refrigerated, the outcome is predictable and certain even in mixed strata.

220

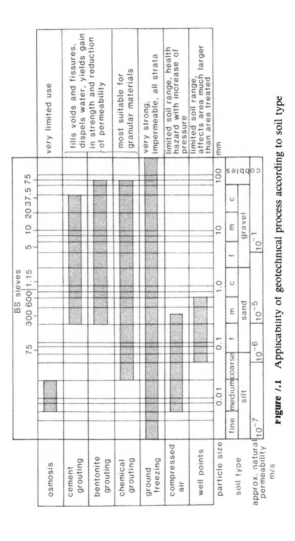

Figure 7.1 Applicability of geotechnical process according to soil type

In the freezing process the pore-water is converted into ice which then bonds the soil particles together, rendering the strata both strong and impermeable. A twentyfold improvement in mechanical strength is common, while the creation of an enclosing frozen membrane offers cofferdam (water-exclusion) performance too.

Frozen conditions are usually created by circulating a cold fluid through a series of coaxial pipes. These freeze-tubes are so disposed that, when the individual ice-cylinders associated with each merge, either a continuous membrane (icewall) is formed around the volume to be protected or excavated, or a supporting column of desired area is created.

The configurations of the site and the existing and/or future structures will determine the shape of the ice-body needed, and therefore the number and orientation of the freeze-tubes. Some examples are given in Figure 7.2 and further referred to in section 7.6. A circular or elliptical shape is preferred if an excavation is to be protected.

Conventional rotary drilling or soft ground boring methods are normally used to form holes into which the sealed freeze-tubes can be introduced.

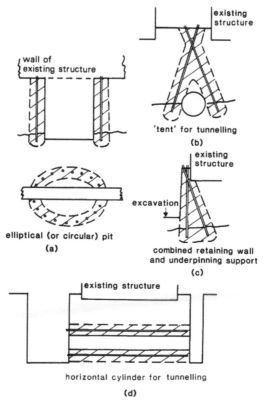

Figure 7.2   Alternative freeze-tube patterns

Alternatively, the freeze-tubes themselves may be drilled or jetted directly into place. It is often necessary to survey the course of the boreholes to ensure that a regular pattern has been maintained over the depth of penetration.

The several attributes of AGF are considered when assessing its relevance as a constructional aid in particular circumstances; once selected, the planner can usually make economies in his temporary works. Thus the significant increase in soil strength enables excavations to proceed without the need for any further support, while the encumbrance of walings and struts is avoided. It should, however, be noted that if the permanent structure/lining is not being constructed immediately following excavation it may be necessary to insulate the exposed icewall to prevent its deterioration.

Similarly, the impermeable nature of frozen ground avoids the need for groundwater lowering over a wide area and, when the total cofferdam has been created, the trapped water can be pumped away to leave a dry stable excavation. Settlement problems that often arise from vibrations during sheet piling, or from the removal of fines during groundwater lowering, are eliminated. The freezing and thawing cycle is simply a physical transformation—a reversible change of state. There is therefore no contamination of aquifers, nor is the level of the groundwater affected. Medical hazards and non-productive time associated with compressed air working do not arise.

Limitations are few, being mainly related to water movement. A flow of water across a zone being refrigerated is a continual source of added heat which at best will require added effort to overcome, or at worst will preclude the use of the method. For this reason extraction of water from nearby wells must be limited or avoided during the currency of refrigeration. It is generally accepted that groundwater velocities should not exceed 2 m per day, particularly if mechanical refrigeration is being used (see below).

7.3 Methods of refrigeration

7.3.1 On-site mechanical plant

The most commonly used plant consists of a reciprocating or screw compressor operating with ammonia or freon. The refrigeration achieved is used to chill a brine-based heat transfer medium (or secondary refrigerant) which is circulated through the freeze tubes before being returned to the plant to be rechilled in a closed-circuit system. The heat removed from the rock or soilwater is dissipated to the atmosphere via a cooling tower or an evaporative condenser or to a convenient water source, e.g. lagoon or river.

The component parts of such a plant—the compressor, motor, heat exchanger(s), expansion valves, pumps, switchgear and instrumentation—are usually all packaged on a common base and installed in a container (see Figure 7.3), or on a trailer, for ease of assembly on site.

Figure 7.3

With two-stage compound compressors, using ammonia as the primary refrigerant and calcium chloride brine, temperatures down to − 35°C or even − 40°C can be achieved when required. A usual design value is in the region of between − 25°C and − 30°C to create the required frozen conditions in a period of only a few weeks.

7.3.2 Off-site-produced expendable refrigerant

Cryogenic liquids—those substances which evaporate at very low temperatures unless stored under pressure in insulated vessels—may be used as refrigerants. Of these, liquid nitrogen and liquid carbon dioxide are produced in bulk for commercial distribution and sale. Both are inert and can be vented safely to atmosphere after their work cycle. Liquid nitrogen is available from many more distribution points than is liquid carbon dioxide, boils directly to gas within the pressure range experienced, and is therefore generally used. With closely-spaced freeze tubes a rapid primary freeze period of only a few days is easily attainable in most situations.

With cryogenic refrigeration site plant requirements are simple, a power supply is not necessary, and the system is silent as well as fast. However, as the refrigerant is exhausted to atmosphere after only one cycle, it being impracticable to recondense it, the cost advantage of fast freezing can rapidly be lost if the frozen state has to be maintained for a long period.

These features often restrict choice of cryogenic refrigeration to remedial works where action is urgently required, or to smaller-scale applications including short tunnels, sealing of damaged sheet piles and underpinning.

7.4 The properties of frozen ground

7.4.1 Strength and creep

The frozen strength of any soil or rock type is dependent on its moisture content and the temperature below freezing point to which it is subject. In general, at a given sub-zero temperature, sand will be stronger than silt, and silt will be stronger than clay. Typical strength/temperature relationships for some generalized soil types have been published from time to time (Figure 7.4).

Frozen sands and soils with large pore spaces develop reasonably high compressive strengths at temperatures only a few degrees below freezing point, but in silty and clayey soils very substantial proportions of the total moisture content may remain unfrozen at temperatures as low as − 10°C. The more clayey the soil the greater the proportion of unfrozen moisture, and this has a marked influence on strength.

Lovell (1957) has shown that the compressive strength of clay–silt soils increases three- to fourfold as the temperature is reduced from − 5°C to

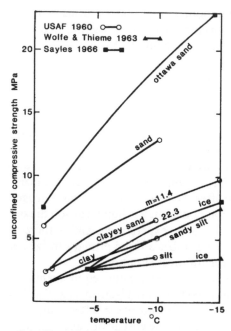

Figure 7.4   Unconfined compressive strength of frozen soils

$- 18°C$. Furthermore the unconfined compressive strength increases exponentially with the relative moisture frozen. As an example, for a silty clay studied, the amount of moisture frozen at $- 18°C$ was only 1.25 times that frozen at $- 5°C$, but the compressive strength at the lower temperature was over four times as high as that at the higher temperature.

As with unfrozen clays, ice and frozen soils behave in an elasto-plastic manner, and under steady load will exhibit creep (time-dependent strain). The mechanical parameters are all both temperature- and time-dependent. The characteristic deformation–time curve at constant temperature and constant load follows firstly a decreasing, then a rapidly increasing, strain-rate with time characteristic (Gardner *et al.*, 1984), the inflection being defined as the failure point (Figure 7.5).

From tests on a sample of organic silty clay Sanger and Kaplar (1963) have shown that, at 80 psi (0.6 MPa) this failure occurred after 17 hours when the sample was maintained at just below $0°C$, yet that failure point had not been reached after 60 hours when tested at $- 2.5°C$.

A combination of high ground pressure and small negative temperatures may, in some frozen soils, lead to measurable creep. A high rate of creep dictates special techniques and monitoring. Consideration of creep does not generally become important in freezings of shallow to medium depth, and is usually confined to formations such as plastic clays or organic silts.

The strength and creep properties are measured in the laboratory on

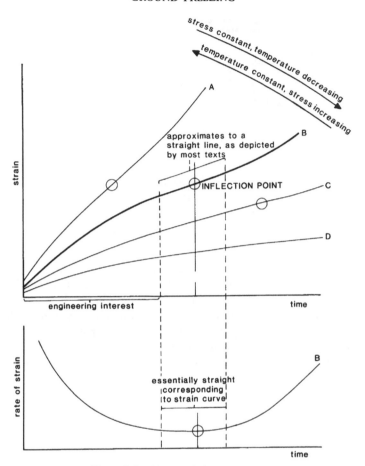

Figure 7.5   Characteristic creep curves

specimens recovered from in-situ frozen locations or on specimens frozen in the laboratory under simulated in-situ conditions.

Specialist testing facilities are required, which include a cold room whose temperature can be controlled to within $\pm 1°C$ for storage of field-frozen samples, creation of laboratory-frozen specimens, and workshop-type activities when the specimens are prepared and conditioned for testing.

In the field freezing does not occur triaxially, but as the freezing front advances. To simulate this in the laboratory, the specimen is frozen uniaxially. Another process associated with freezing is moisture migration, as described later (section 7.4.2). A single apparatus, illustrated in Figure 7.6, satisfies these various conditions: the water contained in the brass pot is well insulated and is therefore the last component to be frozen when the apparatus is placed in the cold room. At this stage the specimen is chilled from the top, the zero isotherm advancing steadily from top to bottom, until totally frozen. During

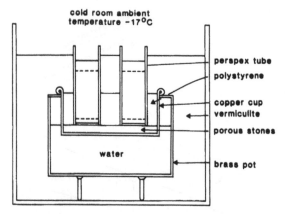

Figure 7.6 Apparatus for frozen sample preparation.

this time the surface of the water is free to find its own level according to any volume changes that may arise. The specimen is then released from the apparatus and stored until the test is to be conducted.

Strength is determined by testing at a constant rate of strain, while creep parameters are measured by tests performed at constant stress. The refrigerated compression equipment illustrated in Figure 7.7 has been built to perform either type of test. A microcomputer program controls the loading

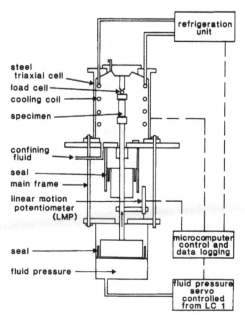

Figure 7.7 Refrigerated triaxial compression apparatus

applied at any time and, in the case of creep tests, takes account of change of cross-sectional area as the sample shortens.

7.4.2 Frost heave

The change in the volume of water as it is chilled is well known. Much of this volume increase is dissipated by expulsion of water ahead of the freezing front, and is a function of the degree of saturation and the permeability of the ground. The remaining volume increase creates a pressure which may result in compression frost shattering of the bounding strata, or ground heave, or remains as a stress, according to the overburden pressure. Such heave as does occur at this stage is very small, and usually insignificant.

Ground heave of an order likely to cause distress occurs at sub-zero temperatures when a so-called frost-susceptible soil has access to a plentiful supply of water. It has been observed, in respect of a surface layer of soil which is subject to climatic freezing, that the soil must have an open system (a condition in which free water in excess of that originally contained in the voids is available to be moved to the freezing front) if segregated ice is to form, i.e. for it to be a frost-susceptible soil (Arvidson and Morgernstern, 1977). Such conditions usually occur in silts or fine sandy silt where the mechanism of upward water movement is attributed to capillary forces acting in the pore spaces of a soil near to the groundwater table within a medium of reasonably high permeability which permits mass movement while freezing conditions apply. The ice lenses so formed displace the ground above and heave is observed at the surface.

Jumikis (1966) conducted a series of experiments to illustrate and explain the upward migration and freezing of moisture which can in some cases result in appreciable rise in ground level. These experiments were directed mainly at the behaviour of road subgrades but the conclusions are relevant in some instances to artificial ground freezing. Freezing as a constructional aid usually involves the creation of vertical boundaries between frozen and unfrozen material, primary water migration therefore being essentially horizontal. Another differing feature is that the temperature gradient across the interface is very much greater than with climatic freezing, thus reducing the conditions which favour bulk migration of water.

In general, heave is not a serious problem, but must be expected and allowed for in unfavourable conditions.

7.4.3 Concreting

When pouring concrete against frozen ground, a principal concern is that the concrete should not itself become frozen before achieving the initial set (Altounyan et al., 1982). Generally the heat of hydration will be sufficient to prevent advance of the freezing isotherm for several days. Increased heat of hydration can be achieved by the use of rapid-setting cements, by using additives such as calcium chloride, by preheating the aggregates and/or by mixing with hot water. Additionally an insulating layer may be placed against the exposed icewall, such that a considerable time can elapse before the temperature of the setting concrete falls to freezing point. During this time a limited thawing of the innermost part of the icewall will occur as the temperature gradient from the hot concrete distorts the location of the zero isotherm. Care must, of course, be exercised to ensure that overheating does not occur.

7.5 Design

7.5.1 Structural

For shallow excavations the calculation of an adequate thickness of a cylindrical or circular icewall to withstand earth and hydrostatic pressures is commonly based on the Domke formula using strength parameters. An extension of the Lamé approach to thick cylinders, the Domke formula was the first to recognize and take some account of the plastic behaviour of frozen ground, and can be usefully expressed in the form

$$S_x = a\left[0.29\frac{P}{Q} + 2.3\left(\frac{P}{Q}\right)^2\right]$$

where S_x = icewall thickness
   a = radius of excavation
   P = external (ground) pressure
   Q = compressive strength of frozen ground.

The failure criterion is simple, and the formula only allows calculation to be carried out on the basis of strength with no check on deformation.

Later work by Vialov (1966) and others based on the Mohr–Coulomb plastic shear criterion, and in some cases taking account of end restraint, has generated many formulae. A useful one for shallow frozen ground structures may be conveniently written as

$$P = \frac{2cN^{1/2}}{N-1}\left[\left(\frac{b}{a}\right)^{N-1} - 1\right]$$

where N = flow factor $\dfrac{1 + \sin\phi}{1 - \sin\phi}$

   ϕ = limiting angle of shear resistance of frozen ground
   b = radius of outer face of icewall.

For small-diameter, shallow pits as required for underpinning purposes, the simplest check by Domke will normally be sufficient to determine the minimum thickness required.

When frozen ground is to be utilized as a loadbearing element it is also necessary to check its bearing capacity. The Terzaghi formula is appropriate (Andersland and Anderson, 1978). But, because the strength is both temperature and time-dependent, the value used in estimating the allowable bearing capacity must take due account of the thermal regime that will apply during the working life of the ice-body. Likewise it is necessary, during the execution of the works, to ensure that the design temperature of the ice-body is attained and preserved.

For tunnels and deep shafts the Finite Element Method (FEM) allows the deformation and stresses to be analysed. This is particularly useful in soils subject to creep, as the changes in deformation and stress with time can also be computed and presented in easily interpreted, diagrammatic and graphical form. Although computer time is expensive, FEM modelling represents the ground behaviour more accurately than elastic or elasto-plastic analysis and is therefore to be preferred for sensitive cases.

7.5.2 Thermal

Having established the volume and average icewall temperature needed structurally, the thermal properties of the formations and the refrigeration capacity are equated with freeze-tube dispositions and freezing periods to establish the scope of the projected freezing operation.

As heat is abstracted from the ground, cooling proceeds in three distinct regimes: ground already frozen is being further cooled, water at the freezing front is being frozen by removal of its latent heat, and the ground beyond the freezing front is being cooled below its natural temperature. The many variables at any given location, which are rarely measured, make a full and rigorous analysis impossible. As with permeability, thermal conductivity varies with the direction of heat flow and with the phase and temperature of the wet material. The moisture content and the velocity of any groundwater flow will affect refrigeration demand.

In practice the designer also correlates theoretical calculations with data collected from previous work in this field. Design methods to assess plant capacity and the time to achieve icewall integrity have been given in the literature by many workers, e.g. Jumikis (1966), Collins and Deacon (1972), Shuster (1972), Sanger and Sayles (1978), but all require refinement based on the experience of the major practitioners, much of which remains their in-house know-how.

The FE method allows detailed prediction of temperature distribution with time, information which is an essential prerequisite to the stress and deformation analyses referred to in section 7.5.1.

7.6 Applications

7.6.1 Open pits and shafts

The common feature of shaft work is that excavation takes place within an icewall—usually circular or elliptical—which encloses the construction volume and isolates it from the groundwater. This principle is also applied, at reduced scale, to gain access beneath foundations of existing buildings to facilitate their underpinning. The position of the freeze circle—the pattern of the freeze-tubes in plan—is located such that the line of the footing is a chord, as shown in Figure 7.2(a). When the enclosed space has been excavated to a stable stratum, the underpinning support which will transfer the load of the building to the stable stratum can be constructed. Once this support is secure the frozen ground is allowed to thaw. During the operation it may be necessary to introduce a bridging beam beneath the original footing to prevent degeneration; if so, the icewall will act as a temporary loadbearing support for that beam, although this has not been shown on the figure (see example v).

The permanent underpinning support may be constructed as a simple extension downwards of the original foundation, i.e. to a similar width, followed by backfilling of the remaining void; alternatively it may be cheaper to erect a single shutter to isolate the underpinning area and the smaller sector and backfill that volume with concrete, utilizing the icewall itself as the remainder of the form.

7.6.2 Tunnels

As with pits and shafts, freeze-tubes can be installed parallel to the tunnel axis, as in Figure 7.2(d). Excavation can then take place within the enclosing icewall although, with time, the ice-front will advance into and across the trapped space; with tunnel diameters of 3 m or less the core is likely to be totally frozen within three weeks of the zero isotherm crossing the excavation line.

Alternatively the freeze-tubes can be installed from the surface to straddle the line of the tunnel (unless there is no cutoff stratum, when additional freeze-tubes may be needed through the tunnel space to ensure ice closure across the base). Where there are space restrictions at the surface, as in Figure 7.2(b), angled freeze-tubes to create a tent-like cross-section may be appropriate.

7.6.3 Structural support/retaining wall

Figure 7.2(c) illustrates a freeze-tube pattern to combine the functions of temporary support to the existing building foundation and a retaining wall to the soil under the building while the neighbouring site is excavated (see example v). It may be necessary, additionally, to install freeze-tubes from within the existing building to create a wider bearing area and thicker icewall, and/or to install ground-anchorages as excavation proceeds, and/or to buttress the icewall if the excavation is particularly deep.

7.7 Examples

(i) At Timmins in Northern Ontario, Canada (1974), exploitation of an ore-body was commenced at the surface by open-pit mining, and also underground from a 900 m-deep shaft sunk within the only outcropping rockhead in the neighbourhood.

There being no further outcropping stratum capable of providing an adequate foundation for the 1500 m-deep number 2 shaft headframe, a deep excavation had to be made through the overburden of fill, muskeg (peat) and soft silty clay. The bedrock dipped from 10 m to 17 m depth across the diagonal of the required 21 m × 16 m headframe (see Figure 7.8).

A frozen enclosure through the overburden, comprising four arcs in plan, was established using a double row of freeze-tubes to 12 m deep, and a triple row at greater depths. Excavation of the unfrozen central core was undertaken by digger/backactor plant and the spoil pushed to one corner by dozer for

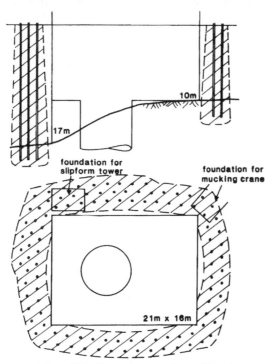

Figure 7.8 Open pit excavation at Timmins

removal to tipper trucks by crane-operated grab. After preparation of the rock surface, headframe erection was commenced using slipform methods. Then shaft sinking could proceed conventionally.

Without ground freezing such an excavation would have needed extensive temporary support by sheet piling, framing and strutting with consequential interference to mucking and construction operations. Besides safeguarding the excavation the icewall also served as the stabilized support for the mucking crane at a shallow corner, and for the Alimak slipforming tower where the icewall was at its thickest.

(ii) In Washington DC (Anon, 1979), excavations for the foundation of a new office tower caused cracking and other damage to neighbouring property. Shoring, which was erected to brace the building, was not completely effective.

The whole length of the exposed sidewall was therefore temporarily supported by a frozen-ground retaining wall extending 11 m in depth, in the manner of Figure 7.2(c). Some 63 freeze-tubes were installed vertically and at various angles to freeze mixed strata comprising soft clay, silt and loose fine sand using a 26 TR trailer-mounted freeze plant, chilling calcium chloride brine.

(iii) At Blackpool (Maishman, 1973), on the other hand, it was feared that excavations to be made within a frame building, supported on shallow square footings to each column, would cause unacceptable distress to that structure.

The ground conditions consisted of 3.5 m of sand underlain by 2 m of peat, then a further 8 m of soft sandy silt to the basal stiff clay formation. Bored cast-in-situ piles were first constructed through the overburden to the stiff clay bearing stratum; the concreting was carried out to the eventual 6 m-below-surface pit floor level only. A frozen-ground icewall was then created around the intended excavation, extending down to the stiff clay, thus ensuring that the weak sandy silt (and the piles it then contained) could not be disturbed, e.g. by heave/buoyancy of the pit floor as the excavation reached its full depth. Once the freezing had been effected, the excavation could proceed over the full 12 m diameter until the prebored piles were exposed in the pit bottom. After casting of the pit structure had been completed, the frozen ground was allowed to thaw. The whole exercise was completed without affecting the existing frame structure.

(iv) In Gibraltar (Harris and Woodhead, 1978) large diesel-driven generating sets supported by shallow foundations and constructed on reclaimed land had suffered differential settlement in service. The requirement, when replacement of the generating sets became due, was to transfer their load through the fill to bedrock some 7 m below floor level. The foundation design called for twin piles, 2.4 m diameter under each 9 m × 4 m foundation block.

The filled ground consisted mainly of limestone rubble from the many tunnels driven in the rock, mixed with sands dredged from the seabed, a material too permeable to achieve dry conditions by pumping methods. The original foundation was first removed to just above the water table (sea level).

Fourteen freeze-tubes were then installed by drilling to 6 m depth, giving 1 m penetration into the new foundation stratum, a mudstone bedrock. Refrigeration for each pile occupied 3 to 4 weeks of the whole 12-week programme.

The same principle could have been applied to create say eight piles of equivalent total load-carrying capacity beneath the original footing, had removal/replacement of the generator sets not been required.

(v) A 15th-century palace/fortress in Burgos (Spain) was to become the HQ of a Spanish bank, and the renovations, alterations and enhancements were designed to include/preserve as much of the fine Gothic–Renaissance period architecture as possible, particularly two of the facades. The enhancements included excavating within the structural boundary to a depth of 11 m to provide 3 sub-surface floors for parking space and safety vaults.

Below the 2 m deep footings lay 4.5 m of gravelly sand, 3 m of limestone, then bedrock of marl with gypsum; the phreatic level was 2.5 m below ground level. Thus, the foundations of the boundary walls had to be extended downwards: (a) to transmit the load directly to the bedrock; and (b) to achieve a cofferdam within which the works could safely proceed. Part of this could be undertaken by conventional diaphragm walling techniques, but to avoid the possibility of distress to the two facades a more refined system was devised. In essence the solution embodied a series or 'chain' of circular frozen cylinders (shallow shafts) as in Figure 7.9. Each 'shaft' was of 4/4.5 m internal diameter, and the procedure was:

(a) to excavate alternate shafts, and construct within them a 3 m long section of wall immediately below the existing footing;
(b) to fill the remaining void with compacted material;
(c) to excavate the remaining 5 m diameter overlapping shafts; and

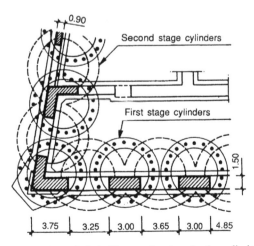

Figure 7.9 Plan of chain-like overlapping circular cylinders.

(d) to construct the intervening sections of underpinning to complete the cofferdam, and again backfill the void with compacted material.

The work was carried out in 1983–84 using two freon freeze-plants each of 160 K kcal/hr capacity, and using 'terpeno', an aromatic hydrocarbon of low viscosity to −60°C, as the secondary refrigerant instead of brine. The primary freeze period for each shaft was 15 days. For safety reasons the frozen boundary was taken to be the −5°C isotherm, to ensure that the icewall thickness at any point was not less than 1 m. The construction works during which the integrity of the icewall was maintained varied between 25 and 30 days.

There were no adverse movements, either of heave or settlement, and use of the ground freezing method ensured that the facades were preserved without distress.

(vi) Near Wrexham, one case is recorded (Gregory and Maishman, 1973) where the process was used to safely abandon an old structure before new construction was commenced over the same site. This arose when a bypass to motorway standards was scheduled to traverse an old colliery site. Twin 300 m-deep mineshafts had stood open since mining ceased there in 1926; one of these shafts lay within the fencelines of the proposed new road. The standing water level in the flooded workings was only a few metres below ground level and coincided with the general groundwater level of the 40 m-thick alluvial overburden deposits.

Thus, to isolate the mineworkings from the effects of any subsequent fluctuation of the groundwater table, it was necessary to construct a reinforced concrete plug within a strong impermeable stratum below the saturated overburden. The shallowest suitable stratum occurred at 60 m below the surface.

To achieve this, the shaft was first backfilled with broken stone to the 60 m level. Twin 0.8 m-diameter sealed steel tubes were then floated into the shaft to rest on the stone filling; after placing freeze-tubes between the twin tubes and the shaft wall, the remaining annular space was filled with sand. Refrigeration of this saturated sand then excluded the groundwater from entering the shaft, and permitted the use of the twin tubes as access, mucking, service and ventilation shafts while excavation for and casting of the reinforced plug took place.

(vii) Metro construction in Antwerp and Brussels involved tunnelling beneath streets, buildings and other structures of various ages and descriptions. At several places the building foundations and the roof of the intended tunnel were in superficial deposits, often of unsaturated sand.

In open locations the sidewalls of the rectangular-section tunnels could usually be constructed as diaphragm walls, followed by cut-and-cover methods to excavate the tunnel space and construct the floor and roof without the need for ground treatment. Where this was not practicable due to the presence of occupied property, the strata above the tunnel roof and flanking the sidewalls were stabilized by ground freezing. From headings constructed in non-vulnerable locations, or within the protection of frozen ground, horizontal freeze-tubes were drilled over the intended roof. One or two arrays of parallel or fanned freeze-tubes were provided according to the thickness of frozen ground canopy required, the thickness necessary being a function of span and loading to be catered for—see Figure 7.10(*a*).

It had been established by tests that the frozen strength of the sand varied according to both its moisture content and the sub-zero freezing temperature. For example, at optimum moisture content the strength at a particular temperature was three times that at its natural moisture content (less than

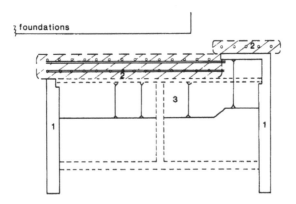

(a) Frozen roof over cónstruction formed in headings

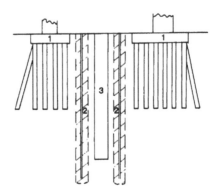

(b) Diaphragm wall constructed between existing sensitive foundations

Figure 7.10   (*a*) Frozen roof over construction formed in headings. 1, flanking walls constructed under bentonite; 2, frozen roof (canopy); 3, excavation in headings with propped supports. (*b*) Diaphragm wall constructed between existing sensitive foundations. 1, piled foundations to existing railway bridge; 2, twin icewalls; 3, new diaphragm wall.

8%), while at optimum moisture content the strength at $-15°C$ was 1.4 times that at $-5°C$.

In view of the low natural moisture content, slow irrigation was applied to the area being stabilized during the primary freeze period to raise the frozen moisture content to a level that would result in the high strength needed to safeguard the structural loads to be supported. Adits were then excavated at regular spacing, and concreted; when this concrete had cured, the intermediate ground could be excavated in its turn, which, when the whole had been concreted, produced a contiguous roof construction.

At one location the metro tunnel had to pass beneath a railway bridge which was supported on piles. The flanking walls of the new tunnel could be constructed under bentonite, but they were required to pass between and very close to the piled footings without causing any disturbance to them—see Figure 7.10(b). This was achieved by forming twin icewalls, one either side of the intended bentonite wall, to act as retaining walls which would prevent movement of the piled foundations.

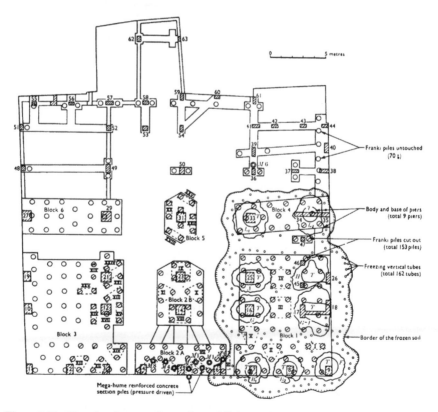

Figure 7.11 Plan showing area frozen for stabilizing, and the new piers (T). From Dumont-Villares (1953) with permission.

(viii) In São Paulo, Brazil (Dumont-Villares, 1956), settlement occurred very rapidly as a 26-storey building was nearing completion in 1942, and urgent measures became necessary to arrest the movement, eliminate tilt and restore the building to its original elevation.

São Paulo is located on a basin of Tertiary sediments, successively partially saturated clays, sands and silty sands which vary in thickness, strength and compressibility, saturated clayey sands of low to medium density, and decomposed then solid gneiss or granite at a depth of some 90 m. Considerable variations occur within short distances both horizontally and vertically, calling for careful investigation at all sites.

In accordance with recognized practice Franki piles were installed with their bulbs founded within the stiff clay layer identified over much of the site (Figure 7.11). Nothing untoward was observed during installation or testing of the piles, and construction of the building then proceeded rapidly without mishap until its full height was reached.

The significance and scale of the settlement was realized some five months after completion of the main structural work, when setting out of the lift guides was started. At that time the total settlement was about 60 mm, and tilt was 1 : 500 towards the front right-hand corner, the whole building moving monolithically. Investigation borings established that the settlements taking place resulted from the presence of a soft silty clay wedge which had not been detected before construction.

The primary task of arresting the tilting movement was achieved by ground freezing under the affected corner. A total of 162 freeze-tubes was placed in holes drilled though the heavily reinforced concrete floor slab and the underlying strata to a depth of 14 m. During placement of the freeze-tubes the rate of settlement increased alarmingly to a peak of 6 mm per day before commencement of refrigeration on 9 February 1942. The last freeze-tube was installed and commissioned on 25 May. During this $3\frac{1}{2}$-month period of increasing refrigerative effort the rate of settlement lessened, until on 19 June settlement ceased and the building became stable. Continuing refrigeration resulted in a small uplift recovery of 10 mm at the point of previous maximum settlement.

Preparations were then made to eliminate the tilt and restore the building as nearly as possible to its original elevation by jacking. A number of freeze-tubes were disconnected to ensure uniform distribution of effort, and to release capacity for additional freeze-tubes while two further columns were built. Several of the original Franki piles within the frozen area were by now under tension and had to be cut, while several columns near the frozen boundary were receiving increased loading. This dictated the provision of additional piles, and the 'Mega Hume' segmental reinforced type was used for this purpose.

Cast-in-situ mass concrete piles in belled pits were installed through the frozen ground into the deep sand to act as the permanent 5 kg/cm² (0.5 MPa) load support when the ground was allowed to thaw. Conventional inverted jacking methods, with steel wedges, were used to level the building on these piles and to take up changes during the thaw period which lasted some 18 months.

7.8 Summary

Ground freezing has played an increasing role in the civil engineering field generally, and underpinning problems specifically, in recent years, and may be expected to contribute there as well as in mining engineering in the future.

The advent of large-scale commercial development of liquid nitrogen as an inert refrigerant is a significant factor in the increasing popularity of freezing for small and emergency excavation problems. Speedy execution, coupled with minimal plant and power requirements at site, are undoubted advantages. The elimination of noise is environmentally attractive. Nevertheless, mechanical refrigeration is more attractive for large-scale applications and for those where the frozen state has to be retained for a long time.

References

1. Altounyan, P.F.R., Bell, M.J., Farmer, I.W. and Happer, C.J. (1982) Temperature, stress and strain measurements during and after construction of concrete linings in frozen sandstone. *Proc. ISGF '82*, Hanover, N.H., 343–348.
2. Andersland, O.B. and Anderson, D.M. (1978) *Geotechnical Engineering for Cold Regions.* McGraw-Hill, New York.
3. Anon. (1979) Soil is frozen to save Tippy Tavern. *Eng. News Record*, 10 May, 12.
4. Arvidson, W.D. and Morgernstern, N.R. (1977) Water flow induced by soil freezing. *Can. Geotech. J.* 14, 237–245.
5. Collins, S.P. and Deacon, W.G. (1972) Shaft sinking by ground freezing for the Ely Ouse–Essex scheme. *Proc. ICE*, May, 129–256.
6. Dumont-Villares, A. (1956) The underpinning of the 26-storey 'Companhia Paulista de Seguros' building, São Paulo, Brazil. *Géotechnique*, 1–14.
7. Gardner, A.R., Jones, R.H. and Harris, J.S. (1982) Strength and creep testing of frozen soils. *Proc. ISGF '82*, Hanover, N.H., 53–60.
8. Gardner, A.R., Jones, R.H. and Harris, J.S. (1984) A new creep equation for frozen soils and ice. *Cold Regions Sci. & Technol.* 9, 271–275.
9. Gregory, O. and Maishman, D. (1973) Motorway construction meets an unusual problem in old shaft treatment. *Proc. IMinE* (Manchester branch), February.
10. Harris, J.S. and Woodhead, F.A. (1978) Ground freezing for large-diameter foundation piers. *Consulting Engineer*, January.
11. Jumikis, A.R. (1966) *Thermal Soil Mechanics.* Chapter 11. Rutgers University Press, New Brunswick.
12. Libad, F.M., Moreno-Barbara, F. and Ortiz, A.U. (1985) Cimentacion para la rehabilitacion de un edificio antiguo. *Inf del Laboratorio de Carreteras y Geotecnia del CEDEX no. 171/PT 46*.
13. Lovell, C.W. (1957) Temperature effects on phase composition and strength of a partially frozen soil. *Highway Research Board Bull. 168*, Washington DC.
14. Maishman, D. (1975) Ground freezing (at Blackpool). In Bell, F.G. (ed.), *Methods of Treatment of Unstable Ground, Proc. Symp. Sheffield 1973*, Newnes-Butterworth, Sevenoaks, 159–171.

15. Sanger, F.J. and Kaplar, C.W. (1963) Plastic deformation of frozen soils. *Proc. Int. Conf. on Permafrost*, Purdue, 1963.
16. Sanger, F.J. and Sayles, F.H. (1978) Thermal and rheological computations for artificially frozen ground construction. In *1st Int. Symp. on Ground Freezing*, Bochum, 1978, 311–338. Reprinted in *Eng. Geol.* 13, 1–4.
17. Shuster, J.A. (1972) Controlled freezing for temporary ground support. *Proc. 1st Int. Conf. on Rapid Excavation and Tunnelling*, 1972, 863–894.
18. Vialov, S.S. (1966) Methods of determining creep, long term strength and compressibility characteristics of frozen soils. *Tech. trans. from Russian no. 1364*, Natl. Research Council of Canada.

8 Underpinning by chemical grouting

G.S. LITTLEJOHN

8.1 Historical introduction

The use of chemicals in grouting evolved logically from cement grouting practice where direct injection of neat cement into fine fissures or small pores was only partially successful. At Thorne in Yorkshire, for example, two shafts started in 1909 came to a standstill at a depth of 150 metres due to heavy water ingress through porous sandstone which contained fine fissures. With a background of proven experience from Hatfield colliery in 1911, the Belgian engineer François employed silicatization at Thorne in 1913, and this commercial success established the process in engineering practice. The technique involved the injection of sodium silicate and aluminium sulphate solution, after which neat cement grout was injected with comparable ease. François concluded that the chemical gel acted simply as a lubricant, which perhaps explains why he did not develop the system for the treatment of alluvium, in spite of the fact that the use of sodium silicate as a grout had been known since 1886 through a patent by Jesiorsky.[1] In reality the chemical gel filled the fine fissures and pores, thereby sealing the walls of the major fractures. Without such a seal and under high injection pressures the water would have been driven from the cement grout into the porous structure of the rock, leaving the grout to stiffen prematurely. Following the early commercial successes in shaft sinking with silicatization, the method was employed in many other countries, particularly South Africa.[2]

As a natural consequence of these successes mining engineers soon turned their attention to the grouting of finer-grained sandstones and sands as a cheaper and quicker alternative to the Poetsch freezing process then in use. The filtering effect of cement in fine-grained materials had been well known for some time through the work of Portier (1905),[3] the inventor of the cementation process for shaft sinking, so the need for a low-viscosity fluid to penetrate the pores and thereafter solidify by chemical reaction was understood. Although Lemaire and Dumont had patented a single-shot process based on dilute silicate and acid solution in 1909,[4] it was not until 1922 that Durnerin,[5] apparently unaware of the chemical grout patents, observed that in the field the reaction would

242

require the use of two reagents, suggested those which would give a gelatinous precipitate of either silica gel or hydrated iron oxide, and then demolished his own proposals on practical grounds. Within three years the Dutch engineer Joosten had solved the practical problems by an ingenious method for the treatment of sands where small volumes of concentrated sodium silicate were injected in stages through a perforated pipe as the pipe was driven to the required depth. Subsequently, as the pipe was drawn back in the same stages a strong brine solution was injected (Figure 8.1). The brine displaced and reacted almost instantaneously with the silicate to form a soft gel in the pores, with the adsorbed film a hard dehydrated gel binding the sand grains together at their point of contact to form a sand mass with crushing strengths of 3 to 5 N mm^{-2}.[6] The high cost of injecting two fluids and the close centres of holes (600 mm in sands) due to high initial viscosities led to a low viscosity system by Guttman who diluted the sodium silicate with sodium carbonate solution, but the main search was for a single-fluid grout of low viscosity which would set after a suitable time. In this respect the majority of commercial successes were based on sodium silicate, popular reagents including lime water, sodium bicarbonate and sodium aluminate. Whilst all gave a soft gel which greatly reduced permeability of the ground there was no appreciable gain in strength. Nevertheless such solutions were used on a large scale, e.g. 4530 tonnes of silicate injected by Rodio at Bou-Hanifia[7] in Algeria under the advice of Terzaghi. This project started in 1933, and it was here that Ischy used his invention, the tube à manchette, which permitted grouts of different properties to be injected in any order and at any interval of time from the same borehole (Figure 8.2).

Starting in 1934 Mayer,[8] at the Laboratoire du Bâtiment et des Travaux Publics, developed successfully a single fluid silicate grout with controlled gelling using hydrochloric acid and copper sulphate, and a cheaper version incorporating stabilized clay whose technical feasibility was confirmed at the Barrage du Sautet on the River Drac. The first large commercial success was in 1936 during the construction of the dam at Genissiat on the River Rhône.[9]

By this time the complex structures of alluvial deposits and methods of measuring permeability were becoming better understood following Terzaghi's early work on soil mechanics published in 1925, and in 1938 a simplified theory of injection into granular material was published by Maag[10] relating factors such as injection pressure, flow rate, density and viscosity of grout, ground porosity and permeability based on the assumption of spherical flow through homogeneous and isotropic material.

Whilst the 1939–45 war naturally hindered practical developments, the activity on patents covering new chemical formulations increased dramatically.[11] During the 1940s phenol–formaldehyde and resorcinol–

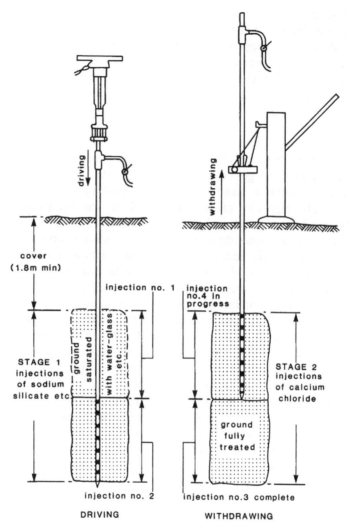

Figure 8.1 Joosten process (after Glossop[1]).

formaldehyde systems evolved (phenoplasts), and by 1953 de Mello, Hauser and Lambe had filed a patent covering acrylate of polyvalent metal (AM-9). This American formulation, although now replaced by a non-toxic version, was unique as a waterproofing grout due to its very low viscosity (1.2 cP), excellent gel time control ranging from a minute to several hours, and its ability to treat fine silts.

On the practical front the early 1950s saw the establishment of chemical grouting as a recognized geotechnical process with particular regard to

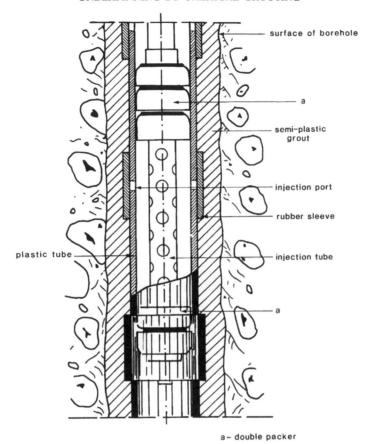

a- double packer

Figure 8.2 Ischy's tube à manchette.

dam cut-offs and tunnel support, the most popular chemicals being silica-based, e.g. aureole grouting at Acif-el-Hammamam, Algeria (Figure 8.3). In this regard Soletanche created a hard silicate gel using an organic ester (ethylacetate)[12] in 1957, capable of producing grouted sand strengths of $2-3 \, N \, mm^{-2}$. Over the same period the process of gelification of ligno-sulphonate using dichromate (chrome-lignins) received attention in England, Sweden and Russia.[12] These grouts provided good gel time control (5–120 min) and grouted sand strengths of $1-2 \, N \, mm^{-2}$, but with a potential dermatitis risk to personnel.

In 1963 the state of the art in grouting was reviewed at the ICE Conference in London[13] and Table 8.1 illustrates the major commercial chemical grout systems considered practicable at that time.

Over the past two decades a great variety of chemical systems has been

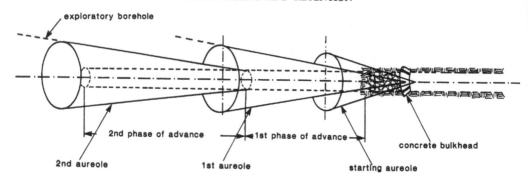

Figure 8.3 Pregrouting by aureoles.

introduced into the grouting market covering a wide range of materials, properties and costs, but the basic objective has been to improve or replace existing grouts. Of particular note are the water reactive materials which gel or polymerize upon contact with water, such as TACSS introduced in Japan in 1967, with initial viscosity ranging ranging from 22 to 300 cP. Another polyurethane is CR250,[15] marketed by 3M in 1979 primarily as a sealant for leaking joints in sewers, in view of its remarkable elastic properties. Bearing in mind these recent innovations and the potential for wetting and drying in sewer sealing the durability of polyurethane grouts is a subject for study. In 1980 two low-toxic systems were introduced as replacements to AM-9. AC-400 grout[16] is an acrylate monomer with the same catalyst system as AM-9, giving a viscosity of 2 cP and grouted coarse sand strength of 0.2–$0.3 \, N \, mm^{-2}$ (10% concentration). Injectite-80 is a polyacrylamide grout[17] which removes the toxicity problem at the expense of viscosity (50 cP) for a 10% concentration, but for short gel time mixes (20 s), as used in sewer sealing, grout strengths of nearly $1 \, N \, mm^{-2}$ are attained in 1 h. Current developments include inorganic reagents for sodium silicate in strengthening applications. There is also a general emphasis towards non-toxic systems which reflects the growing sensitivity to environmental hazards, both in the short-term handling of materials, and the long-term leaching of chemicals from the gel.

In this introduction on the historical development of chemical grouts up to the present time the writer has drawn liberally from the paper by Glossop (1961).[1] For the reader interested in further detail, reference can be made to the Progress Report of the ASCE Task Committee on Chemical Grouting (1957),[11] the ICE Symposium on Grouts and Drilling Muds in Engineering Practice (1963),[13] the ASCE Conference on Grouting in Geotechnical Engineering (1982)[18], the 8th European Conference on

Table 8.1 Classification of grouts (after Skipp and Renner[14]).

Proprietary name (if used)	Basic composition	Type of action	Strength of gel or cement (lb in^{-2})	Strength of treated soil (lb in^{-2})*	Reference
Joosten I	Sodium silicate, calcium chloride (2F)†	a, b, d‡	—	Up to 1000	H.J. Joosten R. Glossop
Joosten II	Sodium silicate, alkali dilution, calcium chloride (2F)	a, b, d	—	600	H.J. Joosten
Joosten III	Sodium silicate, heavy metal salt, ammoniacal colloid (1F)	a, b, d	—	120	H.J. Joosten
Gutman	Similar to Joosten II, sodium carbonate as alkali	a, b, d	—	200–700	I. Guttman R. Glossop
Rodio	Sodium silicate, lime water (1F)	a, b, d	—	100	K. Keil
Langer	Sodium silicate, heavy metal salt, coagulant	a, b, d	—	—	C.F. Kolbrunner
Polivka	Sodium silicate, sodium bicarbonate (1F)	a, b, d		70–100	J.J. Polivka
	Silicate–ethyl acetate (1F)	a, b, d	20–100	Up to 300	
	Resorcinol–formaldehyde (1F)	a, b, d		Up to 300	
	Urea–formaldehyde (1F)	a, b, d	20–200	Up to 500	
AM-9	Acrylamide (1F)	a, b, d		Up to 300	R.L. Shiffman C.R. Wilson
	Calcium acrylate (1F)	a, b, d			J.J. Polivka
	Chrome–lignin (1F)	a, b, d			
Polythixon	Polyester (1F)	a, b, c or b, c		Up to 300	P.H. Cardwell
	Polyurethanes (1F)	a, b, c	20–200	Up to 500	P.H. Cardwell

*Literature gives little data on type and density of sands.

†1F = single-fluid; 2F = two-fluid.

‡a = void filler; b = adhesive; c = single-phase cement; d = two-phase cement.

Soil Mechanics and Foundation Engineering 'Improvement of Ground' (1983),[19] and the ASCE Conference on Grouting, Soil Improvement & Geosynthetics (1992).[48]

8.2 Ground investigation

Prior to any detailed design planning of a chemical grout treatment involving decisions on geometry of injection holes and choice of grouts, a ground investigation should be carried out.

The overall objective of the ground investigation is to provide a detailed geotechnical classification of the different ground types encountered together with their locations and thicknesses.

Where chemical grouting is envisaged, permeability and porosity data for each type of ground should be obtained along with hydraulic gradients and chemical properties of the ground water. The ground water details may influence choice of chemical formulation and extent of treatment, whilst the porosity dictates grout consumption. The coefficient of permeability (k) is the most useful single index of the groutability of soil or rock. For chemical grouts that are free of particles k influences the rate of injection, whilst for particulate grouts, permeability may set practical lower limits for grout injection by permeation, e.g.:

5×10^{-4} m s^{-1} for cement grout
1×10^{-5} m s^{-1} for clay chemical grout
1×10^{-6} m s^{-1} for chemical grout.

Equally important, the measured ratio of ground permeability before and after treatment reflects the effectiveness of the grouting operation. In-situ short cell tests are preferred for permeability assessment of different soil horizons compared with a large scale drawdown measurement unless the latter is augmented by representative grading curves. For ground strengthening applications additional tests are required, e.g. shear strengths from tests on undisturbed samples or in-situ penetrometer or dilatometer values.

Broadly speaking, the data above are invariably lacking in quality and quantity, yet the information is vital when designing a chemical treatment and judging its effectiveness. A ground investigation is satisfactory only when it provides sufficient information to answer the following questions.

1. Can the ground be grouted?
2. For ground treatment what types and amounts of grout are required?
3. Following treatment what strength increase can be anticipated?

8.3 Principles of injection

Whilst ground is invariably irregular in nature, a theoretical appreciation of the injection process based on idealized isotropic conditions in the case of porous ground, nevertheless acts as a useful aid for grout selection and choice of appropriate technique during the assessment of the ground investigation results.

8.3.1 Permeation of porous ground

In permeation grouting chemical grout is injected into the fine pores of soils or rocks at pressures insufficient to disturb the ground structure. Under pressure the grout advances steadily, displacing air and water outwards, the direction of flow being determined by ground permeability, i.e. grout flows most readily into the zones offering least resistance.

In uniform isotropic soils spherical flow is observed and assuming Darcy's law and a Newtonian fluid, Raffle and Greenwood (1961)[20] show that the flow rate Q at a radius of penetration R is related to the hydraulic driving head H as follows:

$$H = \frac{Q}{4\pi k}\left[\mu\left(\frac{1}{r}+\frac{1}{R}\right)+\frac{1}{R}\right] \tag{8.1}$$

where k = ground permeability
μ = grout viscosity in centipoises
r = radius of spherical injection source (for a cylindrical injection source of length L and diameter D, $r = \frac{1}{2}\sqrt{LD}$ approx).

The time for the grout to penetrate to radius R is given by

$$t = \frac{nr^2}{kH}\left[\frac{\mu}{3}\left(\frac{R^3}{r^3}-1\right)-\frac{\mu-1}{2}\left(\frac{R^2}{r^2}-1\right)\right]. \tag{8.2}$$

If the second component inside the main bracket is ignored, the relationship simplifies to the equation proposed by Maag in 1938.

The results of equation 8.2 are embodied in Figure 8.4 which illustrates how the radius of grout penetration increases with time for grout viscosities of 1 cP, 10 cP and 100 cP, respectively. Thus direct estimates can be made of the rate of progress of injection.

If a grout such as a clay-chemical has a shear strength (non-Newtonian fluid) then under constant injection pressure the opposing drag forces on the wetted surfaces of the ground structure gradually increase until the injection pressure is resisted, with no extra available to maintain viscous flow. According to Raffle and Greenwood the pressure gradient (i) required to overcome the Bingham yield strength may be expressed as

$$i = 4\tau_s/d \tag{8.3}$$

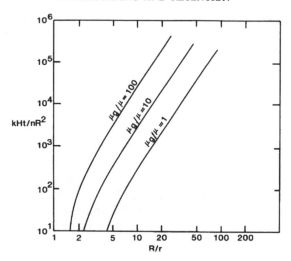

Figure 8.4   Dependence of penetration time on viscosity ratio (after Raffle and Greenwood[20]).

k = soil permeability
H = hydraulic head
t = time
n = porosity of soil
r = radius of source
R = radius of grout at time t
μ = viscosity of water
μ_g = viscosity of grout.

where τ_s = Bingham yield stress
and d = effective diameter of the average pore.

To maintain an advancing flow an additional pressure gradient is required.

With reference to equation 8.3 the average pore diameter can be estimated from the Kozeny equation.

$$d = 2\sqrt{\frac{8\mu k}{\delta_w g n}} \qquad\qquad (8.4)$$

where δ_w = density of water
and g = acceleration due to gravity.

Combining equations 8.3 and 8.4, Table 8.2 indicates typical average pore diameters for different permeabilities assuming a porosity of 25%. For the specific case of silts, experiments by Garcia-Bengochea *et al.* (1978)[21] indicate that the predominant pore size is approximately equal to the effective size (D_{10}) of the soil.

Table 8.2 Relationship between typical average pore diameter and permeability.

k ($m\,s^{-1}$)	d (mm)
1×10^{-2}	0.36
1×10^{-3}	0.114
1×10^{-4}	0.036
1×10^{-5}	0.0114
1×10^{-6}	0.0036

N.B. In ground where the pore or fissure is less than $3\,\mu m$ chemical grouting is generally impracticable and uneconomic.

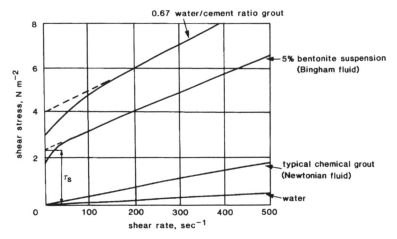

Figure 8.5 Flow properties of typical grouts (after Bell[22]).

For a ground permeability of 1×10^{-5} $m\,s^{-1}$ and given a 5% bentonite solution (Figure 8.5), with a yield of value of $2\,N\,m^{-2}$ then an injection head of 7 bars ($700\,kN\,m^{-2}$) is required for each metre of grout penetration just to overcome the inherent yield strength. For convenience Table 8.3 shows approximate hydraulic gradients to maintain flow in Bingham-type fluids.

Where injection pressures must be limited to avoid ground disturbance and heave then there is a limiting radius of penetration R_L given by

$$R_L = \frac{\delta_w g H d}{4\tau_s} + r. \tag{8.5}$$

Table 8.3 Hydraulic gradient to maintain flow in non-Newtonian grouts (after Scott[23]).

Soil permeability ($m\,s^{-1}$)	Yield value ($N\,m^{-2}$)	Hydraulic gradient
10^{-2}	1	1.2
	10	12
	100	120
	1000	1200
10^{-3}	1	4
	10	40
	100	400
	1000	—
10^{-4}	1	12
	10	120
	100	1200
	1000	—
10^{-5}	1	40
	10	400
	100	4000
	1000	—

Using such expressions design curves may be drawn to create optimum injection hole patterns. For chemical grouting of alluvium, typical final spacings range from 0.5 to 1.5 m on a triangular or rectangular grid.

Whilst chemicals are marketed as pure solutions, they invariably contain particles up to 20 μm say, which may block off fine pores in the ground. Based on empirical rules similar to filter criteria, D_{15} (soil) should be greater than $25D_{85}$ (grout) for successful permeation. In this regard it is noteworthy that silt impurities in commercial bentonite may have particles up to 50 μm. Where particles may affect the efficiency of treatment of fine-grained rocks and soils, a more refined chemical is required, or alternatively the cruder chemical should be clarified by centrifuge.

Bearing in mind that ground is heterogeneous then for permeation of soils grout may penetrate initially the more open structures at the expense of the lower permeability zones. As a result in practice, predetermined quantities related to porosity are normally injected in phases as the hole spacing is gradually reduced, the objective of the subsequent injections being to progressively treat the finer materials and thereby tighten up the ground. In planning a sequence of injections it is normal to commence grouting through holes spaced at intervals 2 to 3 times the final spacing. By this method of 'split spacing' (Figure 8.6) the treatment is less dependent ultimately on the theoretical design and more reliant on the observed effectiveness of the successive phases of treatment.

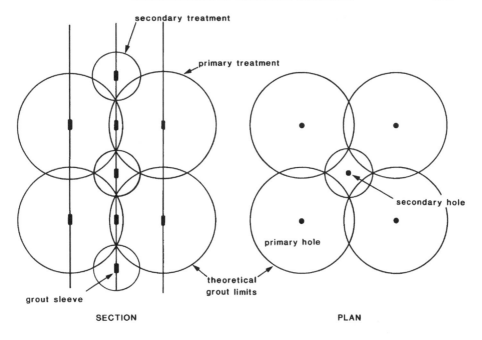

secondary treatment

primary treatment

grout sleeve

theoretical
grout limits

secondary hole

primary hole

SECTION

PLAN

Figure 8.6 Pattern of primary and secondary injections (split-spacing).

8.4 Grout systems

8.4.1 Grout selection

While most of the recent developments in chemical grouting technology have been in the field of single-fluid low-viscosity systems capable of penetrating fine-grained materials, soil and rock formation vary significantly in terms of permeability and porosity, the former being particularly sensitive to small changes in void size. The selection of the most suitable chemical grouts is therefore largely a matter of judgement on the part of the experienced engineer, who should also be aware of the range of grout suspensions available, since considerable economies can result from the use of these cheaper grouts in advance of the more costly and penetrative chemicals.

Table 8.4 contains common grout compositions which are currently in use, and in judging how best to meet a performance specification, the following aspects should be taken into account.

1. Extent and quality of ground investigation with particular reference to permeability and porosity.
2. Optimum injection method and hole pattern.
3. Availability of grout materials.

Table 8.4   Grouts used in alluvial grouting (after Skipp[24]).

Suspensoids	Two-shot solutions	Single-shot solutions
Cement	Sodium silicate–	Chrome–lignin, Sumisol*,
Cement–sand	calcium chloride	T.D.M.
Cement–clay	Hydrochloric acid–	Sodium silicate–sodium
Cement–bentonite	urea formaldehyde	bicarbonate
Cement–bentonite, P.F.A.	monomer	Sodium silicate–sodium
Waste mine slurries		aluminate
Bentonite–gel (with sodium		Sodium silicate–ethyl acetate†
silicate and acid phosphate)		Sodium silicate–mixed esters
		(Durcisseur*)
		Sodium silicate–formamide (base)
		(Siroc*)
		Sodium silicate–oxalate salt
		(Cemex)†
		Resorcinol–formaldehyde (acid
		or alkali catalysis)†
		Polyphenolic–formaldehyde,
		alkali catalyst ($MQA4$*, $MQ5$*,
		Terranier†)
		Acrylamide, $AM9$*, Progil
		R.1295*

*Proprietary grouts on sale.
†Patent protected.

4. Viscosity—time development of grout including sensitivity to temperature, dilution and mix proportioning errors.
5. Stability of grout in-situ.
6. Degree of saturation of ground during service including risk and effect of grout desiccation.
7. Chemical composition of groundwater.
8. Permanence of grout in-situ.
9. Toxicity of grout and chemical components and working environment.
10. Aggressivity of grout and chemical components towards plant and equipment.
11. Residual permeability or strength of grouted ground.
12. Degree of site supervision required including sophistication of systems.
13. Overall cost including materials, mixing and injection.

In practice the in-situ permeability of the ground dictates initially the technical options available in terms of grout selection (Table 8.5, Figure 8.7) but thereafter overall cost dominates final choice of system (Table 8.6).

Table 8.5 Grouting limits of common mixes (after Caron[25]).

Type of soils	Coarse sands and gravels	Medium to fine sands	Silty or clayey sands, silts
Soil characteristics:			
Grain diameter	$D_{10} > 0.5$ mm	$0.02 < D_{10} < 0.5$ mm	$D_{10} < 0.02$ mm
Specific surface	$S < 100$ cm^{-1}	100 cm$^{-1} < S <$ 1000 cm^{-1}	$S > 1000$ cm^{-1}
Permeability	$k > 10^{-3}$ m s^{-1}	$10^{-3} > k > 10^{-5}$ m s^{-1}	$k < 10^{-5}$ m s^{-1}
Type of mix	Bingham suspensions	Colloid solutions (gels)	Pure solutions (resins)
Consolidation grouting	Cement $(k > 10^{-2}$ m s$^{-1})$ Aerated mix	Hard silica gels: double-shot: Joosten (for $k > 10^{-4}$ m s^{-1}) single-shot: Carongel Glyoxol Siroc	Aminoplastic Phenoplastic
Impermeability grouting	Aerated mix Bentonite gel Clay gel Clay/cement	Bentonite gel Lignochromate Light carongel Soft silica gel Vulcanizable oils Others (Terranier)	Acrylamide Aminoplastic Phenoplastic

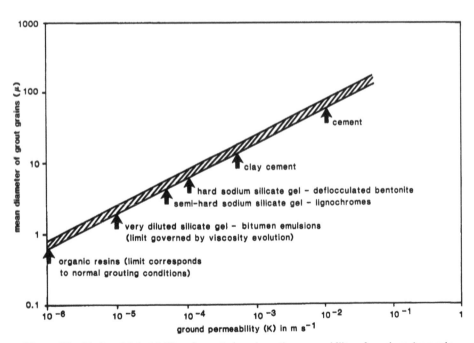

Figure 8.7 Limits of injectability of grouts based on the permeability of sands and gravels (after Cambefort[26]).

Table 8.6 Relative material costs of grout formulations.

Formulation	Relative cost of materials
Cement–bentonite	
w/c = 3, 5% bentonite by wt. of water	1.0
w/c = 2, 3% bentonite by wt. of water	1.3
w/c = 1, 1% bentonite by wt. of water	2.3
Cement	
(w/c = 0.5)	3.4
Silicate–bentonite	
20% bentonite, 7% silicate (by wt. of water)	1.3
Silicate–chloride (Joosten)	4.0
Silicate–ester	
37% silicate, 4.4% ester (by volume)	5.0
47% silicate, 5.6% ester (by volume)	6.5
Silicate–aluminate	
46% silicate, 1.4% aluminate (by weight)	5.0
Phenol–formaldehyde	
13% (by volume)	10.5
19% (by volume)	15.3
Acrylate	
10% (by weight)	18.5
Resorcinol–formaldehyde	
21% (by volume)	23.0
28% (by volume)	31.0
Polyacrylamide	
5% (by volume)	20.0
10% (by volume)	40.0

8.4.2 Viscosity

Bearing in mind the important distinction between suspensions (particulate structures—Bingham fluids) and solutions (Newtonian fluids—see Figure 8.5), then subject to specified limiting injection pressures related to depth of overburden, chemical grout viscosity–time development curves (Figure 8.8) in conjunction with flow equations, determine hole patterns and the time of injection (Figure 8.9).

The viscosity of chemical grouts varies with the concentration of the reactive chemicals (Figure 8.10), and as some of these grouts contain minute particles in suspension it is perhaps more correct to use the term apparent viscosity in such cases. Temperature increases can reduce initial viscosities but the reductions are marginal and quickly compensated by the accelerated gelling process.

Generally speaking, chemical grouts with viscosities less than 2 cP can permeate without trouble ground with a permeability as low as $1 \times 10^{-4}\,m\,s^{-1}$. For higher viscosities of say 20 cP it may be necessary to restrict the grout application to more permeable ground or reduce hole spacing for the same gel time.

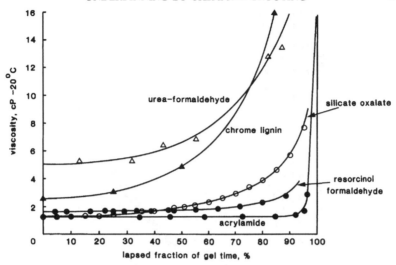

Figure 8.8   Growth of viscosity in period before gelation (modified after James[27]).

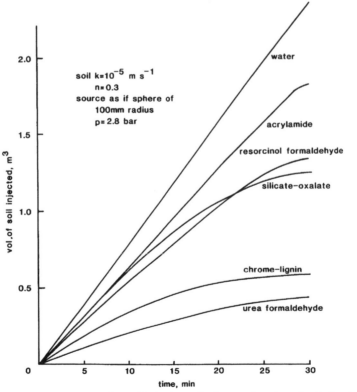

Figure 8.9   Relative volumes of soil filled when injection continued until gelation (modified after James[27]).

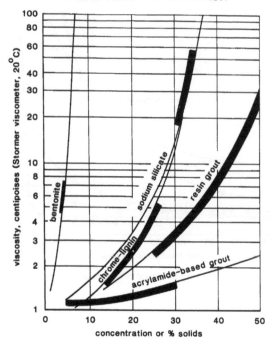

Figure 8.10 Viscosities of various grouts. Heavy lines indicate the solution concentrations normally used in the field (after Karol[4]).

8.4.3 Setting time

Setting or gel time is commonly regarded as the time between initial mixing or addition of the catalyst and the formation of a gel. Between initial mixing and attainment of the hardened state a number of intermediate times may be distinguished on an arbitrary basis.

1. Limiting injection time, at which the apparent viscosity has increased to the extent that injection becomes uneconomically slow.
2. Zero displacement time, when the grout has attained sufficient strength to resist displacement under the imposed hydraulic gradients in the formation.
3. Setting time, defined by some empirical test, e.g. no flow from upturned flask or 100 cP for low viscosity system (2 cP). This time may be used as a quality control for grout mixed on site and on occasions dictates the limiting injection time e.g., 75% of setting time.
4. Fully cured time at which final properties have been developed and the chemical reaction has ceased.

The setting time of most chemical grouts can be varied from minutes to hours, but is temperature-sensitive e.g., gel time halved for temperature increase of 10°C. Traditional control is by concentration of solids although catalysts are also used. For greater concentrations the set is reduced but the effects of dilution in flowing groundwater may extend the set again. Under normal circumstances setting times of 45–90 minutes are employed to give adequate time for mixing, pumping and placement. Where setting times are less than 30 minutes or ambient temperatures are high (>30°C) proportioning pump systems are preferred which delay the mixing of chemical components until the injection point.

In design planning choice of setting time is primarily influenced by grout volume to be injected, ground permeability, groundwater conditions and temperature.

8.4.4 Stability

Many single-fluid chemical grouts are subject to syneresis, i.e. the expulsion of water from the gel. If this property is present to a significant extent new seepage channels can be created even if the voids within a formation are filled initially. In silicate-based grouts exhibiting syneresis, water exudes from the gel within a few hours of setting and the process stabilizes generally after 3 to 4 weeks. Syneresis can however be controlled readily in the mix formulation by increasing the concentration of the silicate and the coefficient of neutralization.

In practice the degree of syneresis for a given grouted material is a function of the ratio of volume to surface area within the grouted structure, since bonding of the gel and solid surface resists internal shrinkage stresses and thereby reduces grout volume change. As a consequence, grouts which may be unsuitable for coarse gravels may be entirely appropriate for fine sands (Figure 8.11). Where grouts exhibiting high syneresis in-situ have been employed the consequence for the split spacing technique is greater grout consumptions in the subsequent injection phases, e.g. secondary and tertiary stages.

8.4.5 Strength of grouted formation

Chemical grouting of a cohesionless soil has the effect of increasing cohesion whilst the angle of internal friction remains sensibly the same (Figure 8.12). The magnitude of the cohesion depends upon the grout type injected, and several researchers over a period of time (Schiffman and Wilson, 1958;[28] Skipp and Renner, 1963;[14] Warner, 1972;[29] Farmer, 1975;[30] Krizek et al., 1982[31]) have published the results of triaxial testing of soils impregnated with various grout systems. In general, strength increases with increasing density and decreasing effective grain size (D_{10}). Well-graded soils have higher strengths than uniform soils of the same effective grain size.

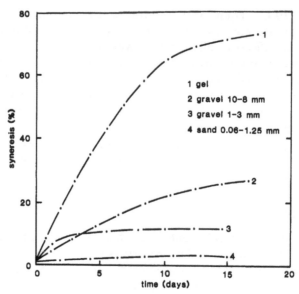

Figure 8.11 Variation of syneresis as a function of grain size; 60% silicate–ethyl acetate gel (after Caron[25]).

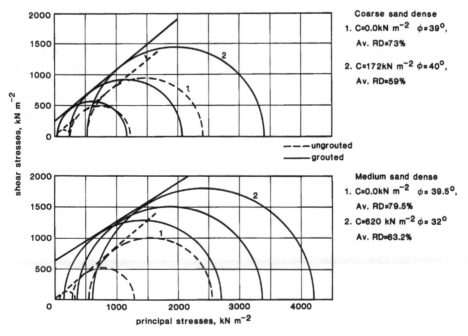

Figure 8.12 Drained triaxial test results for silicate-grouted coarse and medium sands (after Skipp and Renner[14]).

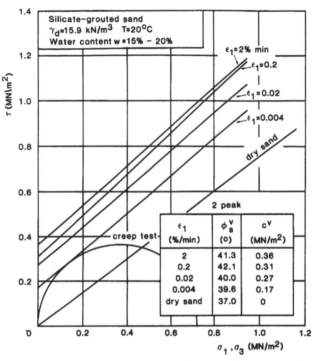

Figure 8.13 Influence of strain rate on shear strength of silicate-grouted sand (after Stetzler[33]).

Most published strength results are taken from unconfined compression tests but there is a lack of standardization on sample size or strain rate. Since the gels are visco-elastic their properties are strain dependent and increase in strain rate increases strength,[32] so that quoted laboratory strengths may be exaggerated when subsequently the treated ground is subjected to a slow rate of loading as in some underpinning applications. Figure 8.13 illustrates the significance of strain rate on the shear strength of silicate grouted sand. Much of the literature also reports dry strengths, where water in the grout is lost through desiccation and the grout matrix shrinks. In such circumstances the intergranular bond may increase and again the stronger results which are recorded may be totally misleading. Since the vast majority of chemical treatments take place in saturated or damp conditions, strength testing should simulate the environment concerned in order to produce relevant data.

In this regard alternating wet and dry cycles in service can be particularly detrimental.[29] In-situ tests or tests on undisturbed samples from full-scale grouting programmes are rare, but a comprehensive field assessment of silicate grouted sands has recently been published by Davidson and Perez[34] covering a variety of engineering properties before and after grouting including in-situ, density, horizontal stress, elastic modulus, shear strength, creep and permeability.

In strong rocks the ultimate set strength of grout is not usually a design consideration since rock is normally self-supporting in applications such as open excavations or tunnels, provided water ingress is controlled.

8.4.6 Creep of grouted formation

Chemically grouted soils are subject to creep, the magnitude depending to some extent upon the ratio of applied stress to inherent strength.[29] According to Karol,[4] limited data suggest that for unconfined tests negligible creep will occur for ratios less than 25%. For confined tests at K_0 'at rest' lateral pressure the ratio may be increased to 50%. These ratios give some guidance on appropriate load safety factors in design, but much more research is required generally on stress–strain–time behaviour to ensure that predictions are based on reproducible test results from representative samples.[35]

8.4.7 Resistance of grout to extrusion

Once injected into pores or fissures a grout must have sufficient strength to resist displacement by hydraulic gradient. If τ_f is the shear strength of the set gel the limiting hydraulic gradient (i) to initiate extrusion can be estimated by reference to equations 8.3 and 8.4.

$$i = \frac{4\tau_f}{d\delta_w g}. \tag{8.7}$$

In the case of a grouted cut-off the hydraulic gradient (i) is calculated as the maximum differential head divided by the width of the curtain. From Table 8.3 it is clear that the demand for high shear strength is greatest in ground of high permeability. Traditionally, grouted cut-offs in alluvium are designed conservatively for a hydraulic gradient of 3 to 5.

Once the grouted cut-off has been designed to resist extrusion, in certain applications involving very high hydraulic gradients e.g. grout sealing around deep mine shafts, the choice of grout must also take account of creep properties.

8.4.8 Permanence

If a grout treatment is required to be permanent, the set grout must not deteriorate with time. It should be resistant to chemical attack and dissolution by groundwater, particularly under an hydraulic gradient.

Chemical grouts based upon an aqueous solution may redissolve on long-term contact with groundwater and the process of dissolution governs the permanence of the grout treatment.

Defining chemical grout permanence as the period of time after which a grout treatment fails to meet its intended function, Hewlett and Hutchinson (1983)[36] suggest a 1–2 order increase in residual permeability for water stopping applications, and a loss of 50% gel forming solids in strengthening work, which for the common formulations is equivalent to a reduction in gel strength of up to 40%.

Slow percolation of groundwater through a gel (assuming typical gel permeabilities of $10^{-10} \, m \, s^{-1}$, or less) causes the leaching water to become progressively more saturated with dissolved gel solids, and this in turn reduces further gel dissolution. However, as the grout dissolves, the permeability and hydraulic gradient increase, both of which will cause the process to accelerate. In general, where water percolates through the grout, e.g. mine shaft lining, the dissolution or disintegration of the gel will be long-term, even at high pressure gradients.

On the other hand where stratified ground is partially treated with grout, e.g. untreated silts interbedded with grouted sands, water flow through the silt can wash the gel surface of the grouted sand. In this case the saturation solubility of dissolving material may not be met and gel removal may be sustained resulting in reduced grout treatment properties.

According to Hewlett and Hutchinson, material residing within the gel network can move through the gel either due to the passage of water resulting from an hydraulic gradient and/or as a result of concentration gradients of dissolved material causing diffusion. The rate at which the gel structure actually dissolves (S) and the rate at which the dissolved material diffuses through the gel structure to the outside (D) can be measured in the laboratory, and if the mechanism of dissolution is diffusion-controlled, the grout retracts from its outer face over a period of time leaving behind untreated ground (Figure 8.14a). Tests to date indicate that weak silicate/oxalate and silicate/bicarbonate gels suffer face retraction, whilst in the case of chrome lignin, phenoplast, aminoplast and acrylamide gels the rate of diffusion exceeds the rate of dissolution, i.e., the mechanism is dissolution-controlled, and the gel structure gradually weakens but retains its overall dimensional structure (Figure 8.14b). In underpinning work the grout treatment is invariably temporary and permanence is not a major concern. As a consequence, cheaper and more dilute systems are often employed in practice.

Overall, the soundest defence is good grouting practice to ensure maximum void filling which in turn precludes access of aggressive water to the grout. Choice of gels which do not suffer shrinkage nor collapse by synesis helps to reduce decay and in the field, completion of each grouting phase with gradually increasing injection pressures is beneficial.

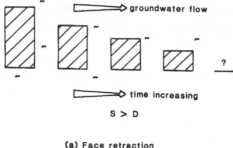

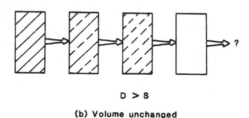

Figure 8.14 Types of grout dissolution (after Hewlett and Hutchinson[36]).

8.4.9 Health and safety aspects

Although there are few accounts of harm or damage, instances of skin reactions have occasionally been reported, always in association with poor hygiene and ineffective clothing. From Japan there have been reports of environmental pollution caused by the use of acrylic and acrylamide grouts.

When presenting the virtues of a product or material, the chemical supplier should at the same time describe as far as possible any hazards likely to arise from its use and the conditions when it will not function properly.

Chemical constituents should be labelled for toxicological, explosive (flashpoint), flammable (boiling point), corrosive, cytoxic (interferes with cell division) and environmental hazards. In this regard the materials used for ground treatment have recently been identified by CIRIA,[37,38] and a guide to the potential hazards to people and the environment is provided.

Generally speaking, toxic hazards from the use of grouts are negligible after gelation since practical hydraulic gradients are incapable of removing significant quantities of grout material from such impervious materials. The greatest toxic hazard is to the workforce and due to spillage of ummixed chemicals and cleaning plant following injection.

8.5 Grouting operations

Once the purpose of grouting, choice of grout and hole pattern have been defined, open grout holes are drilled and prepared for treatment or alternatively the grouting pipes are drilled or jetted to the appropriate depth.

Grout is normally introduced into the ground in one of four ways, choice being dictated by ground conditions and the degree of control required for grout placement into different horizons:

(1) into an open hole in self-supporting ground through pipes caulked at the surface;

(2) through an injection pipe held in place in the hole or casing by a packer (Figure 8.15);

(3) from a pipe driven into the ground and withdrawn as injection proceeds (Figure 8.1);

(4) through a pipe left in place in the ground, as with a tube à manchette (Figure 8.2).

Chemical grouting rates are not large and injection hole diameters generally lie in the range 20 to 100 mm, with hole spacings of 0.5 to 2.5 metres. Where holes are divided into stages for localized treatment of horizons, lengths normally vary from 5 metres in deep rock to 0.25 metres in tube à manchette grouting of highly stratified alluvium.

In terms of grout batching, mixing and pumping, installations can vary from high-speed colloidal mixers[39] which can be used effectively for both powder compositions and single-fluid chemical grouts, in conjunction with in-line grout pumps, to mixing/pump stations holding stock solutions and incorporating automatic dosing systems, based on either volume dosing pumps or flow rate control valves. All these systems are based on in-line pumping of the grout but, as already described, in high ambient temperatures or where short gel times are employed, in-line proportioning

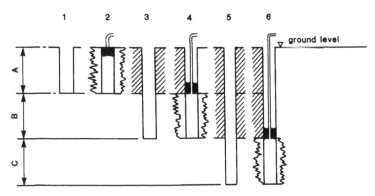

Figure 8.15 Stage and packer grouting in descending stages.

pumps are recommended where the constituent chemical solutions of two-component systems are pumped separately but in the correct proportions to a mixing head at the injection point.

Grout monitoring of flow and pressure ranges from simple visual observations of pressure gauges and pump stroke counters and manual recording, to sophisticated fully automatic continuous recording systems. During grouting operations it is important to maintain on a daily basis the following records for each stage of injection.

Date

Hole number

Depth of injection

Period of injection

Grout type and consumption

Injection pressure at inlet

Remarks e.g., uplift and surface leakage.*

When these records are systematically plotted on sections as the work proceeds, the charts provide an instant picture of progress and tightening-up as the injection sequence progresses (primary→secondary→tertiary). For each major injection phase, pre- and post-treatment water tests should be carried out where reduced permeability is the design objective. For strength improvement, penetrometer tests are most commonly employed (Figure 8.16) and a two-fold increase reflects good grout treatment. Crosshole shear wave velocity measurements have been reported by Davidson and Perez[34] to be effective in determining grout penetration in soil. A back-analysis of grout consumption against ground porosity is also useful for comparison with the design assumptions.

8.6 Applications

To illustrate that chemical grouting is now a well-established geotechnical process for the strengthening of ground in relation to underpinning and improving the loadbearing characteristics of foundations, case records are described which have been drawn from civil engineering practice over the past 25 years.

8.6.1 Ground treatment at Wuppertal, West Germany[40]

Deep excavations were required through 7 metres of water-bearing alluvium immediately adjacent to existing 4- and 8-storey buildings. The alluvium comprised a continuous grading from cobbles to fine sands, with

---

*Surface leakages indicate that further grouting will be ineffective and injection should cease until the grout in-situ has set.

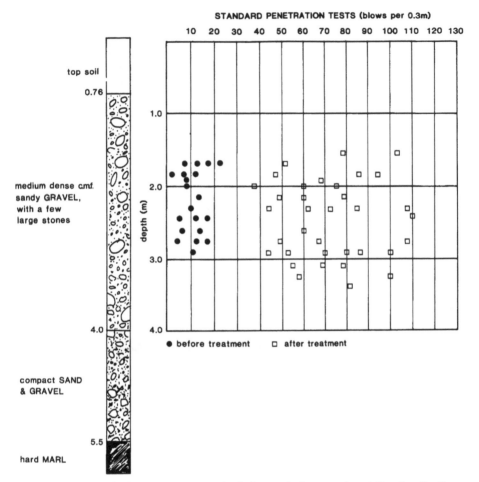

Figure 8.16 Standard penetrometer results before and after grouting at Rugeley Cooling Tower, England.

a permeability in the region of $10^{-4}\,\mathrm{m\,s^{-1}}$, and overlying a hard metamorphic schist.

To protect the properties against loss of ground during excavation, a grouted wall solution (Figure 8.17) was accepted. Initially, clay–cement grout was injected through pipes C to fill the larger voids and limit the spread of the more expensive chemical grout on the excavation side. Thereafter, silicate–acetate grout was injected through pipes A and B. Contact grouting by clay–cement was also undertaken immediately under the foundation to ensure the prior filling of any major voids at this location.

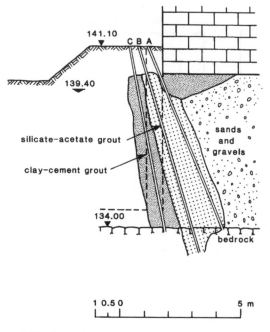

Figure 8.17 Ground treatment at Wuppertal, West Germany.[40]

8.6.2 Increase of formation strength, Minneapolis, Minnesota[4]

On this project, a 50-metre brick chimney was founded on wood piling which had deteriorated above the water table, and reduced the structural support to the foundation slab. Figure 8.18 shows how chemical grouting was employed to strengthen the underlying sands to a level where the structural load from the foundation could be transferred to the sound portion of the piles below the water table. A silicate-based grout was used for the main strengthening and injected within an outer curtain of calcium chloride which acted as a quick-setting hardener and thereby reduced wastage of sodium silicate.

8.6.3 Ground strengthening at East Greenwich sewer, London[41]

In advance of driving a new tunnel in water-bearing sands and gravels under a length of main road with many services overhead and an old brick sewer alongside (Figure 8.19), grouting from the surface was carried out using the tube à manchette method. Initially, bentonite–cement was injected followed by silicate–acetate. Of particular interest is the use of a diluted silicate–acetate in the zone of excavation for the new sewer, to reduce the mechanical strength of the treated ground and thereby ease the

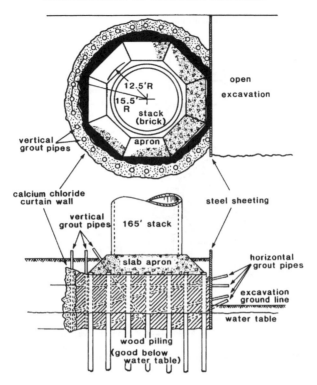

Figure 8.18 Strengthening of sand to increase bearing capacity.[4]

miners' task. As a result of the grouting the required air pressure was reduced which in turn avoided the risk of ground loss or disruption via loss of air or water ingress.

8.6.4 Reduction in pile settlement at Jeddah, Saudi Arabia[42]

Large-diameter piles (960–1200 mm) founded at a depth of 11 metres were designed in end bearing for working loads up to 5800 kN and a specified limiting settlement of 80 mm at 1.5 times working load, but preliminary pile tests (Nos. 190 and 214—see Figure 8.20) indicated excessive settlement caused primarily by ground disturbance beneath the pile during construction. To treat the variable low-permeability silty sand ($k = 1 \times 10^{-5}$ to 5×10^{-7} m s^{-1}), a tube à manchette was installed through the base of each pile (Figure 8.21) after which resorcinol–formaldehyde chemical grout was injected in a controlled sequence to form an enlarged base of strengthened soil.

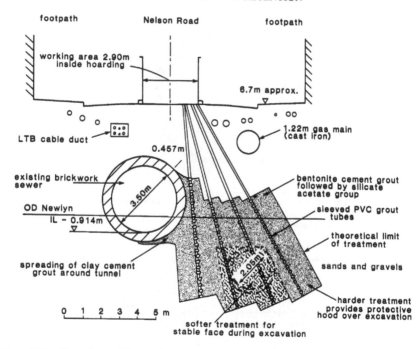

footpath   Nelson Road   footpath

working area 2.90m
inside hoarding

6.7m approx.

LTB cable duct

1.22m gas main
(cast iron)

0.457m

existing brickwork
sewer

bentonite cement grout
followed by silicate
acetate group

OD Newlyn
IL - 0.914m

3.50m

sleeved PVC grout
tubes

theoretical limit
of treatment

spreading of clay cement
grout around tunnel

2.03m

sands and gravels

0 1 2 3 4 5 m

harder treatment
provides protective
hood over excavation

softer treatment for
stable face during excavation

Figure 8.19   East Greenwich sewer: ground treatment. Typical section along Nelson Road.[41]

During production grouting 349 piles were injected using 338 5000 litres of resorcinol–formaldehyde grout. Subsequent loading tests of piles (Figure 8.20), although limited in number, were taken at random (Piles Nos. 18, 114 and 166) and confirmed that high-capacity shallow piles could be constructed economically with grout consolidation compared with much longer friction piles.

8.6.5 Underpinning of multistorey blocks, Paris[43]

At the rue de Monttessuy in Paris, a 3.5 m deep excavation required for a new electric substation was planned with multistorey blocks on three sides. Grouting was employed firstly to create an impermeable curtain round the site extending down to impermeable marls (Figure 8.22) and secondly to strengthen the soil beneath the adjacent foundations thereby underpinning the buildings and reducing risk of settlement during excavation.

Voids which existed beneath the old basement floors were filled with clay–cement, followed by a strong silicate–acetate injection into the sand and gravel immediately adjacent to the excavation. A weaker silicate–aluminate grout was employed in the lower horizon of alluvium to form

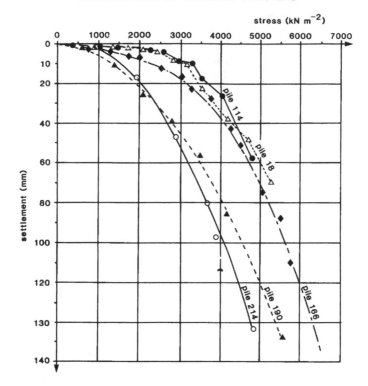

Figure 8.20   Pile test results showing applied stress *v.* head settlement.[42]

the cut-off. Subsequent excavation proceeded without difficulty and the silicate–acetate treatment was so hard that side trimming by pneumatic pick was necessary.

Other successful examples of underpinning include Great Cumberland Place,[44] Rugeley cooling tower (UK), Maryland National Bank (USA),[45] Austrian National Bank (Vienna),[46] and Timisoara Opera (Rumania).[47]

8.7 Conclusions

In fine-grained ground formations chemical grouting is an established engineering expedient for ground impermeabilization and strengthening. In such circumstances, grouting can be indispensable in tunnels and deep excavations as a temporary support and for underpinning buildings and foundations as a permanent ground strengthening.

Nevertheless, after 60 years grouting remains an art where field experience counts. Local variations in the ground cannot be predetermined and a flexible attitude should be maintained at the expense of rigorous

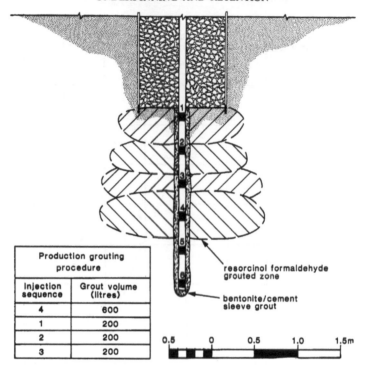

Production grouting procedure	
Injection sequence	Grout volume (litres)
4	600
1	200
2	200
3	200

resorcinol formaldehyde grouted zone

bentonite/cement sleeve grout

0.5   0   0.5   1.0   1.5m

Figure 8.21 Tube à manchette installation and production grouting procedure.[42]

specification of techniques, so that variations can be implemented where appropriate, to improve the engineering result.

Over the past two decades in particular with the introduction of more sophisticated single-fluid grouts and development of injection techniques, there has been a steady reduction in grout material quantity injected per unit volume of ground and a reduction in the number and intensity of grout holes considered necessary. Both these reductions are indicative of a growing confidence in the efficiency and effectiveness of grouting which augurs well for the future.*

---

*However, there is still no relaxation in the quest by Scottish engineers for a single-fluid grout with the viscosity of water and strength of concrete, where water is the most expensive constituent.

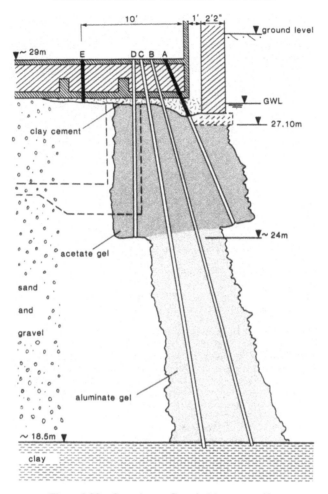

Figure 8.22 Grouting at Rue de Monttessuy.[43]

References

1. Glossop, R. (1961) The invention and development of injection processes. Part II: 1850–1960. *Geotechnique* 11 (4) 255–279.
2. Atherton, F.G. and Garrett, W.S. (1959) The history of cementation in shaft sinking. *Symposium on Shaft Sinking and Tunnelling*, Institution of Mining Engineers, London.
3. Portier (1905) Cimentation des terrains aquifères en vue du creusement des puits. *Congres. Int. Mines*, Liège. Sect. des Mines, Tome 1.
4. Karol, R.H. (1983) *Chemical Grouting*. Marcel Dekker, New York.
5. Durnerin, M. (1922) Le Problème du Foncage des puits sur le Prolonguement du Bassin Houiller de la Sarre en Lorraine. *Congres Scient. Assoc. Ingens. sortie de l'Ecole de Liège*.
6. Joosten, H.J. (1954) *The Joosten process for chemical soil solidification and sealing and its development from 1925 to date*. N.V. Amsterdamsche Ballast Maatschappij.
7. Drouhin, M. (1938) La lutte contre les érosions souterraines au barrage de Bou-Hanifia, etc. *Trans. 2nd Congr. large Dams*, Washington, Vol. 4, 29–49.

8. Mayer, A. (1958) Cement and clay grouting of foundations. French grouting practice. *Proc. ASCE*, Paper 1550 (February).
9. Cambefort, H. (1955) Parafouilles spéciaux en terrains perméables. *Cinquiéme Congr. Grands Barrages*, Paris, Vol. 1, 883–96.
10. Maag, E. (1938) Ueber die Verfestigung und Dichtung das Baugrundes (Injektionen). *Erabaukers der ETH*.
11. ASCE (1957) Chemical Grouting Report. *Proc. ASCE* 83 (SM4), 1–106.
12. Caron, C. (1963) The development of grouts for the injection of fine sands, *Symposium on Grouts and Drilling Muds in Engineering Practice*. Butterworths, London, 136–141.
13. ICE (1963) *Grouts and Drilling Muds in Engineering Practice*. Butterworths, London.
14. Skipp, B.O. and Renner, L. (1963) The improvement of the mechanical properties of sands. *Symposium on Grouts and Drilling Muds in Engineering Practice*. Butterworths, London, 29–35.
15. 3M Company (1981) *3M Sealing Gel System Field Manual*. 3M Company, St. Paul, Minnesota.
16. Clarke, W.J. (1982) Performance characteristics of acrylate polymer grout. *Proc. ASCE Conf. on Grouting in Geotechnical Engineering*, New Orleans, 418–432.
17. Berry, R.M. (1982) Injectite—80 Polyacrylamide Grout. *Proc. ASCE Conf. on Grouting in Geotechnical Engineering*, New Orleans, 394–402.
18. ASCE (1982) *Proceedings of Conference on Grouting in Geotechnical Engineering*. ASCE, 345 East 47th Street, New York.
19. ISSMFE (1983) *Improvement of Ground. Proc. 8th European Conf. on Soil Mech. & Found. Engng.*, Helsinki, Vols. 1 and 2.
20. Raffle, J.F. and Greenwood, D.A. (1961) The relationship between the rheological characteristics of grouts and their capacity to permeate soils. *Proc. 5th Int. Conf. on Soil Mech. & Found. Engng.*, Vol. 2, 789–793.
21. Garcia-Bengochea, I., Lovell, C.W. and Altschaeffli, A.G. (1978) Pore distribution and permeability of silty clays. *Proc. ASCE*, 105 (GT7).
22. Bell, L.A. (1978) *Alluvial Grouting*. M.Sc. Dissertation, University of Durham.
23. Scott, R.A. (1975) Fundamental conditions governing the penetration of grouts. In *Methods of Treatment of Unstable Ground* (ed. F.G. Bell), Newnes-Butterworths, London.
24. Skipp, B.O. (1975) Clay grouting and alluvial grouting. In *Methods of Treatment of Unstable Ground* (ed. F.G. Bell), Newnes-Butterworths, London, 141–158.
25. Caron, C. (1965) Etude physico-chimique des gels de silice. *Annals de l'Institut du Bâtiment et des Travaux Public*, 207–208 (March–April).
26. Cambefort, H. (1977) The principles and applications of grouting. *Q.J.E.G.* 10(2), 57–95.
27. James, A.N. (1963) Discussion to Session 3—*Grouting Symposium on Grouts and Drilling Muds in Engineering Practice*. Butterworths, London, 168–169.
28. Schiffman, R.L. and Wilson, C.R. (1958) The mechanical behaviour of chemically treated granular soils. *Proc. ASTM*, 58, 1218–1244.
29. Warner, J. (1972) Strength properties of chemically solidified soils, *Proc. ASCE* 98, (SM2), 1163–1185.
30. Farmer, I.W. (1975) Undrained stengths of chemically grouted cohesionless soils. *Proc. 2nd Int. Cong. Int. Ass. Eng. Geol.*, Sao Paulo, Vol. 1, 1–6.
31. Krizek, R.J., Benltayf, M.A. and Dimitrios, K. (1982) Effective stress–strain strength behaviour of silicate-grouted sand. *Proc. ASCE Conf. on Grouting in Geotechnical Engineering*, New Orleans, 482–497.
32. Clough, W., Kuck, W. and Kassali, G. (1979) Silicate-stabilized sands. *Proc. ASCE*, 105 (GT1), 65–82.
33. Stetzler, B.U. (1982) Mechanical behaviour of silicate-grouted soils. *Proc. ASCE Conf. on Grouting in Geotechnical Engineering*, New Orleans, 498–514.
34. Davidson, R.R. and Perez, J-Y. (1982) Properties of chemically grouted sand at locks and dam No. 26. *Proc. ASCE Conf. on Grouting in Geotechnical Engineering*, New Orleans, 433–449.
35. Borden, R.H., Krizek, R.J. and Baker, W.H. (1982) Creep behaviour of silicate grouted sand. *Proc. ASCE Conf. on Grouting in Geotechnical Engineering*, New Orleans, 450–469.
36. Hewlett, P.C. and Hutchinson, M.T. (1983) Quantifying chemical grout performance and potential toxicity. *Proc. 8th European Conf. on Soil. Mech. & Found. Engng.* Vol. 1, 361–366.

37. CIRIA (1981) A Guide to the safe use of chemicals in construction. *Special Publication 16*, Construction Industry Research and Information Association, London.
38. CIRIA (1982) Health and safety aspects of ground treatment materials. *Report 95*, Construction Industry Research and Information Association, London.
39. Littlejohn, G.S. (1983) Plant and equipment for cement based grouts. *SAICE Grouting Course*, Geotechnical Division, University of Witwatersrand, Johannesburg.
40. Leonard, M.W. and Mollar, K. (1963) Grouting for support, with particular reference to the use of some chemical grouts. *Symposium on Grouts and Drilling Muds in Engineering Practice*, Butterworths, London, 156–163.
41. Dimpsey, J.A. and Moller, K. (1970). In *Grouting in Ground Engineering (Proc. Conf. Ground Engineering)*, ICE, London, 3–10.
42. Littlejohn, G.S. Ingle, J. and Dadasbilge, K. (1983) Improvement in base resistance of large diameter piles founded in silty sand. *Proc. 8th European Conf. on Soil Mech & Found. Engng.* Vol. 1, 153–156.
43. Ischy, E. and Glossop, R. (1962) An introduction to alluvial grouting. *Proc. I.C.E.* 21, 449–474.
44. Neelands, R.J. and James, A.N. (1963) Formulation and selection of chemical grouts with typical examples of their use. *Symposium on Grouts and Drilling Muds in Engineering Practice*. Butterworths, London, 150–155.
45. Zeigler, E.J. and Wirth, J.L. (1982) Soil stabilisation by grouting in Baltimore Subway. *Proc. ASCE Conf. on Grouting in Geotechnical Engineering*, New Orleans, 576–590.
46. Stadler, G. and Comte, C.H. (1983) Specific grouting projects in the pre-Alpine region. *Proc. 8th European Conf. on Soil & Mech. Found. Engng.* Vol. 1, 163–166.
47. Bally, R.J. and Klein, R. (1983) Some research works and applications of grouting of line grained soils in Romania. *Proc. 8th European Conf. on Soil Mech. & Found. Engng.*, Vol. 1, 127–130.
48. ASCE (1992) *Proc. Conf. on Grouting, Soil Improvement & Geosynthetics*. ASCE, 345 East 47th Street, New York.

9 Lateral shores and strutting

P. LIGHT

9.1 Historical background

From the Roman era until recent times, timber has generally been the principal material used for any necessary temporary supports.

The skills of the 'timbermen' were strengthened over the years by migrants from other industries, such as ship building and mining, to the extent that a gang of such men, under the leadership of an experienced foreman, could be relied upon to carry out, successfully and safely, operations that today require much more analysis and very careful management. Material, especially timber, cannot be wasted, for both economic and environmental reasons, and skilled labour has become progressively scarcer and more expensive.

Less than 30 years ago basements for buildings were still being excavated using the 'trench and dumpling' method. The perimeter trench would first be excavated, strutting introduced progressively downwards, and the permanent retaining wall, generally designed to cantilever and provided with an 'anti-slide' key, would be constructed in the trench. The 'dumpling', or central area of the basement, could then be excavated at a relatively fast rate.

Although this was a labour-intensive and lengthy process, other operations such as underpinning and shoring of adjacent properties could be carried out at the same time, so that the overall time required for the excavation for a typical basement was not necessarily unduly excessive.

Figure 9.1 shows just this type of construction in progress for the Swiss Embassy building about 20 years ago.

Whilst the excavation could be completed reasonably expeditiously, the construction of the permanent reinforced concrete retaining wall in the trench would be impeded by obstructions presented by large timber walings and struts. The wall would have to be constructed either in short lifts or in narrow bays between vertical rows of props. Either way, the large number of construction joints would require significant time and labour, and would increase the risk of water penetration into the finished basement.

The timber raking shores commonly used to support party walls would, of course, have to be erected prior to final demolition of the building to be re-developed and, to avoid undermining, be founded at a level below the future basement foundation level wherever possible. Again, such shores would

276

Figure 9.1 Construction of the Swiss Embassy building by the 'trench and dumpling' method.

present obstructions to the construction of the new building frame, with resulting increases in contract period and costs.

9.2 Types and purposes of strutting

In simple terms, temporary lateral supports to residual structures are required when the removal of adjacent ground or structures would give rise to unstable conditions. These may be considered in three basic categories:

(i)   supports to adjacent buildings;
(ii) supports to adjacent highways and other retained areas; and
(iii) supports to retained elements of structure within the contractor's site.

Figure 9.2 shows part of the temporary supports in place to support an adjacent building and highway on a contract just completed near St James Square in London. A comparison with the methods used at the Swiss Embassy 20 years earlier shows the effects of both the introduction of secant piling techniques to provide the basic perimeter support, and the use of braced steel towers instead of timber rakers to support the adjacent building. The struts were designed to span across the basement and the number of king posts was kept to the minimum possible. The towers were founded in bored in-situ piles constructed using a tripod rig in the basement prior to demolition. The permanent foundation is a reinforced concrete raft.

In the great majority of cases, the building to be demolished is not providing support to the adjacent structure as a whole, but only to the party or dividing wall. If it can be shown without doubt that the wall in question is effectively

Figure 9.2 Construction near St James Square, London showing the introduction of secant piling techniques to provide the basic perimeter support, and the use of braced steel towers instead of timber rakers to support the adjacent building.

tied back to floors and/or crosswalls, and that the pressure exerted on the ground at its foundation is acceptably distributed across its width, then no support is needed. The decision may well be influenced by the possible need to underpin. It will also certainly be influenced by the judgement of the surveyors appointed by the two building owners and/or the engineers advising them.

When supports are required consideration should be given to using 'tie-backs', for example steel waling beams or channels wedged or packed against the wall, with ties a few metres long bolted into the faces of the return walls at each end. An alternative, in the rare cases when access is or can be made available, is to tie back the walings to the top of the floor joists or slabs within the adjoining owner's building. Such methods, which assist the economic construction of the new building, require a degree of co-operation on the part of the adjoining owners and their professional advisers, which is rarely forthcoming—usually because of legal constraints, or understandable reluctance to disturb tenants.

At the Standard Chartered Bank in Bishopsgate, completed in 1985, the perimeter of the excavation was supported generally by a reinforced concrete skin wall, constructed progressively downwards as the excavation progressed, and propped by a system of steel struts supported on king posts. The king

Figure 9.3 Construction of the Standard Chartered Bank in Bishopsgate.

Figure 9.4 Horizontal beams and girders.

posts had been cast into a number of the permanent large-diameter piles before general excavation commenced (see Figure 9.3).

An interesting feature of this project is that the Consulting Engineers, Pell Frischmann & Partners, had decided to avoid conventional underpinning of adjoining buildings (which could have led to differential settlement) by designing the new retaining wall so that it could be constructed in a sequence of bays varying in width from about 2 to 3 m. The technique involved grouting of the sand/gravel (overlying the London Clay) beneath the adjoining foundations, and maintaining measured loads—using flatjacks—in the struts used to frame the excavations and in those used to re-strut the new wall.

Retention problems, such as upholding facades, are dealt with in some detail elsewhere.[1] The types of supports include horizontal beams and girders designed to transfer loads to shear walls or braced towers (Figure 9.4), vertical cantilevered frames designed to transfer loads to the ground, and additional bracing to the permanent structure to make the retained elements self-supporting.

9.3 Responsibilities and relationships

All parties involved in construction have particular duties and responsibilities, under both statutory and common law. The client, as everyone, has the duty of care to all who may be affected by his activities, and also has various obligations to serve proper notices in good time under The Building Regulations.[2,3]

Until the commencement of the contract period, when the contractor takes over the site, the client shares responsibility for the health and safety of people on his premises with the employers of those people.[4] Accidents have occurred as a result of unguarded openings being left in floors during investigation and survey operations. Such preliminary works require the same attention to planning and supervision as do later operations.

During the contract period the prime responsibility shifts towards the contractor as controller of the premises but, as always, all persons—including employees—have to ensure, in so far as is reasonably practicable, the health and safety of themselves and all others who may be affected by their operations.

A common cause of delay, and sometimes confusion, at the start of a city development, is the lack or lateness of the party wall agreement. It is the responsibility of the client, and his professional advisers, to obtain the agreement of any adjoining owners whose buildings and/or areas may be affected, not only regarding the extent of the works but, even more importantly, the method used to maintain structural safety. The contractor may well be given the opportunity to propose detailed methods of underpinning and provide associated temporary supports but, unless he has specific design responsibilities, he remains responsible for materials and workmanship only. Under-pinning is normally permanent, and is designed by the consulting engineer.

The responsibility for lateral strutting to party walls is somewhat less clear. The shores will often have been designed by the contractor, and although checked and accepted, with the drawing possibly forming part of the legal agreement between building owners, a share of responsibility for design is attached to the contractor. The contractor is advised, in any case, to assume that he remains responsible in all respects, as he is indeed completely liable for any failures that may occur as a result of wrong sequence of working, poor materials, or lack of properly skilled and trained labour and supervision. The precise cause of failure is rarely clear-cut.

For the majority of cases, outside party wall agreements, the responsibility for design, construction and removal of temporary supports rests clearly with the contractor.

Sub-contractors involved in this type of work are usually primarily con-cerned with demolition and/or excavation, and may well not have the skills in-company necessary to design and execute the strutting operations. In such cases, and whether or not proprietary items of falsework are hired from specialist suppliers, or design is sub-let to outside consultants, the main contractor must ensure that both the design and construction are checked and supervised by competent people. It is essential to make proper provision for this in tender prices, as failure to do so could cost dearly all those concerned.

Other parties affected by the works, who have duties to their own 'customers' (tenants, employers, users of water, gas, etc., and the travelling public), include:

(i) adjoining building owners;
(ii) highway authorities;
(iii) statutory authorities (water, electricity, etc.); and
(iv) local authorities (e.g. building controls).

Finally, the digging activities of archaeological investigators on sites of historic interest must either be completed under the overall supervision of the owner's consulting engineer—before the site is handed over to a contractor—or must be carried out under the latter's supervision during the contract period.

9.4 Design considerations

Typically, the cost of lateral supports to a basement excavation in London can amount to 25% of the cost of sub-structure, assuming that the perimeter is formed with contiguous or secant bored piling, or similar construction that may be considered as part of the permanent works.

This particular cost may be avoided if the permanent works are used to maintain support to the perimeter using 'top-down' construction techniques,[1] but the 'value engineering' exercise must be carried out thoroughly by the design and construction teams before any conclusion is drawn regarding the best economic solution. These techniques will probably generate, for instance, higher excavation costs because of restricted headroom and more difficult access. However, if construction above ground can start whilst the basement is still being excavated and built, an overriding advantage in reduced contract period, with attendant costs, may be won. As always, adequate provision must be made to ensure reasonably healthy and safe conditions, with particular attention to noise, dust and fumes.

Factors contributing to the assessment of design loads for basement excavation are:

  (i) safety (consequences of failure);
 (ii) the nature of the ground, including groundwater;
(iii) the surcharge;
(iv) permitted ground movements;
 (v) thermal strain in the propping system;
(vi) overall dimensions of excavation; and
(vii) period required for temporary support.

The first of these must remain paramount, as human life is at stake, and also the costs resulting from significant failure are usually considerable. CIRIA Report 104[5] recommends that a factor of safety of at least 2 should be applied to any prop or anchorage system. This does not mean, however, that safety factors should be cumulative to the extent that the designer protects himself unreasonably at the expense of his client. What the designer must do is to take adequate measures to allow for any shortcomings in his knowledge of the actual physical conditions. Costs of timely and well-directed site investigations are usually balanced by savings resulting from more economic design and less delays in construction.

The magnitude and rate of development of soil pressures have provided a contentious subject for discussion for many years. In general, designers, possibly increasingly influenced by fear of expensive litigation, have become more and more cautious, and this understandable attitude is likely to prevail until more meaningful knowledge is gained from site measurements and observations.

Figure 9.5 shows a comparison of some design loadings derived from various sources.[6-9] It may be of interest to note that when the author was first introduced to the construction of deep basements in London Clay the lateral load was normally calculated using a rectangular (approx.) pressure diagram, where $P = 24H$, lb/ft^2 (equivalent to $P = 1.64H$, kN/m^2). This was applied successfully to all timber strut design.

Surcharges, when considering fully supported excavations, may be imposed by adjacent structures and highways (as both dead and live loads),

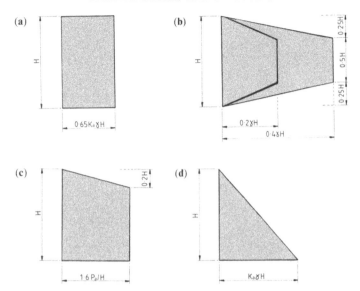

Figure 9.5 Comparison of design loadings from cohesionless: (*a*), (*b*) and (*d*) and cohesive (*c*) soils. Surcharge loads have been omitted for clarity. H = depth of excavation; γ = unit weight density of soil; K_a = coefficient of active earth pressure; P_a = thrust due to active earth pressure ($P_a = K_a \gamma H^2/2$). (a) Tomlinson (1976)—after Terzaghi and Peck (1967); (b) Tomlinson (1976)—after Terzaghi and Peck (1967)—use lower value when movements are minimal and construction period is short; (c) CP2 earth-retaining structures; (d) Rankine.

by stacked materials or by construction plant. The last can include large mobile cranes, used to lift into place other plant such as piling equipment or sections of tower cranes.

As it is unrealistic to assume that propping systems can be designed to prevent any movement whatsoever, the designer must make a judgment, often by agreement with other parties' engineers, on the order of deformation to be permitted. This will vary from a few millimetres in the case of adjacent buildings on strip or raft footings, to the order of 20 mm where there are areas or roads, with 'non-sensitive' services, to be supported. Sophisticated calculations are seldom justified, as natural soils never comply with the necessarily simplified assumptions made regarding the extent and nature of the ground, its homogeneity and its elasticity.[10]

Changes in temperature induce loads in steel strutting systems, which can amount to a significant proportion of the total load. The designer must allow for such effects, whether by increasing strut sizes (which again increases temperature loads!) or by providing constant load jacks or heat shielding measures. The costs of end connections to struts, however, could well be the determining factor in deciding which solution to adopt.

The shape and extent of the excavation, and the time it is required to be temporarily supported, are factors that undoubtedly influence the magnitude

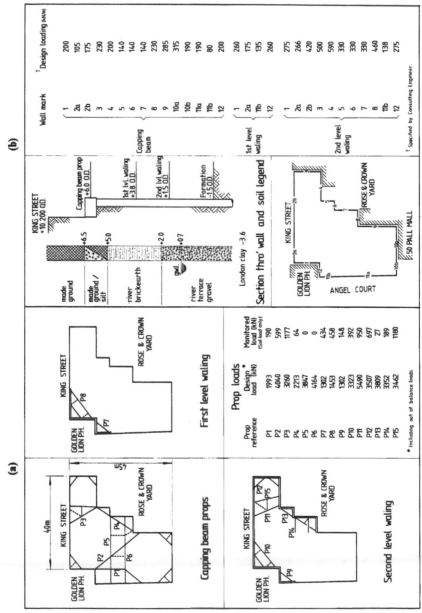

Figure 9.6 Comparison between (*a*) actual prop loads and (*b*) design waling loads for a recent basement excavation in London.

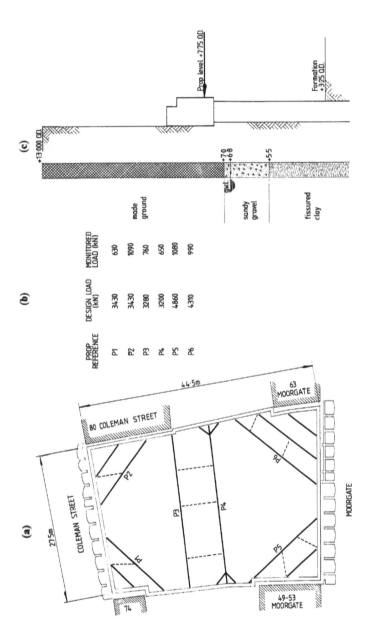

Figure 9.7 Measured loads and design loads in props provided to support a basement excavation in London Clay: (*a*) plan showing propping layout; (*b*) prop loads; (*c*) section through wall and soil legend. Loads include allowance for both soil and temperature difference of 15°C.

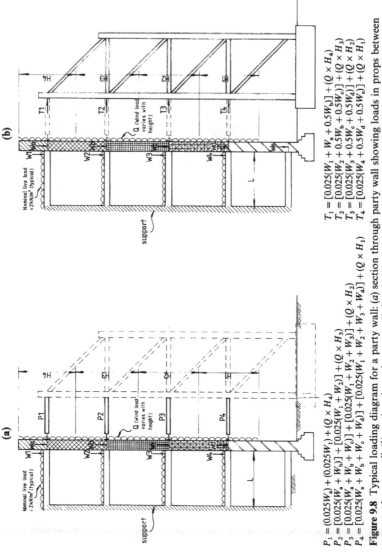

$P_1 = (0.025W_2) + (0.025W_1) + (Q \times H_4)$
$P_2 = [0.025(W_a + W_b)] + [0.025(W_1 + W_2)] + (Q \times H_3)$
$P_3 = [0.025(W_a + W_b + W_c)] + [0.025(W_1 + W_2 + W_3)] + (Q \times H_2)$
$P_4 = [0.025(W_a + W_b + W_c + W_d)] + [0.025(W_1 + W_2 + W_3 + W_4)] + (Q \times H_1)$

$T_1 = [0.025(W_1 + W_a + 0.5W_b)] + (Q \times H_4)$
$T_2 = [0.025(W_2 + 0.5W_b + 0.5W_c)] + (Q \times H_3)$
$T_3 = [0.025(W_3 + 0.5W_c + 0.5W_d)] + (Q \times H_2)$
$T_4 = [0.025(W_4 + 0.5W_d + 0.5W_e)] + (Q \times H_1)$

Figure 9.8 Typical loading diagram for a party wall: (a) section through tower and party wall showing loads in props between tower and party wall; (b) section through party wall showing loads in tower. $W_1 - W_4$ = dead load and live load from adjacent building floor at each level; $W_a - W_e$ = dead load from party wall between floor levels (hence total weight of wall = $W_a + W_b + W_c + W_d$, etc.); Q = wind load on party wall; L = span between party wall and support; $P_1 - P_4$ = loads in props between tower and party wall; $T_1 - T_4$ = loads in tower.

of active soil pressures developed, but insufficient data appear to have been gathered to enable designers to take these into account, other than on a judgmental basis.

Figure 9.6 shows a comparison between actual loads and design loads for a recent basement excavation in London. It may be seen that the actual loads due to soil pressure fell considerably short of the design loads and in some props failed to materialize at all. The high degree of redundancy built into the temporary support structure, including the contribution made by the perimeter piling, would probably account for some of these apparent anomalies. In this case it was not permitted to transfer load into the lines of piles, so that the 'out-of-balance' forces had to be transferred back into the prop and waling system.

Measured loads and design loads in props provided to support a basement excavation in London Clay are shown in Figure 9.7. In this case allowance was made for a temperature-induced increase of load. The props were in use over a period of about five months.

Above ground, the design problems are somewhat easier to solve, but again the quality of the information available to the designer is of critical importance. The strength and stability of the structure to be supported must be assessed[11] after carrying out comprehensive investigations, within the limitations of available access. If access is not available, or the cost of investigation is considered to be unacceptable, then the designer must make appropriate allowances and assumptions, which may in turn lead to a re-appraisal of the need to spend money on investigation!

Figure 9.8 shows a typical loading diagram for a party wall. The assumed percentage of dead load acting horizontally, and its distribution,[12] is critical to the design of the supports, as is the assumed permitted deflection. The last is a matter for judgment and agreement between the parties concerned.

For a strutting scheme to be economic, the effects on the construction of the permanent works have to be provided for as part of the design process. The main objective is usually to facilitate access so as to provide the opportunity to build fast without loss of safety or quality. An understanding of construction processes such as formwork fixing and striking, concrete placing and finishing, and steelwork erection is vital to this process.

Finally, the method of dismantling the temporary supports must be considered, not only as a separate operation, but also in relation to the sequence of building the permanent structure and the latter's development of strength and stability.

9.5 Construction considerations

The costs of an operation are largely influenced by the method of construction, and have three basic components—labour, plant and materials.

Material costs depend in turn not only on cost of manufacture, but also on the labour and plant elements of handling, storage, transport and delivery to site. Whether or not the final cost of, for instance, steel in temporary supports can be reduced by designing for re-use rather than for scrapping depends on these factors. The hiring of proprietary systems may well be less costly, in any case, if the period of use is relatively short and this possibility should be considered even if a small amount of tailoring is needed to enable system propping to be used.

The total cost per tonne of steelwork can vary considerably depending on the above factors and also, importantly, on the simplicity of connections and the access available for erection. Dismantling and removal costs must be added finally to complete the equation.

Failures in lateral movement systems, whether involving retained facades, larger elements of buildings, or highways, are frequently the results of inadequate co-ordination between the parties involved. The guidance and advice given in 1975, by the Advisory Committee on Falsework (The Bragg Report),[13] and later in the Code of Practice for falsework[14] includes a strong recommendation that the contractor appoints a temporary works (or falsework) co-ordinator. This does not lead to additional costs, as the duties involved are, in any case, necessary as part of efficient management of the process.

The duties of the temporary works co-ordinator are set out in the above named documents (albeit somewhat narrowly in the Code of Practice, where falsework by definition does not include earth work supports), and include:

(i)   ensuring that the design brief is adequate and that it accords with actual conditions on site;
(ii) ensuring that the design is properly checked by a competent person;
(iii) ensuring that the construction of the propping system is completed in accordance with the drawings and is ready to be loaded; and
(iv) ensuring that the temporary supports are not removed until adequate permanent support is provided.

Failures can also result, just as accidents in general, from lack of time available to complete properly all of the steps necessary to the process. Network planning techniques, covering the period from the client's decision to issue enquiry documents to the removal of the temporary supports, should be used to ascertain latest possible key dates for decisions or approvals and earliest dates for completion. Activities critical to the progress of the works can then be identified and pursued in a controlled manner.

Construction cannot be managed efficiently without feedback information. This applies to productivity data necessary to check and modify assumptions made in programming, the amount and quality of labour and plant, and the choice of method. It applies also to the planning of strutting systems in the sense that over-design leads to unnecessarily heavy, or too many, props and walings, which often restrict access to other operations, and hence cause time to be added to the contract period, as well as increased costs.

Monitoring of actual loads is thus a worthwhile investment, for the benefit not only of temporary works designers in the longer term, but also of each contract where instruments are installed. For example, it may be possible to remove selected struts for a limited time in order to facilitate access for permanent works construction.

9.6 Quality management

Quality may be defined as continually satisfying customer requirements. Within the context of temporary strutting, the designer's main customer is the contract manager who is responsible for the method of construction and structural safety. In turn, the manager's customer is the client, as defined by the contract, and he also has other customers in the sense that the contractor has duties of care and responsibilities as discussed previously (see section 9.3).

This 'customer satisfaction' can only be assured by giving attention to detail. In lateral support systems above ground, designed to retain facades or parts of buildings, important details include the connections with the old brickwork or masonry and the interconnections within the retained elements themselves. The strength of the existing structure must be carefully assessed.[11]

Simple pull-out tests are well worthwhile to determine the design loads for expanding bolt or chemical fixings. Provision may have to be made for worst cases, such as fixings being made into mortar joints, as realistic limits must be set on the amount of close site supervision it is economic to provide.

Below ground level, where unexpected conditions often prevail, the contractor must always be prepared to deal quickly with problems such as removing large obstructions or coping with water ingress. Lateral support, except for faces generally about 1.5 m deep and a few metres wide, must be maintained at all times. If ground water is present steel sheeting will probably need to be driven, if possible to a depth adequate to cut off the flow.

In general, detailing should take account of potential lack of fit,[15] no surfaces ever being truly vertical or horizontal, and of eccentricities in loading paths. On-site cutting and welding provides an expedient method of overcoming these problems, but is costly, especially as very close inspection and testing is necessary to maintain required quality standards, unless connections are over-designed to allow for a high percentage of bad workmanship.

In the author's experience, quality can best be assured by applying the following principles:

(i) ensure good communications by appointing a trained temporary works co-ordinator;[13,14]

(ii) design for simple, fast and safe erection and dismantling;
(iii) spend money on obtaining adequate knowledge of site conditions and the nature and strength of structures to be supported;
(iv) design with understanding of the limitations of the labour, plant and supervision to be employed, and of the method of construction to be adopted;
(v) use quality management procedures to maintain checking and approval disciplines.

9.7 Possible future developments

Results of load monitoring in propping systems for excavations on a number of sites in various ground conditions indicate that design methods need to be reviewed and improved. This will necessarily be a fairly slow and continuing process, but may lead to improvements in techniques for basement construction. Design of basement walls for 'top-down' construction may also be influenced.

Recently developed materials such as geotextiles, and techniques such as the use of sprayed concrete and soil nails[1] have not so far been used to any great extent in support systems for excavation, possibly because initial cost studies indicate no economic gain. In the long term, however, there may well be potential for reducing or eliminating lateral struts and hence improving access for permanent construction.

In the shorter term, careful site investigation and the application of value engineering techniques could lead to overall economies without increasing risk of failure. More use of site instrumentation could provide the opportunity to install readily adaptable systems, provided that suitable connections are designed and that permitted tolerances are realistic.

Finally, as in all other aspects of construction, real progress towards more economic yet still safe methods of support can only be made if all members of the team, from the client and his advisers to the sub-contractor, are working together in a less adversarial framework than has so far prevailed. Team work within companies should perhaps evolve towards 'partnering' between them.

Acknowledgements

The site measurements of prop loads referred to in this chapter were made by H. T. Yu, B.Sc, M.Sc, Ph.D of Trafalgar House Technology Ltd. Figures 9.5, 9.6, 9.7 and 9.8 were provided by T. Balakumar, B.Sc (Eng) of Trollope & Colls Engineering Department.

References

1. Institution of Civil Engineers (1988) *Proc. Conf. on Economic Construction Techniques: Temporary Works and Their Interaction with Permanent Works*; 16 November, 1988.

2. HMSO (1985) Manual to the Building Regulation.
3. HMSO (1986) The Building (Application to Inner London) Regulations.
4. HMSO (1974) The Health and Safety at Work, etc, Act.
5. CIRIA (1984) Design of retaining walls embedded in stiff clays. *CIRIA Report No. 104.*
6. TRADA (1981) *Timber in Excavations (Appendix 2).*
7. Institution of Structural Engineers (1951) *Civil Engineering Code of Practice No. 2—Earth Retaining Structures.*
8. British Steel Corporation (1979) *Piling Handbook.* 2nd ed.
9. Tomlinson, M. J. (1976) *Foundation Design and Construction.* Longmans, London.
10. Capper, P. L., Cassie, W. F. and Geddes, J. D. (1980) *Problems in Engineering Soils.* 3rd ed. E. & F.N. Spon Ltd.. London.
11. Institution of Structural Engineers (1980) *Appraisal of Existing Structures.*
12. BS 5628: Part I: 1978, Code of Practice for Use of Masonry. British Standards Institution, London.
13. HMSO (1975) *Final Report of the Advisory Committee on Falsework.*
14. BS 5975: 1982, Code of Practice for Falsework. British Standards Institution, London.
15. CIRIA (1981) Lack of fit in steel structures. *CIRIA Report No. 87.*

10 Soil anchorages

G. S. LITTLEJOHN

10.1 Introduction

The earliest reports of commercial anchoring into soil to provide lateral support to a vertical excavated face date from 1958, when Bauer grouted bars into a heavy alluvium for an excavation in Cologne. Since this date there has been a startling increase in the use of anchorages throughout the world, and millions of soil anchorages have been installed. These range from bars of a few tonnes capacity in soil reinforcement, to prestressed tendons of several hundred tonnes to provide lateral restraint to deep excavations and unstable natural slopes.

Today, anchorages may be employed to solve problems involving direct tension, sliding, overturning, dynamic load and ground prestressing, and the range of applications includes retaining walls, dry docks and highway pavements subjected to hydrostatic uplift, tall buildings, pile testing, pipe jacking, caisson sinking, slope stabilization, and tension nets for sports stadia.[1] This wide range of applications involving a variety of engineering disciplines perhaps best explains why soil anchorage development and usage have been so dramatic over the past 50 years.

If reliable performances are to be maintained, a technical understanding and appraisal of soil anchorage systems is required by the practising engineer. The purpose of this chapter is to provide guidance on site and ground investigation requirements, design methods, corrosion protection measures, construction techniques and testing procedures for grout injected soil anchorages.

10.2 Definition

A soil anchorage is considered to be an installation that is capable of transmitting an applied tensile load to a loadbearing soil stratum. The installation consists basically of an anchor head, free anchor length and fixed anchor (see Figure 10.1). The term anchor is used exclusively to denote a component of the anchorage, e.g. anchor head and free anchor length. The anchor head is the component of a soil anchorage that is capable of transmitting the tensile load from the tendon to the surface of the ground or structure requiring

292

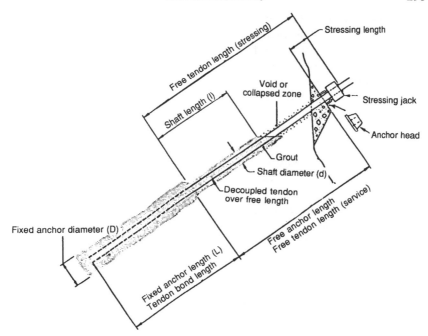

Figure 10.1 Soil anchorage nomenclature.

support. The fixed anchor length is the designed length of the anchorage over which the tensile load is capable of being transmitted to the surrounding soil. A tendon, which usually consists of steel bar, strand or wire either singly or in groups, is the part of an anchorage that is capable of transmitting the tensile load from the fixed anchor to the anchor head.

10.3 Site investigation

10.3.1 General

The ground is one structural component of the anchorage system, and the importance of a good quality site investigation cannot be over-emphasized. Investigations are most satisfactorily undertaken in a number of stages: (i) initial desk and field study; (ii) main field and laboratory investigation; and (iii) investigation during construction. The work required at any stage will be dependent on the nature of the project and, for a major retaining wall, could require that the main field and laboratory investigation stages be split into preliminary and detailed phases. The planning and general requirements of site investigations are well understood and the purpose of this section is to highlight particular features that are relevant to soil anchorages.

The geometry of a soil anchorage and its mode of operation require, in particular, a detailed knowledge of ground conditions local to the fixed anchor. Whilst there may be adequate data to indicate both the feasibility and advantages of an anchorage system, it is usual to find that there is insufficient detailed information to permit its economic design or construction. In this regard the data required for the safe design of temporary anchorages are often similar to those necessary for permanent works.

The broad aim of the investigation should be to determine, by the most economic means, the nature of the block of ground that is influenced by, or influences, the installation and behaviour of the anchorages. Since soil anchorages are installed at various inclinations, lateral variations in ground properties should be investigated as thoroughly as the vertical variations. For structures such as an anchored retaining wall it is recommended that the maximum centres of investigation locations, e.g. boreholes, pits, etc., should not exceed 20 m unless the geology and the characteristics of the relevant strata are well known. In addition, the plan dimensions of the site need to be carefully defined so as to include the probable fixed anchor zone. Too often in practice, particularly for deep excavations, anchorages are installed beyond the site perimeter, where there is a dearth of ground data.

10.3.2 Field sampling

Whilst the available sampling techniques are described in national standards, e.g. BS 5930,[2] particular emphasis should be placed on obtaining samples that can identify the fabric/structure of the stratum in which the fixed anchor length may be installed. In view of the relatively small diameter of most anchorage systems, minor differences in ground characteristics can be of significant influence in the behaviour of the anchorage/ground complex.

In soils, samples of a size and type suitable for examination and laboratory tests relevant to the nature of the soil should be taken from each stratum, and at maximum intervals of 1.5 m in thick strata. Intermediate disturbed samples, suitable for simple classification tests, should also be obtained, so providing a specimen of the ground at a maximum of 0.75 m intervals of depth. In variable strata, continuous undisturbed samples may be necessary in the probable vicinity of the fixed anchor zone.

Determination of the groundwater conditions on the site will be essential for the overall design of the project, particularly where excavations are proposed. This area of investigation, particularly the recording of long-term groundwater conditions, is too often given scant attention during routine site investigation works.

It has to be appreciated that the speed of boring, together with the addition of water—either to stabilize a borehole or as circulation fluid during drilling—usually precludes the measurement of equilibrium groundwater conditions during the investigation period. Nevertheless, all observations of water conditions during drilling should be carefully recorded throughout the progress of the investigation, as these often permit an initial assessment of true groundwater conditions. For example, the percentage of circulation water return during rotary drilling may assist in a first appreciation of groundwater level.

The long-term groundwater conditions can only be measured satisfactorily by the installation of standpipes or piezometers. Where groundwater may be contained in several aquifers separated by impervious strata, it may be necessary to install piezometers at different levels to record the head in each aquifer.

Chemical analysis of the ground and groundwater, e.g. sulphate·content, is important in determining the appropriate type of cement for the tendon bonding grout.

10.3.3 Field testing

Standard penetration tests (SPTs) should be made at regular intervals of depth in all granular strata in order to obtain an indication of their in-situ density classification. In the probable zone of the fixed anchor length, tests should be made at least at 1 m intervals of depth in each borehole.

The in-situ strength of cohesive soils can also be assessed by the SPT. The results provide a valuable check on laboratory data when made alternately with undisturbed sampling intervals.[3] The test is particularly applicable where the soils possess a structure or fabric that may preclude representative undisturbed sampling, e.g. clays containing partings of water-bearing sand.

The value of any empirical in-situ test depends critically on its being performed and interpreted in a standard manner as prescribed by its originator. Specialist engineering supervision is essential to obtain reliable results on which technical and commercial decisions are to be based. When making standard penetration tests, particular attention must be paid to maintaining stability of the bottom of the borehole by attention to size of bore, boring methods, and hydrostatic balance.

The radial stress/strain characteristics of the ground mass can be obtained in granular and cohesive soils by a pressuremeter test. Recent developments in pressuremeter design include a self boring probe.[4]

Where the ground investigation proves strata that may lead to potential grout loss, then permeability tests supplemented by fabric description may be required to quantify this problem. At low injection pressures loss of cement grout is unlikely where the permeability is less than 5×10^{-4} m/sec.

A generalized measure of redox potential and soil resistivity can be obtained to assess the potential ground corrosiveness to embedded metals. For example, the redox potential provides guidance on the risk of micro-biological corrosion, frequently characterized by pitting attack and most commonly encountered in heavy clay soils.

10.3.4 *Laboratory testing*

Tests for the classification of soils are described in national codes e.g. BS 1377.[5] It is recommended that the grading of granular soils, and the liquid and plastic limits of cohesive soils be determined for every stratum encountered in the investigation.

The determination of low plasticity indices in a cohesive soil, even in very localized zones, can influence both the type of anchorage and the method of drilling. For example, the drilling technique adopted may have to be that used for sands rather than clays, i.e. cased as opposed to uncased.

For granular soils, it is only possible to estimate the in-situ shear strength characteristics on the basis of in-situ tests, combined with grading and particle shape assessments from laboratory classification tests.

In granular soils of mixed grading, it may be justifiable to determine the peak shear strengths of samples prepared to a series of densities from loose to dense. The in-situ strength may then be inferred by interpolation at the density estimated from field tests. It should be noted that the shear strength is dependent on stress level. Hence, all laboratory tests should be made at a stress level approximating to that in-situ.

For cohesive soils, the shear strengths should be obtained by triaxial compression tests. The type to be employed will depend on the design method, the mass permeability of the soil and the probable rate of stressing of the anchorage.

Where stressing of the anchorage is rapid, the undrained shear parameters should be used in the necessary total stress analysis. For a slow rate of stressing or where the long-term anchorage behaviour is required, the shear strength in terms of effective stresses should be obtained from a consolidated drained test. Alternatively, and particularly where the behaviour of the pore water pressure during shear is of interest, a consolidated undrained test with pore water pressure measurement should be made. The rate of strain employed should be sufficiently slow to ensure equalization of pore water pressure throughout the specimen.

Where a cohesive soil possesses a relatively high mass permeability, e.g. clayey silts, clay possessing a permeable fabric, and marl, both undrained and effective shear parameters should be determined to permit a study of the influence of this property under different rates of loading of the anchorage.

In all tests, the stress/strain characteristics of the soil during shear should be recorded. In soils exhibiting a marked structure or fabric, e.g. laminated clays, it may be necessary to study the influence of these features on the shear strength of soil in the plane of shear induced around the fixed anchor length.

For systems that will apply a high average stress to a clay lying between the fixed anchor length and the structure, the compressibility characteristics should be determined by testing in a suitable type of oedometer. The rate of consolidation determined by such tests will also assist in assessing the mass permeability of the soil, particularly where tests can be made on large diameter specimens in a hydraulic oedometer.[6] Such data may influence the choice of design method, i.e. whether a total or effective strength analysis is relevant and also provide guidance as to the probable loss of prestress through case history comparisons.

10.3.5 Presentation of data

Of particular importance in investigations for soil anchorages is the recording of the structure or fabric of the soil. Relatively minor features in the strata can have a significant influence on anchorage behaviour in view of the limited volume of the stressed zone. For example, the presence of thin, even single grain, partings of silt or sand within a clay can have a marked effect on the behaviour of the soil and, in particular, on the action of drilling water on the soil prior to grouting. This, in turn, can severely limit the load that can be placed on clay anchorages, particularly of the underreamed type. Similarly, the presence of even small pockets of granular material within a generally cohesive glacial drift can have a major influence on the installation technique. Such minor features within an apparently uniform soil can significantly affect the cost of a project.

10.3.6 Investigation during construction

During the course of anchorage installation, further evidence on the nature of the ground conditions is revealed by excavation or in the boreholes of the individual anchorages. This information should be recorded and appraised as an extension of the engineering geology work. No site investigation can explore ground conditions as frequently as the anchorage installation process can, and any major variations in the ground conditions from those anticipated have to be recorded and their significance assessed. It should be emphasized, however, that production drilling associated with anchorage installation is not geared to investigate the ground in detail. All ground data obtained during anchorage drilling should be recorded and subjected to daily analysis. Such a system can act as an early warning device, should variation in strata levels or ground type require changes in the design or installation method.

Any adjacent activities on the site that may influence anchorage behaviour should also be carefully monitored and recorded, and their possible influence assessed at an early stage in the work. Construction activities include groundwater lowering, piling, blasting and freezing. In addition, if the site is adjacent to an electrical installation the ground should be checked for stray electrical currents, since these can cause corrosion of a steel tendon if this becomes the anode in a galvanitic process.

10.4 Design

10.4.1 General

A grouted soil anchorage may fail in one or more of the following modes:

(i) failure within the soil mass;
(ii) failure of the soil/grout bond;
(iii) failure of the grout/tendon bond; and
(iv) failure of the steel tendon or anchor head.

For the design of a soil anchorage each mode of failure must be considered in order to ensure an adequate load factor of safety, having regard to magnitude and mode of loading, period of service and consequences of failure.

10.4.2 Overall stabillity

10.4.2.1 General. In assessing overall stability the designer should take account of:

(i) location of critical failure planes to ensure that sufficient fixed anchor length exists beyond such planes;
(ii) building and planning constraints, which can restrict or deny the use of anchorages outside the working area of the project; and
(iii) physical constraints—related to ground conditions, the presence of underground services, abandoned mine workings, etc.

To assist the anchorage contractor, the designer should detail working load of the anchorages, minimum free anchor lengths and general layout of the anchorages. The designer should also maintain some flexibility and be prepared to permit changes of design to accommodate obstructions or changes in physical conditions.

10.4.2.2 Excavations. In deep excavations constrained by some form of retaining wall, assessment of the overall stability has to consider the interaction between the ground, structure and anchorages as a complete system.

For cohesionless soils where the wall is retained by a single line of anchorages, analysis can be attempted using a sliding block method (see Figure 10.2). The earth pressure P_n on the vertical cut through the proximal end of the fixed anchor is calculated with a nominal angle of shearing resistance (ϕ'_n), and the resultant force R_n on the inclined plane of the sliding block forms the same angle ϕ'_n with the normal to the sliding plane. ϕ'_n has been correctly assumed if the weight of the soil block (W) and the forces P_n and R_n are in equilibrium. If this is not the case then ϕ'_n has to be altered.

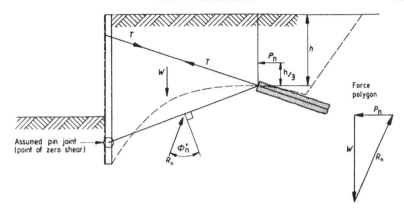

Figure 10.2 Sliding block method of analysis (after Littlejohn[7]).

When equilibrium is achieved the factor of safety is given by:

$$S_f = \frac{\tan \phi'}{\tan \phi'_n} \tag{10.1}$$

where ϕ' is the actual angle of shearing resistance. A minimum factor of safety of 1.5 is recommended for overall stability.

The main attraction of the sliding block method is its simplicity and, although the forces acting are assumed to be concurrent, it is considered that the value S_f is a safe estimate since the stabilizing passive resistance available from the embedded depth of soil within the excavation is ignored in the calculation. Where support systems involve several rows of anchorages the shape of the sliding block is not known with any certainty from experiments and the overall stability should be evaluated using a logarithmic spiral method.[7,8] A logarithmic spiral has the property that the radius from the spiral centre to any point on the curve forms a constant angle ϕ' with the normal line to the curve (see Figure 10.3). None of the forces along the sliding plane will therefore create a moment about the spiral centre, and they can therefore be neglected when considering the equilibrium of moments about this point.

By choosing an appropriate factor of safety S_f, then ϕ'_n can be calculated from equation (10.1). Using a ϕ'_n spiral, the location of this curve is adjusted around the retaining wall until the overturning and stabilizing moments balance. Since passive resistance beneath the excavation is ignored, a conservative solution is again achieved when fixed anchors are positioned beyond the ϕ'_n spiral.

In both stability analyses described, the basic assumption is that anchorage prestress increases the shear strength of the cohesionless soil sufficiently to displace the potential failure plane beyond the proximal end of the fixed anchor. Care should therefore be taken not to apply these methods outside the range of cohesionless soils.

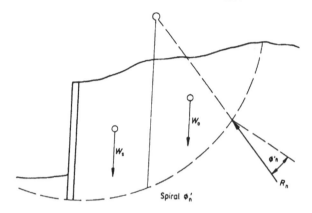

Figure 10.3 Stability analysis: spiral shaped sliding surfaces (after Littlejohn[7]).

$$S_f = \frac{\tan \phi'}{\tan \phi'_n}. \text{ For equilibrium, } \frac{\text{moment due to } W_s}{\text{moment due to } W_o} = 1.$$

In cohesive soils it is clear that anchorage prestress will only increase the shear strength of the soil gradually as consolidation occurs. Consequently, in this situation a conventional analysis of the overall stability should be carried out neglecting the presence of the soil anchorages. The fixed anchors must then be located some distance, typically 2 to 3 m, beyond the potential slip zone to ensure that excessive pressures are not transmitted across this zone, which might lead to premature failure. A variety of methods of stability analysis for slopes can be applied to deep excavations.[9]

10.4.2.3 Uplift capacity. For vertical or downward inclined anchorages subjected to external loads, e.g. basement slabs or highway pavements subjected to hydrostatic uplift, the individual anchorages must be installed at a depth sufficient to resist safely the applied working load without failure occurring in the soil mass. There is little experimental evidence to substantiate the methods, but current practical experience with vertical anchorages in soil indicates that general failure with accompanying surface heave does not occur for slenderness ratios (h/D) in excess of 15, where h is the depth to the top of the fixed anchor and D is the diameter of the fixed anchor. For slender anchorages $(h/D > 15)$ the failure mechanism in the ground tends to be local to the fixed anchor zone. In current practice a depth of 5 m to the top of the fixed anchor is also commonly assumed as a prudent minimum depth of embedment for any soil anchorage application.

10.4.3 Soil/grout interface

10.4.3.1 General. Design rule predictions of ultimate load-holding capacity are invariably created by assuming that the ground has failed along a shear surface, postulating a failure mechanism and then examining the relevant forces in a stability analysis. Using simple practical terms, there are basically two load transfer mechanisms by which ground restraint is mobilized locally as the fixed anchor is withdrawn, namely end-bearing and side shear. Fixed anchors fail in local shear via one of these mechanisms or by a combination of both, provided that sufficient constraint is available from the surrounding ground. In this context, general failure is defined as the full mobilization of slip lines or the generation of significant deformations extending into the exposed surface.

The ultimate load-holding capacity of the anchorage (T_f) is dependent on the factors listed below, although, due to lack of knowledge, item (e) is not generally isolated in design calculations, and the design model is therefore imperfect and does not represent accurately physical states at the ground/ grout interface.

(a) Definition of failure.
(b) Mechanism of failure.
(c) Area of failure interface.
(d) Ground properties mobilized at the failure interface.
(e) Stress conditions acting on the failure interface at the moment of failure.

The design rules described in sections 4.3.3 and 4.3.4 for soils apply to individual anchorages and no allowance is made for group effects or interference.

It should be noted that all anchorages are subject to a testing procedure (see section 4.7.2), and the calculated ultimate load-holding capacities are therefore for guidance only and may have to be modified depending on the results of such testing.

10.4.3.2 Anchorage types

General. Anchorage pull-out capacity for a given ground condition is dictated primarily by anchorage geometry. The transfer of stresses from the fixed anchor to the surrounding ground is also influenced by construction technique, particularly the grouting procedure, and to a lesser extent the methods of drilling and flushing. Accordingly, the types of anchorage to which the design rules are applicable will now be described (see Figure 10.4).

Type A anchorages. Type A anchorages consist of tremie (gravity displacement), packer or cartridge-grouted straight-shaft boreholes, which may be temporarily lined or unlined depending on hole stability. This type is most commonly employed in rock and very stiff to hard cohesive deposits, and resistance to withdrawal is dependent on side shear at the ground/grout interface.

Type B anchorages. Type B anchorages consist of low pressure (typically grout injection pressure $p_i < 1000 \, \text{kN, m}^{-2}$) grouted boreholes, via a lining

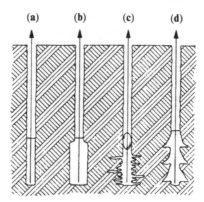

Figure 10.4 Main types of cement grout injection anchorages. (*a*) Type A; (*b*) Type B; (*c*) Type C; (*d*) Type D.

tube or in-situ packer, where the diameter of the fixed anchor is increased with minimal disturbance as the grout permeates through the pores or natural fractures of the ground. This type is most commonly employed in coarse granular alluvium, but the method is also popular in fine-grained cohesion-less soils. Here cement based grouts cannot permeate the small pores but, under pressure the grout compacts the soil locally to increase the effective diameter of the fixed anchor and enhance the shearing resistance of the soil. In practice, resistance to withdrawal is dependent primarily on side shear but, when calculating the ultimate load-holding capacity, an end-bearing component may be included.

Type C anchorages.   Type C anchorages consist of boreholes grouted to high pressure (typically $p_i > 2000\,\mathrm{kN\,m^{-2}}$), via a lining tube or in-situ packer. The fixed anchor length is enlarged by hydrofracturing of the ground mass to give a grout root or fissure system beyond the core diameter of the bore-hole. Often pressure is applied during a secondary injection after initial stiffening of primary grout placed as for Type B anchorages. Secondary injections are usually made via either a tube à manchette system (see Figure 10.5) or miniature grout tubes incorporated within the fixed anchor length; the former is advantageous if several injections are envisaged.

Whilst this anchorage type is commonly applied in fine cohesionless soils, some success has also been achieved in stiff cohesive deposits. Design is based on the assumption of uniform shear along the fixed anchor.

Type D anchorages.   Type D anchorages consist of tremie grouted boreholes in which a series of enlargements, either bells or underreams, have previously been formed. This type is employed most commonly in firm to hard cohesive deposits. Resistance to withdrawal is dependent on side shear and end-bearing,

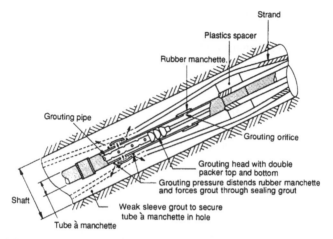

Figure 10.5 Details of tube à manchette for pressure grouting control (after Mitchell[10]).

although, for single or widely spaced underreams (see equation 10.8), the ground restraint may be mobilized primarily by end-bearing.

Although not common, this type can be used in cohesionless soils in conjunction with some form of side-wall stabilization over the enlargement length. Typically this may be by preinjection of cement or chemical grout in the ground around the fixed anchor or by pumping polymer drilling fluid into the borehole during the drilling and underreaming operation.

10.4.3.3 Fixed anchor design in cohesionless soils

Type B anchorages.   For low pressure grouted anchorages of Type B, design equations for the estimation of the ultimate load-holding capacity are based primarily on piling design technology.[7,11-14] For guidance, T_f in kN, may be estimated from equation (10.2):

$$T_f = Ln \tan \phi' \qquad (10.2)$$

where L is the fixed anchor length (in m)
   ϕ' is the effective angle of shearing resistance (in degrees)
   n is the factor that takes account of the drilling technique (rotary percussive with water flush), depth of overburden, fixed anchor diameter, grouting pressure in the range 30 to 1000 kN m^{-2}, in-situ stress field and dilation characteristics.

Field experience[7] indicates that for coarse sands and gravels (permeability $k_w > 10^{-4}$ m s^{-1}), n ranges from 400 to 600 kN m^{-1}, whilst in fine to medium sands ($k_w = 10^{-4}$ to 10^{-6} m s^{-1}), n reduces to 130 to 165 kN m^{-1}. These figures were initially measured in normally consolidated materials for borehole anchor diameters of 0.1 m approximately, and where the enlarged diameter D varies significantly, n is modified in the same proportion.

As an alternative, equation (10.3) may be used where anchorage ultimate load-holding capacity is related to anchorage dimensions and soil properties:[7]

$$T_f = A\sigma'_v \pi DL \tan \phi' + B\gamma h \frac{\pi}{4}(D^2 - d^2) \qquad (10.3)$$

$$= (\text{side shear}) + (\text{end-bearing}).$$

where A is the ratio of contact pressure at the fixed anchor/soil interface to the average effective overburden pressure

σ'_v is the average effective overburden pressure adjacent to the fixed anchor (in $kN\,m^{-2}$)

D is the diameter of the fixed anchor (in m)

L is the length of the fixed anchor (in m)

ϕ' is the effective angle of shearing resistance (in degrees)

B is the bearing capacity factor equivalent to $N_q/1.4$ (see Figure 10.6)

γ is the unit weight of soil overburden (submerged unit weight beneath the water table (in $kN\,m^{-3}$))

h is the depth of overburden to top of fixed anchor (in m)

d is the diameter of grout shaft above fixed anchor (in m).

Type C anchorages. It is a feature of Type C anchorages that calculations are based on design curves created from field experience in a range of soils rather than relying on a theoretical or empirical equation using the mechanical properties of a particular soil. In alluvium, for example, test results[16] have indicated for 0.1 to 0.15 m diameter boreholes, ultimate load-holding capaci-

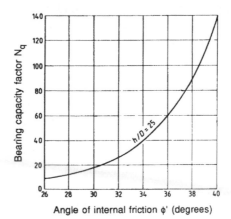

Figure 10.6 Relationship between bearing capacity factor N_q and effective angle of shearing resistance ϕ' (after Berezantzev *et al.*[15]).

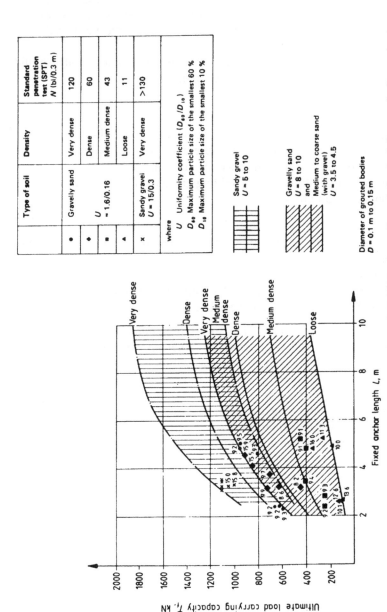

Figure 10.7 Ultimate load-holding capacity of anchorages in sandy gravels and gravelly sand showing influence of soil type, density and fixed anchor length (after Ostermayer and Scheele[18]). Note that: (i) field evidence is limited to a fixed anchor length of 2 to 5 m approximately; and (ii) the relationships between soil density and standard penetration test (SPT) values are not in accordance with BS 5930.

ties of 90 to $130\,\mathrm{kN\,m^{-1}}$ of fixed anchor at a p_i of $1000\,\mathrm{kN\,m^{-2}}$, and 190 to $240\,\mathrm{kN\,m^{-1}}$ at a p_i of $2500\,\mathrm{kN\,m^{-2}}$ where p_i is the grout injection pressure.

In more recent years, design curves for Type C anchorages have been extended through proving tests in Germany.[17,18] For sandy gravels and gravelly sands Figure 10.7 shows that the ultimate load increases with density and uniformity coefficient.

The fixed anchor length for anchorages in cohesionless soils should normally not be less than 3 m, nor more than 10 m.

10.4.3.4 Fixed anchor design in cohesive soils

Type A anchorages. For tremie or gravity grouted straight-shaft anchorages of Type A, design rules are similar to those developed for bored piles[7,19,20] and are based on the use of undrained shear strengths.

For guidance, the ultimate load-holding capacity T_f in kN, may be estimated from equation (10.4):

$$T_f = \pi D L \alpha C_u \qquad (10.4)$$

where C_u is the average undrained shear strength over the fixed anchor length (in $\mathrm{kN\,m^{-2}}$)
α is the adhesion factor
D is the diameter of the fixed anchor (borehole diameter) (in m).
L is the length of the fixed anchor (in m).

The actions of drilling and grouting cause stress changes within the ground, which cannot be accurately modelled by either an effective stress or total stress analysis. An effective stress analysis will indicate a higher calculated ultimate load-holding capacity, whilst a total stress analysis yields results more closely resembling actual ultimate capacities. Therefore, bearing in mind the short duration of the test on an anchorage and the fact that effective stress analysis implies deformation that is accompanied by loss of prestress, a total stress analysis is considered more appropriate.

Type C anchorages.   Where high grout pressures can be safely permitted, Type C anchorages, with or without post-grouting, can be used.

The results of a large number of fundamental tests are shown in Figure 10.8, which can be used as a design guide for borehole diameters of 0.08 to 0.16 m. Skin friction (τ_M) increases with increasing consistency and decreasing plasticity. The consistency index, I_c, is given by the following equation:

$$I_c = \frac{L_L - m}{L_L - P_L} \qquad (10.5)$$

where L_L is the liquid limit (in %)
P_L is the plastic limit (in %)
m is the natural moisture content (in %).

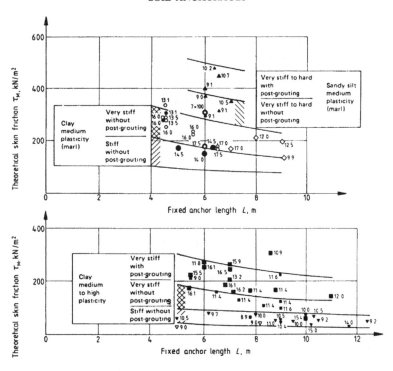

Failure load was reached	Failure load was not reached	Post-grouting	Type of soil	$L_L\%$	$P_L\%$	$l_c\%$
▲	△	Without	Silt, very sandy (marl)	~45	~22	~1.25
▲	△	With	Medium plasticity			
●	○	Without		32 to 45	14 to 25	1.03 to 1.14
○	○	With	Clay (marl) Medium plasticity			
◖	◔	Without		36 to 45	14 to 17	1.3 to 1.5
◗	◓	With				
	◇	Without	Silt Medium plasticity	23 to 28	5 to 11	0.7 to 0.85
■		Without		48 to 58	23 to 35	1.1 to 1.2
■		With	Clay Medium to high plasticity			
▼		Without		45 to 59	16 to 32	0.8 to 1.0

Figure 10.8 Skin friction in cohesive soils for various fixed anchor lengths, with and without post-grouting (after Ostermayer[17]).

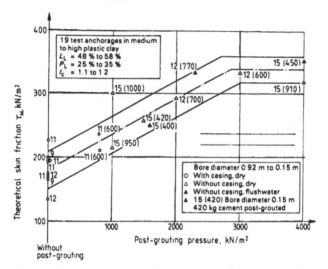

Figure 10.9 Influence of post-grouting pressure on skin friction in a cohesive soil (after Ostermayer[17]). Note that the theoretical skin friction is calculated from the borehole diameter and designed fixed anchor length.

In stiff clays (i.e. a consistency index, $I_c = 0.8$ to 1.0) with medium to high plasticity, skin frictions of 30 to 80 kN m^{-2} are the lowest recorded, whilst the highest values ($\tau_M > 400$ kN m^{-2}) are obtained in sandy silts of medium plasticity and very stiff to hard consistency ($I_c = 1.25$). The technique of post-grouting is also shown to generally increase the skin friction of very stiff clays by some 25 to 50%, although greater improvements (from 120 kN m^{-2} up to about 300 kN m^{-2}) are claimed for stiff clay of medium to high plasticity. From Figure 10.9, the influence of post-grouting pressure on skin friction is quantified showing a steady increase in τ_M with increase in post-grouting pressure.

Type D anchorages. For multi-underreamed anchorages of Type D, a variety of design rules are available.[7,13,19,21] For guidance, the ultimate load-holding capacity T_f, in kN, may be estimated from equation (10.6).

$$T_f = \pi DL\, C_u + \frac{\pi}{4}(D^2 - d^2)N_c C_{ub} + \pi d\ell\, C_a \qquad (10.6)$$

$$= \text{(side shear)} + \text{(end-bearing)} + \text{(shaft resistance)}$$

where D is the diameter of underream (see Figure 10.10) (in m)
   L is the length of fixed anchor (in m)
   C_u is the average undrained shear strength over fixed anchor length (in kN m^{-2})
   d is the diameter of shaft (in m)
   N_c is the bearing capacity factor (value of 9 commonly assumed)

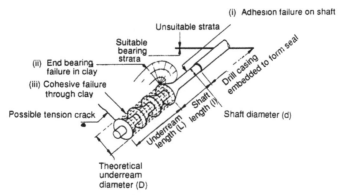

Figure 10.10 Diagram of multi-underream anchorage at ultimate capacity (after Bassett[13]).

C_{ub} is the undrained shear strength at proximal end of fixed anchor (in $kN\,m^{-2}$)

ℓ is the length of shaft (in m)

C_a is the shaft adhesion (in $N\,m^{-2}$) (a value for C_a of $0.3\,C_u$ to $0.35\,C_u$ is commonly assumed).

In the absence of results from trial anchorages in the field, empirical multiplier reduction coefficients ranging from 0.75 to 0.95 are sometimes applied to the side shear and end-bearing components of equation (10.6) to allow for factors such as construction techniques and underream geometry.[13,22]

In the particular case where the clay adjacent to the fixed anchor contains open or sand-filled fissures, a reduction coefficient of 0.5 is recommended for the side shear and end-bearing components.[7]

Of vital importance in cohesive deposits is the time during which drilling, underreaming and grouting take place. This should be kept to a minimum in view of the softening effect of water on the clay. The consequences of delays of only a few hours include reduced load-holding capacity and significant short-term losses of prestress. In the case of sand-filled fissures, for example, where water flushing is employed, a period of only 3 to 4 h may be sufficient to reduce the C_u strength to near the fully softened value.

Underreaming is ideally suited to clays of C_u greater than $90\,kN\,m^{-2}$. Some difficulties in the form of local collapse or breakdown of the neck portion between the underreams should be expected where C_u values of 60 to $70\,kN\,m^{-2}$ are recorded. Underreaming is virtually impracticable below a C_u of $50\,kN\,m^{-2}$. Underreaming is also difficult in soils of low plasticity, e.g. plasticity index <20.

With regard to spacing of underreams (δ_u) in type D anchorages, equations (10.7) and (10.8) can be used to estimate the maximum allowable spacing to give failure along a cylindrical surface, which requires less fixed anchor displacement to develop load. Where fixed anchor displacement is unimportant, more widely spaced underreams may be employed when they will act independently.

$$\pi D \delta_u C_u < \frac{\pi}{4}(D^2 - d^2)N_c C_u \tag{10.7}$$

from which

$$\delta_u < \frac{(D^2 - d^2)}{4D} N_c \tag{10.8}$$

The fixed anchor length should not normally be less than 3 m, nor more than 10 m. In this regard, where the fixed anchor comprises several decoupled strands of different lengths in order to distribute the stress more uniformly to the ground, longer fixed anchor lengths may be acceptable subject to proving tests.

10.4.3.5 Fixed anchor interaction.   To limit interaction between fixed anchors, a spacing of not less than four times the diameter (D) centre-to-centre is assumed (where D is the largest diameter within the fixed anchor) and, in practice, a minimum spacing in the range 1.5 to 2 m is commonly adopted.

The distance between a fixed anchor and adjacent foundation or underground services should exceed 3 m, bearing in mind that for surface foundations it may be necessary to increase the fixed anchor depth to 5 m or more, either to improve the ultimate resistance to withdrawal or to encourage the possibility of local shear failure around the fixed anchor, as opposed to general shear failure, under ultimate conditions (see section 10.4.2.3).

10.4.4 Grout/tendon interface

Three mechanisms of bond, namely adhesion, friction and mechanical interlock, are widely recognized but, in practice, recommendations pertaining to grout/tendon bond values commonly take no account of the length or type of tendon, or the strength and cover of the surrounding grout. For these reasons it is still advisable to measure experimentally the embedment length for untried field conditions.

As an initial guide for cement bonded tendons, the ultimate bond stress assumed to be uniform over the tendon bond length (see Figure 10.1) should not exceed:

(i) $1.0\,\mathrm{N\,mm^{-2}}$ for clean plain wire or plain bar;
(ii) $1.5\,\mathrm{N\,mm^{-2}}$ for clean crimped wire;
(iii) $2.0\,\mathrm{N\,mm^{-2}}$ for clean strand or deformed bar; and
(iv) $3.0\,\mathrm{N\,mm^{-2}}$ for locally noded strands.

The above values are based on a minimum grout compressive strength of $30\,\mathrm{N\,mm^{-2}}$ prior to stressing. They may be applied to single-unit, parallel, or multi-unit tendons, provided that the clear spacing is not less than 5 mm.[23,24] For noded strands and tendons that can mobilize mechanical interlock or the shear strength of the grout, the minimum spacing criterion does not apply. For resinous grouts, ultimate bond values should be obtained from proving tests in the absence of relevant documented test data.

The tendon bond length should not be less than the following, unless full-scale tests confirm that shorter bond lengths are acceptable.

(i) 3 m where the tendon is homed and bonded in-situ; and
(ii) 2 m where the tendon is bonded under factory controlled conditions.

Bond strength can be significantly affected by the surface condition of the tendon, particularly when loose or lubricant materials are present at the interface. The surface of the tendons should therefore be free from loose rust, soil, paint, grease, soap or other lubricants. A light film of rust on the tendon is not harmful and may improve bond. On the other hand, tendons showing signs of pitting should not be used.

In order to minimize debonding[25] it is recommended that, for all anchorages, the tendon area should not exceed 15% of the borehole area for parallel multi-unit tendons and 20% of the borehole area for single-unit tendons or suitably noded multistrand tendons.

10.4.5 Materials and components

10.4.5.1 Cementitious grouts. All conventional hydraulic cements, namely ordinary, rapid hardening, sulphate resisting and low heat Portland, are acceptable but, to avoid stress corrosion of the steel tendon, the total chloride content of the grout derived from all sources should not exceed 0.1% by weight of cement.

To ensure that the cement grout has good bond and shear strength, the mix should attain an unconfined compressive strength of $40 \, N \, mm^{-2}$ at 28 days. The bleeding of the tendon bonding grout should generally not exceed 2% of the volume 3 h after mixing. Higher values may be permitted in the case of permeable ground where the bleed water is filtered from the grout during injection under pressure. Given these design considerations the water/cement ratio typically ranges from 0.4 to 0.7 for soil anchorages.

Admixtures should only be used if tests have shown that their addition enhances the properties of the grout, e.g. by improving workability or durability, reducing bleed or shrinkage, or increasing rate of strength development. Detailed guidance on the design of cement based grouts is available elsewhere.[26]

10.4.5.2 Resinous grouts. Epoxy and polyester resins are most commonly used in tendon encapsulations for the protection of the bond length. For anchorages, ultimate compressive and tensile strengths in excess of

$75\,\text{N}\,\text{mm}^{-2}$ and $15\,\text{N}\,\text{mm}^{-2}$, respectively, are recommended traditionally for efficient load transfer, but post-gelation shrinkage, elastic modulus and percentage extension at failure are also important in relation to crack elimination in protection systems.

Post-gelation shrinkage should preferably be nil and not more than 5%, otherwise debonding can occur, which, in turn, creates a potential corrosion hazard through the formation of leakage paths. To match the ductility of steel tendon it would appear that percentage extension of the resin at failure can be in the range 1 to 1.5%, but currently a minimum elastic modulus that will guarantee negligible creep under high service loads ($>500\,\text{kN}$) cannot be recommended. As a consequence, where resin grout is proposed for tendon bonding, full-scale tests (including sample sectioning) should be carried out prior to anchorage installation to prove the efficiency of the grout mix.

Given that the resin/hardener reaction is highly exothermic it can be beneficial both technically and economically to use inert fillers. Many crushed minerals are suitable but, for capsule or encapsulation grouts, fillers should be graded with 100% passing a $200\mu m$ sieve.

10.4.5.3 Tendon. Tendons usually consist of steel bar, strand or wire either singly or in groups. For soil anchorages Table 10.1 includes typical data for prestressing steel that may be used in tendon design. For such high strength steels the loss of prestress due to relaxation is small.

Under normal circumstances working loads should not exceed 62.5% and 50% of the characteristic strength of the tendon for temporary and permanent works, respectively.

To distribute load to the soil more uniformly, strands of different length are sometimes used within the fixed anchor zone. When these strands are stressed simultaneously, displacements at the anchor head are the same for all strands, and thus the strains, and hence stresses, differ in individual strands. In such cases the stress in the shortest strand should limit the acceptable working load. If the design requires uniform stresses within the tendon, mono-strand stressing is essential.

Centralizers should be provided on all tendons to ensure that the tendon is centred in the grout column. Centralizers should provide within the borehole a minimum grout cover of 10 mm at the centralizer, and should be fitted at centres according to the angle of the soil anchorage and the possible sag between points of support in order to provide a minimum grout cover of 5 mm to the tendon.

Spacers should be provided in the fixed anchor length of all parallel multi-unit tendons to ensure separation of not less than 5 mm between the individual components of the tendon, and thus the effective penetration of grout to provide adequate bond. The spacer should not be compressible nor cause decoupling, and a minimum of three spacers should be provided in each fixed anchor length.

Table 10.1 Typical sizes and specified characteristics strengths for prestressing tendon design (reproduced from BS 8081:1989[27] with permission of British Standards Institution).

Types of steel	Nominal diameter (mm)	Specified characteristic strength (kN)	Nominal steel area (mm²)
Non-alloy			
Wire	7.0	60.4	38.5
7-Wire strand	12.9	186	100
	15.2	232	139
	15.7	265	150
7-Wire drawn strand	12.7	209	112
	15.2	300	165
	18.0	380	223
Low alloy steel bar			
Grade 1030/835	26.5	568	552
	32	830	804
	36	1048	1018
	40	1300	1257
Grade 1230/1080	25	600	491
	32	990	804
	36	1252	1018
Stainless steel			
Wire	7	44.3	38.5
Bar	25	491	491
	32	804	804
	40	1257	1257

10.4.5.4 Anchor head. The anchor head normally consists of a stressing head in which the tendon is anchored, and a bearing plate by which the tendon force is transferred to the structure or excavation.

The stressing head should be designed to permit the tendon to be stressed and anchored at any force up to 80% of the characteristic tendon strength, and should permit force adjustment up or down during the initial stressing phase. Monitoring requirements during service will dictate the need for a normal, restressable (load adjustments of ±10% possible), or detensionable head. The stressing head should also permit an angular deviation of ±5° from the axial position of the tendon without having an adverse influence on the ultimate load carrying capacity of the anchor head.

Bearing plates for high capacity anchorages are normally designed to a national structural code.

10.4.6 Safety factors

The traditional aim in design is to make a structure equally strong in all its parts, so that when purposely overloaded to cause failure each part will collapse simultaneously.

Table 10.2 Minimum safety factors recommended for design of individual anchorages (reproduced from BS 8081:1989[27] with permission of British Standards Institution).

Anchorage category	Minimum safety factor			
	Tendon	Ground/grout interface	Grout/tendon or grout/ encapsulation interface	Proof load factor
Temporary anchorages where a service life is less than six months and failure would have no serious consequences and would not endanger public safety, e.g. short-term pile test loading using anchorages as a reaction system.	1.40	2.0	2.0	1.10
Temporary anchorages with a service life of say up to two years where, although the consequences of failure are quite serious, there is no danger to public safety without adequate warning, e.g. retaining wall tie-back.	1.60	2.5*	2.5*	1.25
Permanent anchorages and temporary anchorages where corrosion risk is high and/or the consequences of failure are serious, e.g. main cables of a suspension bridge or as a reaction for lifting heavy structural members.	2.00	3.0†	3.0*	1.50

* Minimum value of 2.0 may be used if full-scale field tests are available.
†May need to be raised to 4.0 to limit ground creep.
Note 1   In current practice the safety factor of an anchorage is the ratio of the ultimate load to design load. Table 10.2 defines minimum safety factors at all the major component interfaces of an anchorage system.
Note 2   Minimum safety factors for the ground/grout interface generally lie between 2.5 and 4.0. However, it is permissible to vary these, should full-scale field tests (trial anchorage tests) provide sufficient additional information to permit a reduction.
Note 3   The safety factors applied to the ground/grout interface are invariably higher compared with the tendon values, the additional magnitude representing a margin of uncertainty.

...Have you heard of the wonderful one hoss shay,
That was built in such a logical way?
It ran for a hundred years to a day,
And then, of a sudden it...
...went to pieces all at once,—
All at once, and NOTHING FIRST,—
Just as bubbles do when they burst

The Deacon's Masterpiece,
by Dr. Oliver Wendell Holmes

Thus for each potential failure mechanism a safety factor must be chosen having regard to how accurately the relevant characteristics are known, whether the system is temporary or permanent, i.e. service life, and the consequences if failure occurs, i.e. danger to public safety and cost of structural damage.

Since the minimum safety factor is applied to those anchorage components known with the greatest degree of accuracy, the minimum values used in practice invariably apply to the characteristic strength of the tendon or anchor head and thereby encourage a ductile failure. Recommended safety factors for design are listed in Table 10.2.

With regard to failure at the soil/grout and grout/tendon interface of the fixed anchor, load safety factors (S_f) generally range from 2 to 4, where S_f is defined as the ultimate load (T_f) divided by the working load (T_w). T_f may be regarded as the maximum load attained when the fixed anchor can be withdrawn steadily, e.g. creep in a plastic clay, or the maximum load attained prior to a sudden failure and loss of load, e.g. loss of bond in a very dense gravel. As more poor-quality soil has been exploited by anchorages, so safety factors have steadily increased in value. It is also fair to say that engineers today are less tolerant of individual anchorage failures, and whereas a 5% failure rate was common in the 1960s, concern is quickly expressed today whenever the figure exceeds 1%.

With reference to failure within the soil where the overall stability or uplift capacity is being assessed, the load safety factor generally ranges from 1.5 to 2 for slope stability and 2 to 3 for uplift capacity, but lower values may be employed where the analysis is judged to be conservative, e.g. where shear resistance of the soil is known to act but the uplift capacity is based solely on the weight of soil mobilized at failure.

10.5 Corrosion protection

10.5.1 General

Out of millions of prestressed ground anchorages that have been installed around the world, 35 case histories of failure by tendon corrosion have been recorded,[28] some of which were protected only by cement grout cover.

Table 10.3 Proposed classes of protection for soil anchorages (after Fédération Internationale de la Précontrainte[28]).

Anchorage category	Class of protection
Temporary	Temporary without protection
	Temporary with single protection
	Temporary with double protection
Permanent	Permanent with single protection
	Permanent with double protection

Invariably the corrosion has been localized and failures have occurred after service of only a few weeks to many years. As a consequence, it is considered that all permanent anchorages and temporary anchorages exposed to aggressive conditions should be protected, the degree of protection depending primarily on factors such as consequence of failure, aggressivity of the environment and cost of protection.

The object of design against corrosion is to ensure that during the design life of the soil anchorage the probability of unacceptable corrosion occurring is small. Various degrees of protection are possible and, for corrosion resistance, the anchorage should be protected overall, as partial protection of the tendon may only induce more severe corrosion on the unprotected part.

Choice of class of protection (see Table 10.3) should be the responsibility of the designer. By definition, single protection implies that one physical barrier against corrosion is provided for the tendon prior to installation. Double protection implies the supply of two barriers where the purpose of the outer second barrier is to protect the inner barrier against the possibility of damage during tendon handling and placement.

10.5.2 Principles of protection

Protective systems should aim to exclude a moist gaseous atmosphere around the metal by totally enclosing it within an impervious covering or sheath.

Cement grout injected in-situ to bond the tendon to the soil does not constitute a part of a protective system because the grout quality and integrity cannot be assured. Furthermore, fluid materials that become brittle on hardening crack in service as the structure suffers differential strains, the onset of cracking depending upon tensile strength and ductility.

Non-hardening fluid materials such as greases also have limitations as corrosion protection media. Reasons include:

(i) fluids are susceptible to drying out—usually accompanied by shrinkage and change in chemical properties;

(ii) fluids are liable to leakage if even slight damage is sustained by their containment sheaths;

(iii) fluids having virtually no shear strength are easily displaced and removed from the metal they are meant to protect; and

(iv) even in ideal conditions the long-term chemical stability, e.g. susceptibility to oxidation, is not known with confidence.

These aspects require that non-hardening materials are themselves protected or contained by a moisture-proof, robust form of sheathing, which must itself be resistant to corrosion.

Nevertheless, non-hardening fluids such as grease fulfil an essential role in corrosion protection systems, in that they act as a filler to exclude the atmosphere from the surface of a steel tendon, create the correct electrochemical environment and reduce friction in the free length. Whilst a layer of grease is not considered acceptable as one of the physical barriers required in the decoupled free length of a double corrosion protection system, grease is acceptable as a protective barrier in a restressable or detensionable anchor head, since the grease can be replaced or replenished.

Use of thicker metal sections for the tendon, with sacrificial area in lieu of physical barriers, gives no effective protection, as corrosion is rarely uniform and extends most rapidly and preferentially at localized pits or surface irregularities. Non-corrodible metals may be used for anchorage components, subject to verifying their electrochemical behaviour relative to other components, and stress corrosion characteristics in appropriate environments.

10.5.3 Protective systems

There is a variety of protective coatings or coverings. The principles of protection are the same for all parts of the anchorage, but different detailed treatments are necessary for the tendon bond length, tendon free length and anchor head.

In the free length, protection is achieved generally by: (i) injection of solidifying fluids to enclose the tendon; (ii) the use of pre-applied coatings; or (iii) a combination of both, depending on circumstances. The protective system should permit reasonably uninhibited extension of the tendon during stressing, and thereafter, if the anchorage is restressable. Greased and sheathed tendons are a popular solution in such circumstances (see Figure 10.11).

Sacrificial metallic coatings for high strength steel ($>1040 \, N \, mm^{-2}$) should not be used when such coatings can cause part of the steel tendon to act as a cathode in an uncontrolled manner in a galvanitic process.

The bond length requires the same degree of protection as the free length. In addition, the protective elements have all to be capable of transmitting high tendon stresses to the ground. This requires strength and deformability characteristics that have to be checked structurally.

Drilled hole grouted solid
plastics binding tape

Tendons comprise 10 strands each
greased and then sheathed in
polypropylene. Minimum thickness
of plastics coating = 0.8 mm

Figure 10.11 Typical free length detail for single protection of strand tendon (after Littlejohn[7]).

The deformation of individual elements of the corrosion protection system should not be such as to allow continuing creep nor expose the tendon bond length through cracking. The requirements of no creep and no cracking are in conflict and few materials are available that can comply with them under the intensity of stress around the fixed anchor.

Certain materials, notably epoxy or polyester resins, have appropriate strength, ductility and resistance to corrosion. They may be substituted for cementitious grouts but are more expensive.

When used to encapsulate bond lengths of tendon in combination with plastics ducts, compatibility of elastic properties of the anchorage components has to be examined to minimize decoupling or debonding of the resin from the duct.

To ensure effective load transfer between duct and grout, ducts are corrugated. The pitch of corrugations should be within six and twelve times the duct wall thickness, and the amplitude of corrugation should be not less than three times the wall thickness. The minimum wall thickness is 0.8 mm, but consideration of material type, method of installation and service required may demand a greater thickness. Duct material should be impervious to fluids. Typical examples of double protection arrangements for the bond length of strand and bar tendons are shown in Figures 10.12 and 10.13.

Unlike fixed anchors, anchor heads cannot be wholly prefabricated. Because of the strain in the tendon associated with prestressing, friction grips for strand and locking nuts on bars cannot fix the tendon until extension has been achieved. All existing locking arrangements require bare wire, strand or bar on which to grip, and any preformed corrosion protection of the tendon has to be removed. This leaves two sections of the tendon, above and below the bearing plate (outer head and inner head, respectively), that require separate protective measures in addition to the protection of the bearing plate itself. If the environment is aggressive, early protection of the anchor head is recommended for both temporary and permanent anchorages.

The essence of inner head protection is to provide an effective overlap with the free length protection, to protect the short exposed length of tendon

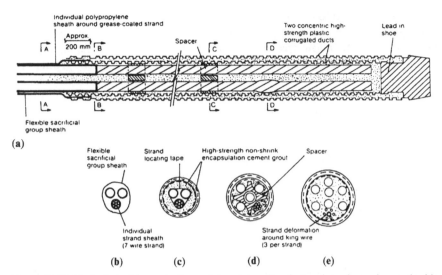

Figure 10.12 Typical double protection of bond length of strand tendon using a double corrugated sheath and cement grout (after Fédération Internationale de la Précontrainte[28]). (*a*) Typical longitudinal section through encapsulation (showing two strands only); (*b*) three-strand system, section A–A; (*c*) three-strand system, section B–B; (*d*) five-strand system, section C–C; (*e*) eight-strand system, section D–D.

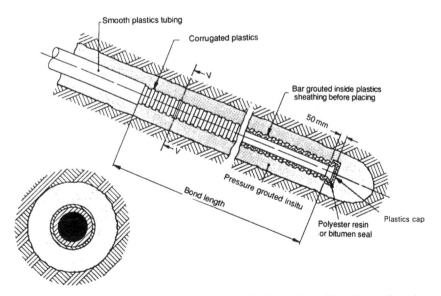

Figure 10.13 Typical double protection of bond length of smooth or ribbed bar tendon using a double corrugated sheath (reproduced from BS 8081: 1989[27] with permission of British Standards Institution). Enlarged view V–V is shown at bottom left.

below the plate and to isolate the short section of the exposed tendon passing through the plate. In satisfying these recommendations, the protective measures have to allow free movement of the tendon, which, in certain instances, may be solved by the use of a telescopic duct. Cement grouts are generally considered unsuitable for inner head protection. Primary grout should not be in contact with the structure and, where a weak, low bleed secondary grout is required to fill the void above the primary grout, it may be subject to cracking during structural movement. Grease-based corrosion protection compounds or similar ductile materials immiscible with water may be required. They may be preplaced or injected and should be fully contained within surrounding ducts and retained by an end seal.

Outer head protection of the bare tendon, the friction grips or the locking nuts above the bearing plate generally falls into two categories, controlled by whether the anchorage is restressable or not. Where restressability is called for, both the anchor head cap and the contents should be removable to allow access to an adequate length of tendon for restressing. Clearly these requirements will vary depending on the stressing and locking system employed. Grease is the most commonly used material within plastics or steel caps. Alternatives include corrosion-resistant, grease-impregnated tape and heat shrink sleeving.

Where restressability is not a requirement of the anchorage, then the cap and its contents are not required to be removable. Consequently, resins or other setting sealants may be used and a mechanical coupling between the cap and the bearing plate is not essential. Where the anchor head is to be

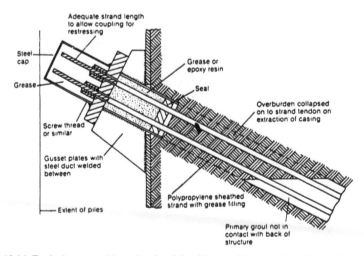

Figure 10.14 Typical restressable anchor head detail for double protection of strand tendon (after Fédération Internationale de la Précontrainte[28]). All steel components of gussets, ducts, bearing plates and caps coated with two coats of pitch epoxy. Free length classified as single protection, since grease is discounted as a protective barrier.

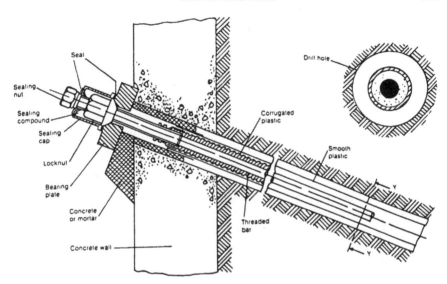

Figure 10.15 Typical detensionable anchor head for double protection of bar tendon (after Fédération Internationale de la Précontrainte[28]). Enlarged view Y–Y is shown at upper right.

totally enclosed by the structure, the outer head components may be encased in dense concrete as an alternative protection, given adequate cover.

The bearing plate and other essential exposed steel components at the anchor head should be painted with bitumastic or other protective materials, prior to being brought to site. Steel surfaces should be cleaned of all rust and deleterious matter prior to priming, e.g. by blast cleaning. The coatings should be compatible with the materials selected for both inner head and outer head protection. Bearing plates on concrete structures may be set in a seating formed of concrete, cement, epoxy or polyester mortar, or alternatively may be seated direct onto a cast-in-steel plate.

Typical examples of double protection arrangements for the anchor head are illustrated in Figures 10.14 and 10.15.

10.6 Construction

10.6.1 General

During soil anchorage construction the method of drilling, with or without flushing, the tendon installation, the grouting system and the time period of these operations may influence the capacity of the anchorage. Anchorage construction should be carried out in a manner whereby the validity of design assumptions is maintained and a method statement detailing all operations, including drilling and grouting plant information, should be prepared prior to site anchorage work. Anchorage work is specialized and should always be carried out under the supervision of experienced personnel.

10.6.2 Drilling

Any drilling procedure capable of supplying a stable hole that is within the permitted tolerances and free of obstructions in order to accommodate easily the tendon may be employed. Drilling necessarily disturbs the ground and the method should be chosen relative to the ground conditions to cause either the minimum of disturbance or the disturbance most beneficial to the anchorage capacity.

Care should be taken not to use high pressures with any flushing media, in order to minimize the risk of hydrofracture of the surrounding soil, particularly in built-up areas. In this connection, an open return within the borehole is desirable to limit pressures, and also permits the driller to monitor major changes in soil type from the drill cuttings or flush.

Unless otherwise specified, the drill hole entry point should be positioned within a tolerance of ±75mm. The drilled hole should have a diameter not less than the specified diameter, and allowances for swelling may be necessary if the hole is open for several hours in, for example, overconsolidated clays. For a specified alignment at entry point, the hole should be drilled to an angle tolerance of ±2.5°, unless, for closely spaced anchorages, such a tolerance could lead to interference of fixed anchor zones in this case the inclination of alternate anchorages should be staggered. Soil anchorages should have a minimum inclination of approximately 10° to the horizontal to facilitate grouting.

Assuming an acceptable initial alignment, overall drill hole deviations of 1 in 30 should be anticipated. On occasions, ground conditions may dictate a relaxation of this tolerance and, for downward inclined holes, it is probable that the vertical deviations will be higher than lateral deviations.

After each hole has been drilled its full length and thoroughly flushed out to remove any loose material, the hole should be probed to ascertain whether collapse of material has occurred and whether it will prevent the tendon being installed completely. In unstable soil conditions where the hole has been cased during drilling, the hole should still be probed to check if soil has flowed up inside the casing, e.g. in sandy silt beneath the water table. For downward inclined holes, up to 1 m of overdrill may be added to cater for detritus that cannot be removed.

Tendon installation and grouting should be carried out on the same day as drilling of the fixed anchor length, since a delay between completion of drilling and grouting can have serious consequences due to ground deterioration, particularly in overconsolidated, fissured clays and marls.

During the drilling operations, all changes in soil type should be recorded together with notes on water levels encountered, drilling rates, flushing losses or gains and stoppages.

10.6.3 Tendon

Ideally, tendon steel in the bare condition should be stored indoors in clean dry conditions. If left outdoors such steel should be stacked off the ground and be completely covered by a waterproof tarpaulin that is supported and fastened clear of the stack so as to permit circulation of air and avoid condensation.

Bare or coated tendons should not be dragged across abrasive surfaces or through surface soil, and only fibre rope or webbing slings should be used for lifting coated tendons. In the event of damage, tendon that is kinked or

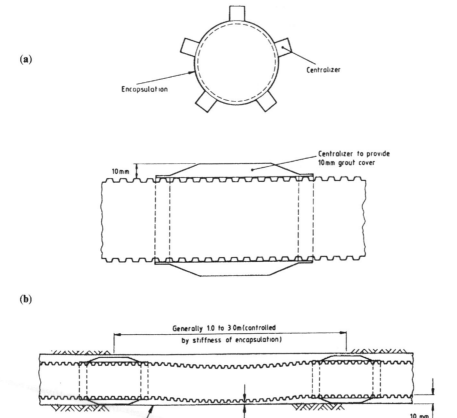

Figure 10.16 Centralizers: (a) cross-section; (b) longitudinal section.

sharply bent should be rejected because load-extension characteristics may be adversely affected.

Over the bond length, bar tendon, multi-unit tendons and encapsulations should be centralized in the borehole to ensure a minimum grout cover to the tendon, or encapsulation of 5mm between centralized locations and 10 mm at centralizer locations (see Figure 10.16).

For multi-unit tendons where the applied tensile load is transferred by bond, spacers should ensure a minimum clear spacing of 5mm. Given tendons with local or general nodes that provide mechanical interlock, occasional contact between tendon units is permissible.

A minimum of three spacers should be provided in each fixed anchor length, and both centralizers and spacers should be provided at centres according to the inclination and stiffness of the tendon, in order to provide the minimum clear spacing of cover (see also section 10.4.5.3). At the bottom of the tendon, use of a sleeve or nose cone will minimize the risk of tendon or borehole damage during homing.

Immediately prior to installation the tendon should be carefully inspected for damage to components and corrosion, after which it should be lowered at a steady controlled rate. For heavy tendons weighing in excess of 200 kg, approximately, mechanical handling equipment should be employed, as manual operations can be difficult and hazardous. The use of a funnelled entry pipe at the top of a cased hole is also recommended to avoid tendon damage as it is installed past the sharp edge of the top of the casing.

On occasion, particularly at the start of a contract, the tendon may be withdrawn after the installation operation, in order to judge the efficiency of the centralizer and spacer units, and also to observe damage, distortion or the presence of smear, e.g. in clay. Where significant distortion or smear is observed, improvements in relation to the fixing or design of the centralizers, or the borehole drilling method may be necessary.

10.6.4 Grouting

Grouting performs one or more of the following functions:

(i) to form the fixed anchor in order that the applied load may be transferred from the tendon to the surrounding soil;
(ii) to augment the protection of the tendon against corrosion; and
(iii) to strengthen the soil immediately adjacent to the fixed anchor in order to enhance anchorage capacity.

The need for function (iii) should be highlighted by the ground investigation and/or as a result of pregrouting. To check that the loss of grout over the fixed anchor length is insignificant during injection for anchorages in permeable soil, it is normally adequate to observe a controlled grout flow rate coupled with a back pressure. The efficiency of fixed anchor grouting can be finally checked by monitoring the response of the soil to further injection when the back pressure should be quickly restored.

Where pressure grouting is not carried out as part of routine anchorage construction, a falling head grout test can be used where the borehole is prefilled with grout typically having $w/c = 0.4$–0.6, and the grout level observed until it becomes steady. If the level continues to fall it should be topped up and, after sufficient stiffening of the grout (but prior to hardening), the borehole should be redrilled and retested. The test may be applied to the entire borehole or restricted to the fixed anchor length by packer or casing over the free anchor length.

In general, if the grout volume exceeds three times the borehole volume for injection pressures less than total overburden pressure, then general void filling, which is beyond routine anchorage construction, is indicated. This extra grout merits additional payment.

For the preparation of cement grout, batching of the dry materials should be by mass, and mixing should be carried out mechanically for at least 2 min in order to obtain a homogeneous mix. Thereafter, the grout should be kept in continuous movement, e.g. slow agitation in a storage tank. As soon as practicable after mixing, the grout should be pumped to its final position, and it is undesirable to use the grout after a period equivalent to its initial setting time (see Figure 10.17).

High speed colloidal mixers (1000 r/min minimum) and paddle mixers (150 r/min minimum) are permissible for mixing neat cement grouts, although the former mixer is preferred in water-bearing ground conditions since dilution is minimized.

Pumps should be of the positive displacement type, capable of exerting discharge pressures of at least $1000 \, \text{kN m}^{-2}$, and rotary screw (constant pressure) or reciprocating ram and piston (fluctuating pressure) pumps are acceptable in practice.

Before grouting, all air in the pump and line should be expelled, and the suction circuit of the pump should be airtight. During grouting, the level of

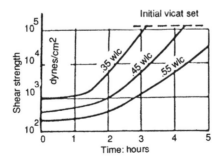

Figure 10.17 Setting times for Ordinary Portland (Type 1) cement grouts at 18°C (after Littlejohn[26]).

grout in the supply tank should not be drawn down below the crown of the exit pipe, as otherwise air will be injected.

An injection pressure of $20 \, \text{kN m}^{-2}$ per metre depth of ground is common in practice. Where high pressures that could hydrofracture the ground are permitted, careful monitoring of grout pressure and quantity over the fixed anchor length is recommended. If, on completion of grouting, the fluid grout remains adjacent to the anchored structure then the shaft grout should be flushed back 1 to 2 m to avoid a strut effect during stressing.

In regard to quality controls, emphasis should be placed on those tests that permit grout to be assessed prior to injection. As a routine, initial fluidity by flow cone or flow trough, density by mud balance and bleed by 1000 ml graduated cylinder (75 mm diameter) should be measured daily along with 100 mm cube samples for later crushing at, for example, 7 and 28 days. These quality controls relate to grout batching and mixing, and the tests do not attempt to simulate the properties of the grout in-situ. For example, water loss from grout, when injected under pressure into fine- to medium-sized sand, creates an in-situ strength greater than the cube strength for similar curing conditions.

Records relating to each grouting operation should be compiled, e.g. age of constituents, air temperature, grouting pressure, quantity of grout injected, and details of samples and tests, as appropriate.

10.6.5 Anchor head

The stressing head and bearing plate should be assembled concentrically with the tendon, with an accuracy of ± 10 mm, and should be positioned not more than 5° from the tendon axis.

After final grouting or satisfactory testing, cutting of the tendon should be done without heat, e.g. by a disc-cutter, in which case the cut should not be closer than one tendon unit diameter from the face of the holding wedge or nut.

Projecting tendons, whether stressed or not, should be protected against accidental damage. This protection is not common in practice and if individual tendon components are mechanically damaged, e.g. kinking of strand, then, when assessing a safe anchorage capacity, these components should be considered redundant, unless tests confirm adequacy.

10.6.6 Stressing

Stressing is required to fulfil two functions, namely:

(i) to tension the tendon and to anchor it at its secure load; and
(ii) to ascertain and record the behaviour of the anchorage so that it can be compared with the behaviour of control anchorages, subjected to on-site suitability tests.

A stressing operation means an activity involving the fitting of the jack assembly on to the anchor head, the loading or unloading of the anchorage including cyclic loading where specified, followed by the complete removal of the jack assembly from the anchor head. Stressing and recording should be carried out by experienced personnel under the control of a suitably qualified supervisor, since any significant variation in procedure can invalidate comparison with control anchorages.

At the present time equipment calibration is not carried out regularly, and discrepancies between jack and load cell readings are not uncommon on site. Jacks should be calibrated at least every year, using properly designed test equipment with an absolute accuracy not exceeding 0.5%. The calibration should cover the load rising and load falling modes over the full working range of the jack, so that the friction hysteresis can be known when repeated loading cycles are being carried out on the tendon. Load cells should be calibrated after every 200 stressings or after every 60 days in use, whichever is the more frequent, unless complementary pressure gauges used simultaneously indicate no significant variation, in which case the interval between calibrations may be extended upto a maximum of one year. Pressure gauges should also be calibrated regularly, e.g. after every 100 stressings or after every 30 days, whichever is the more frequent, against properly maintained master gauges, or whenever the field gauges have been subjected to shock. If a group of three gauges is employed this frequency does not apply since a malfunction in one gauge can be cross-checked against the remaining two.

On every contract the method of tensioning to be used and the sequence of stressing should be specified at the planning stage. In general, no tendon should be stressed at any time beyond either 80% of the characteristic strength (equivalent to 80% GUTS in the USA) or 95% of the characteristic 0.1% proof strength. In addition, for cement grouted fixed anchors, stressing should not commence until the grout has attained a crushing strength of at least $30 \, \text{N} \, \text{mm}^{-2}$. However, in sensitive soil, e.g. clay or marl, which may be weakened by water softening or disturbance during anchorage construction, it may be necessary to stipulate a minimum number of days before stressing.

Details of all forces, displacements, seating and other losses observed during stressing, and the times at which the data were monitored should be recorded for every anchorage.

Finally, it is worth noting that when a stressing operation is the start point for future time-related load measurements, stressing should be concluded with a check-lift load measurement.

During stressing, safety precautions are essential, and operatives and observers should stand to one side of the tensioning equipment and never pass behind when it is under load. Notices should also be displayed stating 'DANGER—Tensioning in Progress' or similar wording.

10.7 Testing

10.7.1 General

There are three classes of tests for all anchorages:

(i) proving tests;
(ii) on-site suitability tests; and
(iii) on-site acceptance tests.

Proving tests may be required to demonstrate or investigate, in advance of the installation of working anchorages, the quality and adequacy of the design in relation to soil conditions and materials used, and the levels of safety that the design provides. The tests may be more rigorous than on-site suitability tests and the results, therefore, cannot always be directly compared, e.g. where short fixed anchors of different lengths are installed and tested, ideally to failure.

On-site suitability tests are carried out on anchorages constructed under identical conditions to the working anchorages, and loaded in the same way and to the same level. These tests may be carried out in advance of the main contract or on selected working anchorages during the course of construction. The period of monitoring should be sufficient to ensure that prestress or creep fluctuations stabilize within tolerable limits. The tests indicate the results that should be obtained from the working anchorages.

On-site acceptance tests are carried out on all anchorages and demonstrate the short-term ability of the anchorage to support a load that is greater than the design working load and the efficiency of load transmission to the fixed anchor zone. A proper comparison of the short-term service results with those of the on-site suitability tests provides a guide to longer term behaviour.

10.7.2 On-site acceptance tests

Every anchorage used on a contract should be subjected to an acceptance test. As a principle, acceptance testing should comprise standard procedures and acceptance criteria that are independent of ground type, and be of short duration. In this regard the maximum proof loads are dictated by Table 10.2, but acceptable load increments and minimum periods of observation have gradually been reduced over the years to save time and money (see Table 10.4). At each stage of loading, the displacement should be recorded at the beginning and end of each period, and for proof loads the minimum period of 1 min is extended to at least 15 min with an intermediate displacement at 5 min, so that any tendency to creep can be monitored.

In order to establish the seat of load transfer within the anchorage, the apparent free length of the tendon may be calculated from the load–elastic displacement curve over the range of 10% T_w to 125 T_w (temporary

Table 10.4 Recommended load increments and minimum periods of observation for on-site acceptance tests (after Littlejohn[29]).

Temporary anchorages load increment ($\% T_w$)		Permanent anchorages load increment ($\% T_w$)		Minimum period of observation (min)
1st load cycle* (%)	2nd load cycle (%)	1st load cycle* (%)	2nd load cycle (%)	
10	10	10	10	1
50	50	50	50	1
100	100	100	100	1
125	125	125	125	15
100	100	100	100	1
50	50	50	50	1
10	10	10	10	1

* For this load cycle, which often includes extraneous non-recoverable movements such as wedge 'pull-in', bearing plate settlement and initial fixed anchor displacement, there is no pause other than that necessary for the recording to displacement data.

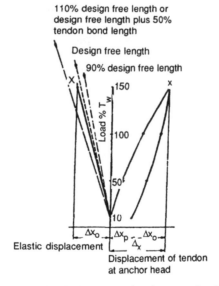

Figure 10.18 Acceptence criteria for displacement of tendon at anchor head (after Littlejohn[29]).

anchorages) or 10% T_w to 150% T_w (permanent anchorages), using the manufacturer's value of elastic modulus and allowing for such effects as temperature and bedding of the anchor head. It is normally adequate simply to record the ambient temperature during the test, unless the monitoring equipment or anchored structure is known or observed to be temperature-sensitive. The free length analysis should be based on the results obtained during the destressing stage of the second or any subsequent unloading cycle, otherwise extraneous non-recoverable movements may mask the reproducible behaviour of the anchorage in service (see Figure 10.18).

For simplicity in practice the following equation is employed:

$$\text{Apparent free tendon length} = \frac{A_t E_s \Delta X_e}{T} \qquad (10.9)$$

where A_t is the cross-section of the tendon, E_s is the manufacturer's elastic modulus for the tendon unit, ΔX_e is the elastic displacement of the tendon (ΔX_e is equated to the displacement monitored at proof load minus the displacement at datum load, i.e. $10\% T_w$ say, after allowing for structural movement) and T is the proof load minus datum load.

On completion of the second cycle, the anchorage is reloaded in one operation to $110\% T_w$ say, and locked-off, after which the load is re-read to establish the initial residual load. This moment represents zero time for monitoring load/displacement-time behaviour during service. Where loss of load is monitored accurately using load cells with a relative accuracy of 0.5%, readings can be attempted within the first 50 min. Where monitoring involves a stressing operation, e.g. lift-off check without load cell, an accuracy of less than 5% is unlikely and longer observation periods of one day and beyond are required. Where displacement-time data are required, a dial gauge/tripod system (see Figure 10.19) is suitable for short duration testing, given that the tripod base should be surveyed accurately for movement. In practice, dial gauges reading to 0.01 mm are commonly used during the test and, where movement of the tripod base is anticipated, its position is checked before and after the test to an accuracy of 1 mm.

For the testing procedures outlined above, acceptance criteria based on proof load–time data, apparent free tendon length, and short-term service behaviour are proposed for temporary and permanent anchorages. These criteria are discussed in the following section.

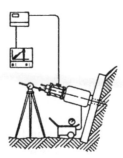

Figure 10.19 Typical method of measuring tendon displacement using a dial gauge.

10.7.3 Proof load–time data

If the proof load has not reduced during the 15 min observation period by more than 5% after allowing for any movement of the anchored structure, the anchorage may be deemed satisfactory. If a greater loss of prestress is recorded the anchorage should be subject to two further proof load cycles and the behaviour recorded. If the 5% criterion is not exceeded on both occasions the anchorage may be deemed satisfactory. If the 5% criterion is exceeded on either cycle the proof load should be reduced to a value at which compliance with the 5% criterion can be achieved. Thereafter, the anchorage may be accepted at a derated proof load, if appropriate.

As an alternative to these recommendations, the proof load can be maintained by jacking and the anchor head monitored after 15 min, in which case the creep criterion is 5% ΔX_e.

For anchorages that have failed a proof load criterion, tendon unit stressing may help to ascertain location of failure, e.g. for a temporary anchorage, pull-out of individual tendon units may indicate debonding at the grout/tendon interface, whereas, if all tendon units hold their individual proof loads, attention is directed towards failure of the fixed anchor at the soil/grout interface.

10.7.4 Apparent free tendon length

The apparent free tendon length should be not less than 90% of the free length intended in the design, nor more than the intended free length plus 50% of tendon bond length or 110% of the intended free tendon length (see Figure 10.18). The latter upper limit takes account of relatively short encapsulated tendon bond lengths and fully decoupled tendons with an end plate or nut.

Where the observed free tendon length falls outside the limits, a further two load cycles up to proof load should be carried out in order to gauge reproducibility of the load–displacement data. If the anchorage behaves consistently in an elastic manner, the anchorage need not be abandoned, provided the reason can be diagnosed and accepted. In this regard it is noteworthy that the E value of a long multi-strand tendon may be less than the manufacturer's E value for a single strand, which has been measured over a short gauge length between rigid platens. A reduction in the manufacturer's E value of up to 10% should be allowed in any field diagnosis.

10.7.5 Short-term service behaviour

Using accurate load cell and logging equipment, the residual load may be monitored at 5, 15 and 50 min. If the rate of load loss reduces to 1% or less per time interval for these specific observation periods after allowing for

Table 10.5 Acceptance criteria for service behaviour at residual load (after Littlejohn[29]).

Period of observation (min)	Permissible loss of load (% initial residual load) (%)	Permissible displacement (% of elastic extension Δe of tendon at initial residual load) (%)
5	1	1
15	2	2
50	3	3
150	4	4
500	5	5
1500 ($\sim$1 day)	6	6
5000 ($\sim$3 days)	7	7
15000 ($\sim$10 days)	8	8

temperature (where necessary), structural movements and relaxation of the tendon, the anchorage may be deemed satisfactory. If the rate of load loss exceeds 1%, further readings may be taken at observation periods up to 10 days (see Table 10.5).

If, after 10 days, the anchorage fails to hold its load as given in Table 10.5, the anchorage is not satisfactory and, following an investigation as to the cause of failure, the anchorage should be: (i) abandoned and replaced; (ii) reduced in capacity; or (iii) subjected to a remedial stressing programme.

Where prestress gains are recorded, monitoring should continue to ensure stabilization of prestress within a load increment of $10\% T_w$. Should the gain exceed $10\% T_w$, a careful analysis is required and it will be prudent to monitor the overall structure/ground/anchorage system. If, for example, overloading progressively increases due to insufficient anchorage capacity in design or failure of a slope, then additional support is required to stabilize the overall anchorage system. Destressing to working loads should be carried out as prestress values approach proof loads, accepting that movement may continue until additional support is provided.

As an alternative to load monitoring, displacement-time data at the residual load may be obtained at the specific observation periods in Table 10.5, in which case the rate of displacement should reduce to 1% or less per time interval. This value is the displacement equivalent to the amount of tendon shortening caused by a prestress loss of 1% initial residual load, i.e.

$$\Delta e = \frac{\text{initial residual load} \times \text{apparent free tendon length}}{\text{area of tendon} \times \text{elastic modulus of tendon}} \qquad (10.10)$$

If the anchorages are to be used in the work and, on completion of the on-site acceptance test, the cumulative relaxation or creep has exceeded 5% initial residual load or 5% Δe, respectively, the anchorage should be restressed and locked-off at, for example, $110\% T_w$. This procedure ensures that a contingency overload is locked into the soil anchorage at the start of its service.

As a general guide, either acceptance criterion for short-term service, i.e. rate of prestress loss or rate of displacement, may be applied quite independently for the common range of free tendon lengths. For short free tendon lengths ($< 5\,\text{m}$), loss of prestress becomes the more appropriate criterion, while for long free tendon lengths ($> 30\,\text{m}$) it is clear that creep displacement may be more important to limit, and therefore more appropriate as an acceptance criterion.

10.7.6. Monitoring service behaviour

As for buildings, bridges and dams, monitoring of structure/soil/anchorage systems will be appropriate on occasions. In general monitoring is recommended for important structures where the following circumstances apply:

(i) wherever the behaviour of anchorages can be ascertained safely by monitoring the behaviour of the structure as a whole, e.g. by precise surveying of movements;

(ii) wherever the malfunctioning of anchorages could endanger the structure and cause it to become a hazard to life or property, and where problems would not be detected before the structure became unserviceable other than by monitoring;

(iii) wherever, due to the nature of the soil and/or the protective system, tendons cannot be bonded to the walls of their holes, so that breakage of a tendon at any point renders it ineffective throughout its length;

(iv) where anchorages are of a pattern that has not been proved adequately in advance, either by rigorous laboratory tests or by site performance under similar circumstances; and

(v) where anchorages are in soil liable to creep.

Two methods of monitoring are in common use: (i) measurement of loads on individual anchorages; and (ii) measurement of the performance of structures or excavations as a whole. The second method is preferable.

When monitoring individual anchorages, the maximum loss or gain of prestress that can be tolerated during service should be indicated, taking into account the design of the works. Variations up to 10% of working load do not generally cause concern. Prestress losses greater than 10% should be investigated to ascertain cause and consequence, and for prestress gain, remedial action—which may involve partial destressing or additional anchorages—is recommended when the increases exceed $20\% T_w$ and $40\% T_w$ for temporary and permanent anchorages, respectively.

In general, monitoring should initially be at short intervals of three to six months, with later tests at longer intervals depending on results. The number of anchorages to be monitored should be indicated by the designer of the works; 5 to 10% of the total is typical in current practice.

Whilst anchorage technology has developed rapidly, there is still a reluctance to invest in performance studies during service. Although an absence of problems may be the reason, the following section on service performance is included to illustrate the benefits that can be gained, together with examples of permanent anchorages with significant periods of service.

10.7.7 Service performance

The advantages of monitoring include: (i) the engineer being able to feed back performance observations into future designs and thereby optimize such parameters as overload allowances and load safety factors; and (ii) the prospective client being accurately and confidently informed of how anchorages installed at his expense will perform after installation. Furthermore, the data collection permits all parties to judge at the earliest possible stage whether anchorages being monitored are, in fact, acting satisfactorily. On a more general front, this form of monitoring may permit correlation of anchorage load and structural movement, and thereby lead to a better understanding of anchorage/ground/structure interaction.

Generally speaking, short-term monitoring, over three to six months, of anchorages installed in cohesionless soils has shown a rapid stabilization of load after initial post-tensioning.[7] Where significant overall ground movements are mobilized during excavation, such stabilization of anchorage loads usually occurs shortly after completion of the excavation. No long-term prestress losses due to creep have been noted in cohesionless soils and there appears to be little concern in practice over the ability of the anchorages to maintain their load-holding capacity in the long term, given adequate corrosion protection. Early examples of permanent anchorages installed in sands and gravels in the UK are included in Table 10.6.

In cohesive soils, such as clays that are known to be susceptible to creep, the dearth of monitoring has left some engineers concerned about the long-term behaviour of anchorages installed in clays. Again, to the author's knowledge, there are no incidents of adverse service behaviour due to creep,

Table 10.6 Early examples of permanent anchorages installed in cohesionless soils in the UK.

Location	Number of anchorages	Working load (kN)	Types of soil	Date of installation
Tilbury, Essex	52	300	Gravels	1968
Grosvenor Road, London	44	300 and 360	Gravels	1969
Ponders End	18	300 and 400	Gravels	1969
Bromley Theatre, Kent	10	630	Sands and gravels	1970
Solihull, Birmingham	14	250	Gravels	1970

Table 10.7   Early examples of permanent anchorages installed in cohesive soils in the UK.

Location	Number of anchorages	Working load (kN)	Types of soil	Date of installation
New Pithay, Bristol	26	500	Marl	1964
Kilburn Square, London	18	300	London Clay	1968
Coventry	102	450 and 900	Marl	1969
Scarborough	23	400	Sandy clay	1969
Neasden Underpass, London	580	100–500	London Clay	1969
Derby Underpass	40	650	Marl	1970

and it should be noted that modern national codes demand a stabilizing trend for prestress loss with time, coupled with an appropriate overload allowance in routine on-site acceptance testing (see section 10.7.5). Early examples of permanent anchorages installed in cohesive soils in the UK are included in Table 10.7.

In the case of Kilburn Square, satisfactory behaviour for all 18 anchorages was confirmed by lift-off tests after 11 years of service, when residual loads ranged from 108% to 93% of the initial lock-off load of 312 kN.

10.8 Applications

In many countries anchorages have established a permanent place in construction practice. However, for engineers not yet fully familiar with modern anchoring technology, this section outlines practical applications in order to provide some perspective and encourage further exploitation.

The formation of excavations in city centres for deep basements, underground car parks and metro stations often entails construction of a retaining wall around the site. Temporary anchorages installed through these walls can reduce cost and allow better use of heavy mechanical plant because construction work can proceed unrestricted. Noteworthy examples include the multi-tied 34 m deep excavation for the Security Pacific National Bank, Los Angeles; the Lok Fu Mass Transit Railway Station, Hong Kong; the Central Bank of Brazil, Brasilia; and the rather unusual inclined wall for the Munich subway (see Figure 10.20).

Present confidence is perhaps best reflected by Figure 10.21 showing the deep cut for New York's World Trade Centre, where stability depended on over 1000 prestressed soil anchorages, and the New York Metro—seen straddling the excavation—maintained regular services throughout the construction period.

Permanent anchorages are often required for motorways or road improvement schemes to tie back retaining walls forming cuttings and underpasses, or to safeguard existing buildings, as in the case of the anchored wall adjacent to Paganini's house in Genoa (see Figure 10.22).

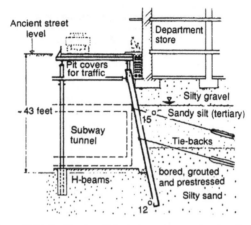

Figure 10.20 Inclined wall for subway excavation in Munich.

Figure 10.21 Basement excavation for the World Trade Centre in New York.

In deep highway cuts or basements beneath the groundwater table, resistance to hydrostatic uplift can be provided by prestressed anchorages. The Kuwait City Ring Road is currently the largest contract of this kind with over 10 000 anchorages in weakly cemented silty sands. Swelling or heave of an excavation floor can be resisted in the same way, e.g. the Seelisberg tunnel on the Swiss National Highway N2.

Anchorages may also be used as a remedial measure to arrest wall or ground movement, e.g. slope stabilization at Grimwith Dam, England.

With the increase in height of office and residential tower blocks, often associated with large-diameter piles, traditional methods of pile and plate load testing using kentledge can be uneconomic or impracticable, particularly where test loads are high or the space available is restricted. In compact or

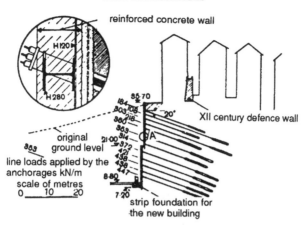

Figure 10.22 Cross-section of anchored wall in Genoa, Italy (after Barla and Mascardi[30]). Detail of section A is shown on upper left.

stiff soil, standard anchorage systems can mobilize loads up to 5000 kN via ring reaction beams.

A wide range of applications now exists for both temporary and permanent anchorages and it is clear that soil anchorage technology is well established in many parts of the world.

10.9 Final remarks

Experience indicates that higher quality and more-detailed ground investigations are required at the planning stage of many anchorage projects to permit their economic design and construction. In addition, there is a need to define clearly in contract documents the design responsibilities of the designer and specialist anchorage contractor in order to minimize contractual problems.

In the field of permanent anchorages, corrosion protection ranges from double protection (implying two physical barriers) to simple cement grout cover. The latter solution is not considered acceptable when the safety of people and property in the event of anchorage failure is balanced against the cost of providing protection. The required degree of protection should be specified at the time of bid, and single protection should represent the minimum standard for permanent anchorages.

Given the specialized nature of soil anchorage work and the wide variety of anchorage types and construction procedures, coupled with the variability of ground, more reliance should be placed on performance specifications related to choice of materials and acceptance testing of all anchorages, compared with control of construction. Such testing should involve proof loading to show a margin of safety, load–displacement analysis to confirm that the resistance to withdrawal is mobilized correctly in the fixed anchor zone, and short-term monitoring of the service behaviour to confirm a stabilizing trend and ensure reliable performance in the long term.

If reliable performances are to be maintained, a technical appraisal of anchorage systems is required by the practising engineer, in addition to routine comparisons on the basis of cost and duration of contract. Modern national codes facilitate these technical appraisals but, more importantly, the adoption of code recommendations should ensure both the safety, and the satisfactory performance of the anchorage system.

Millions of anchorages have been installed successfully for temporary and permanent works throughout the world, the development in anchoring techniques having been dramatic over the past 40 years. With an absence of serious failures, there is a strong base upon which anchorage specialists can build and expand their market with confidence. There is no room for complacency however; engineers must rigorously apply high standards, and much field development remains to be tackled.

Acknowledgement

Extracts from BS 8081:1989 are reproduced with the permission of BSI. Complete copies can be obtained through national standard bodies.

References

1. Littlejohn, G.S. (1982) The practical applications of ground anchorages. *Proc. 9th FIP Congress*, Stockholm.
2. BS 5930:1981 Code of Practice of Site Investigations. British Standards Institution, London.
3. Stroud, M.A. and Butler, F.G. (1975) The Standard Penetration Test and the engineering properties of glacial materials. *Proc. Symp. on the Engineering Behaviour of Glacial Materials*, University of Birmingham, England.
4. Mair, R.J. and Wood, D.M. (1987) Pressuremeter testing —methods and interpretation. *CIRIA Ground Engineering Report B3 : In-situ Testing*. CIRIA/Butterworths, London.
5. BS 1377:1990 Methods of Test for Soil for Civil Engineering Purposes. British Standards Institution, London.
6. Rowe, P.W. (1972) The relevance of soil fabric to site investigation practice. *Géotechnique*, 27, 195–300.
7. Littlejohn, G.S. (1970) Soil anchors. *ICE Conf. on Ground Engineering*, London, 33–44.
8. Anderson,W.F., Hanna, T.H. and Abdel-Malek, M.N. (1983) Overall stability of anchored retaining walls. *Proc. ASCE* 109 (11), 1416–1433; (12), 1817–1818.
9. GCO (1984) Geotechnical manual for slopes. Geotechnical Control Office, Government of Hong Kong (295 pp.).
10. Mitchell, J.M. (1975) Some experiences with ground anchors in London. *Proc. of Conf. on Diaphragm Walls and Anchorages*, ICE, London, 129–133.
11. Lundahl, B. and Adding, L. (1966) Dragförankringer i flytbenägen mo under grundvattenytan. *Byggmästaren*, 44, 145–152.
12. Robinson, K.E. (1969) Grouted rod and multi-helix anchors. *Proc. 7th Conf. on Soil Mech. Found. Eng.*, Mexico, Speciality Session No. 15, 126–130.
13. Bassett, R.H. (1970) Discussion to paper on soil anchors. *ICE Conference on Ground Engineering*, London, 89–94.
14. Oosterbaan, M.D. and Gifford, D.G. (1972) A case study of the Bauer earth anchor. *Proc.*

    of Speciality Conf. on Performance of Earth and Earth Supported Structures, ASCE, Purdue University, Pt. 2, 1391–1401.

15. Berezantzev, V.G., Khristoforov, V.S. and Golubkov, V.N. (1961) Load bearing capacity and deformation of piled foundations. *Proc. 5th Int. Conf. on Soil Mech. Found. Eng.* Paris, 2, 11–15.

16. Jorge, G.R. (1969) The regroutable IRP anchorage for soft soils or low capacity karstic rocks. *Proc. 7th Conf. on Soil Mech. and Found. Eng.*, Mexico, Speciality Session No. 15, 159–163.

17. Ostermayer, H. (1974) Construction, carrying behaviour and creep characteristics of ground anchors. *ICE Conference on Diaphragm Walls and Anchorages.* London, 141–151.

18. Ostermayer, H. and Scheele, F. (1978) Research and ground anchors in non-cohesive soils. *Revue Française de Géotechnique* No. 3, 92–97.

19. Neely, W.J. and Montague-Jones, M. (1974) Pull-out capacity of straight-shafted and under-reamed ground anchors. *Die Siviele Ingenieur in Suid-Africa Jaargang,* 16(4), 131–134.

20. Sapio, G. (1975) Comportamento di Tiranti de Ancoraggio in Formazioni de Argile Preconsolidate. *Atti XII Convegno Nazionale de Geotecnica,* Cosenze.

21. Bastaple, A.D. (1974) Multibell ground anchors in London Clay. *Proc. 7th FIP Congress,* New York, Tech. Session on Prestressed Concrete Foundations and Ground Anchors, 33–37.

22. Buttling, S. (1977) Report on discussion to session IV by C. Truman-Davies. *Review of Diaphragm Walls,* ICE, London, 76.

23. Bruce, D.A. (1976) The design and performance of prestressed rock anchors with particular reference to load transfer mechanisms. *Ph.D. Thesis,* Dept. of Engineering, University of Aberdeen, Scotland.

24. Barley, A.D. (1978) A study and investigation of underreamed anchors and associated load transfer mechanisms. *M.Sc. Thesis,* Dept. of Engineering, University of Aberdeen, Scotland.

25. Littlejohn, G.S. and Bruce, D.A. (1977) Rock Anchors: State-of-the-Art. Foundation Publications Ltd., Brentwood, Essex, England.

26. Littlejohn, G.S. (1982) Design of cement based grouts. *Proc. Geotechnical Engineering Speciality Conference, Grouting in Geotechnical Engineering,* ASCE, New Orleans, 35–48, plus disc. 1–6.

27. BS 8081:1989 Code of Practice for Ground Anchorages. British Standards Institution, London.

28. Fédération Internationale de la Précontrainte (1986) *Corrosion and Corrosion Protection of Prestressed Ground Anchorages.* Thomas Telford Ltd., London.

29. Littlejohn, G.S. (1991) Routine on-site acceptance tests for ground anchorages. *Ground Engineering* 24(2), 37–43.

30. Barla, G. and Mascardi, C. (1974) High anchored wall in Genoa. *ICE Conference on Diaphragm Walls and Anchorages,* London, 123–128.

11　In-situ earth reinforcing by soil nailing

D.A. BRUCE

11.1 Introduction

Since the late 1960s, engineers in Europe, Japan and North America have been exploiting the special advantages of the technique of soil nailing.[1] This geotechnical engineering process comprises the in-situ reinforcement of soils and has a wide range of applications for stabilizing excavations such as are associated with deep foundations or cut and cover tunnelling schemes. It has been researched with large budgets since 1975 by collaborations of contractors, universities and government organizations. It has been the subject of international conferences, symposia and seminars since 1979, and has given rise to a rapidly expanding literature of technical papers and articles worldwide. There are abundant successful case histories to cite in a wide variety of ground conditions and applications, and 'first uses' have been reported recently in such diverse locations as South Africa,[2] New Zealand,[3] and Hungary.[4]

11.2 Characteristic features

Soil nailing is a practical and proven technique used in constructing excavations by reinforcing the ground in-situ with relatively short, fully bonded inclusions—usually steel bars. These are introduced into the soil mass as staged excavation proceeds, and act to produce a zone of reinforced ground. This zone then performs as a homogeneous and resistant unit to support the unreinforced ground behind, in a manner similar to a conventional gravity retaining wall (Figure 11.1).

11.2.1 In-situ reinforcement techniques

There are three main categories of in-situ reinforcement used to stabilize slopes, namely nailing, reticulated micropiling, and dowelling.

In soil nailing, the reinforcement is installed horizontally or subhorizontally so that it improves the shearing resistance of the soil by acting in tension (Figure 11.2(a)). One major variant, the 'Hurpinoise' method,[7] involves simply driving an angle section into the soil, without predrilling and grouting.

Figure 11.1 The analogy between a gravity retaining wall and a soil-nailed wall.[5]

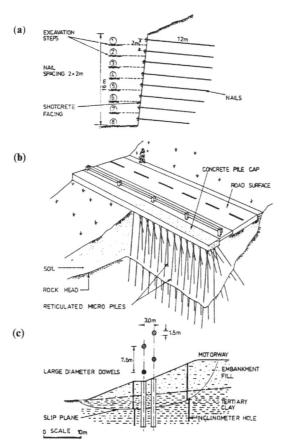

Figure 11.2 The family of in-situ soil reinforcement techniques: (*a*) Soil nailing (after Schlosser[6]); (*b*) reticulated micropiling (after Boley and Crayne[8]); and (*c*) soil dowelling (after Gudehus and Schwarz[11]).

This chapter, however, focuses on the drilled and grouted version—by far the most popular, widely used, and usable version, given practical concerns with installation and corrosion protection.

Reticulated micropiles are steeply inclined in the soil at various angles both perpendicular and parallel to the face (Figure 11.2(b)). The overall aim is similar to soil nailing, namely to provide a stable block of reinforced soil, which supports the unreinforced soil by acting like a gravity retaining structure. In this technique the soil is held together by the multiplicity of reinforcement members acting to resist compression and tension, bending, and shearing forces. Fondedile's pali radice system was the original form of this construction[9] whilst, more recently, Nicholson Construction has applied a similar technique in the USA under the trade name *Type A INSERT Wall*.[10]

Soil dowelling is applied to reduce or halt downslope movements on well-defined shear surfaces (Figure 11.2(c)). Slopes treated by dowelling are typically much flatter than those in soil nailing or reticulated micropile applications. Gudehus[12] has shown that the most efficient way to improve mechanically the shearing resistance on a weakened shear surface through the soil is to use relatively large-diameter piles, which combine a large surface area with high bending stiffness. Thus, the diameter of a soil dowel is generally far greater than that of a soil nail or micropile.

11.2.2 Selecting in-situ reinforcement

Although there are fundamental differences in the mechanical action of these three in-situ reinforcement techniques, there are circumstances where more than one may be applied to slope stabilization as illustrated in Figure 11.3. The following points merit consideration when choosing the appropriate in-situ reinforcement technique.

Laboratory experiments (e.g. Jewell[13]) have shown the influence of the inclination and properties of reinforcing members on the shearing resistance of reinforced soil. These indicate that the reinforcement gives the best increase in strength when it is angled across the potential rupture surface in soil so that the reinforcement is loaded in tension. At other orientations in the soil the reinforcement provides less benefit, and can even reduce the shearing resistance of the soil mass if it acts in compression.

The conclusion, therefore, is that in applications where a steep slope is to be excavated in a homogeneous granular soil, it is most efficient to place the reinforcement through the face in a direction close to the horizontal, as in Figure 11.3(a). To stabilize the soil with reinforcement placed in substantially vertical directions (Figure 11.3(b)) will require a much higher density of reinforcement. For this type of application soil nailing is likely to be more cost-effective than reticulated micropiling.

In marginally stable granular or scree slopes when stability must be improved, but where excavation is not foreseen, then either soil nailing or Type A INSERT Walls would be appropriate. Where drilling equipment cannot be placed on the slope, Type A walls would be best (Figure 11.3(c)). Where access is not problematical, either technique could be applied (Figures 11.3(c) and 11.3(d)), with economic considerations being decisive.

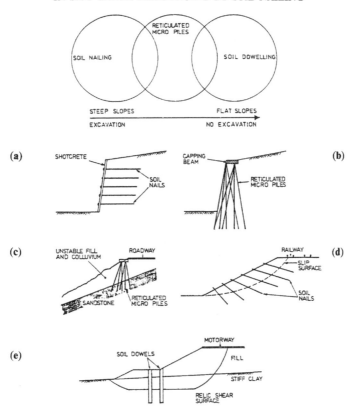

Figure 11.3 Overlap of in-situ soil reinforcement applications: (*a*) and (*b*) in excavations; (*c*) and (*d*) for general slope stabilization; and (*e*) to stabilize residual slips in clay.[1]

In flatter clay slopes where stability is governed by a well-defined shear zone, larger diameter soil dowels would be most appropriate (Figure 11.3(e)), possibly in concert with a RODREN[14] type deep well concept (Figure 11.4).

Type A walls and soil dowelling are not described further in this chapter. The former are described in various publications.[8–10,15–17] Soil dowelling is described in references 12 and 18–20.

11.2.3 Fundamental design considerations

Just as in the design of a gravity retaining wall, the stability of a nailed structure must be checked against both external and internal forces. Regarding external forces:

(i) the reinforced zone must be able to resist the outward thrust from the unreinforced interior, without sliding;
(ii) the combined loading from the reinforced zone self weight, and the

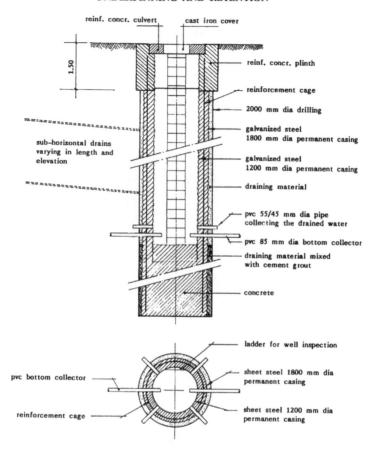

reinf. concr. culvert   cast iron cover

reinf. concr. plinth

reinforcement cage

2000 mm dia drilling

galvanized steel
1800 mm dia permanent casing

sub-horizontal drains
varying in length and
elevation

galvanized steel
1200 mm dia permanent casing

draining material

pvc 55/45 mm dia pipe
collecting the drained water

pvc 85 mm dia bottom collector

draining material mixed
with cement grout

concrete

ladder for well inspection

pvc bottom collector

sheet steel 1800 mm dia
permanent casing

reinforcement cage

sheet steel 1200 mm dia
permanent casing

Figure 11.4 Structural RODREN drainage well.[14]

lateral soil thrust it is resisting must not cause a foundation bearing failure; and

(iii) the stability of the retaining structure must be checked against the deeper seated overall failure mechanisms.

With respect to internal stability, the reinforcing elements must be installed in a pattern dense enough to ensure an effective interaction with the soil in the reinforced zone. The reinforcement elements must also have sufficient length and capacity to ensure a stable reinforced zone. In particular:

(i) each individual reinforcement should be capable of holding the soil immediately surrounding it in equilibrium. This local stability aspect dictates the spacing of the reinforcement; and

(ii) overall slip failure in the reinforced zone must also be considered to ensure against failure by insufficient bond, or breaking of the reinforcement.

These criteria govern the required length of the reinforcement. Each of these aspects of design is referred to in subsequent sections.

11.2.4 Comparison with prestressed ground anchorages

Superficially, there would appear to be a number of similarities between nails and prestressed ground anchorages when used for slope or excavation stability. Indeed, it is tempting to regard nails merely as 'passive' small-scale anchorages. However, there are major functional distinctions to be drawn, which will favor the choice of the one over the other:

(i) Ground anchorages are stressed after installation so that in service they ideally prevent any structural movement occurring. In contrast, soil nails are not prestressed and require a finite (albeit very small) soil deformation to cause them to work.

(ii) Nails are in direct, bonded contact with the ground over most of their length (typically 3–10 m), whereas ground anchorages transfer load only along the distal, fixed anchorage length. A direct consequence of this is that the distribution of stresses in the retained soil mass is different for each type.

(iii) Since nails are installed at a far higher density (typically one per 2–3 m^2 of face) the consequences of a one-unit failure are not necessarily so severe. In addition, the constructional tolerances of installation need not be so high, given their overall, interactive mode of operation.

(iv) As high loads have to be preapplied to anchorages, appropriate bearing facilities must be provided at the head to eliminate the possibility of 'punching' through the facing of the retained structure. Substantial bearing arrangements are not necessary with nails, as the low individual head loadings are easily accommodated on small steel-bearing plates placed on the shotcreted surface.

(v) Individual anchorages tend to be longer (say 15–45 m) and so may necessitate larger scale construction equipment. Also, an anchorage system is often provided to stabilize a substantial retaining structure, such as a diaphragm wall or bored pile wall, which will itself necessitate large-scale equipment.

As is noted in section 11.2.6, certain soil conditions are not suited to nailing, whereas anchoring is applied more widely. In addition, nailing appears more limited in terms of excavated depth potential. If the overall stability calculations show the problem to be deep seated, then ground anchorages will most probably be required in place of, or together with, nails. Conversely, for vertical excavations, soil nailing has frequently proved preferable to other methods of lateral support incorporating prestressed ground anchorages (such as Berlin, or diaphragm walls), in appropriate geological, geometrical and performance conditions.

11.2.5 Comparison with reinforced earth walls

Although soil nailing shares certain features with the older and more widely known technique of reinforced earth for retaining wall construction,[21] there are also some fundamental differences.[6]

The main similarities are:

(i) The reinforcement is placed in the soil unstressed and is not then prestressed: the reinforcement forces are mobilized by subsequent deformation of the soil.

(ii) The reinforcement forces are sustained by frictional bond between the soil and the reinforcing element. The reinforced zone is stable and resists the thrust from the unreinforced soil it supports, like a gravity retaining structure.

(iii) The facing of the retained structure is thin—prefabricated elements in the case of reinforced earth, and, usually, shotcrete in soil nailing—and does not play a major role in the overall structural stability.

The main dissimilarities are:

(i) Although at the end of construction the two structures may look similar, the construction sequence is radically different. Soil nailing is constructed by staged excavations from 'top down' while reinforced earth is constructed 'bottom up' (Figure 11.5). This has an important influence on the distribution of the forces that develop in the reinforcement, particularly during the construction period.

(ii) Soil nailing is an in-situ reinforcement technique exploiting natural ground, the properties of which cannot be preselected and controlled as they are for reinforced earth fills.

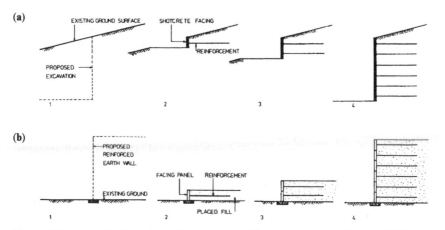

Figure 11.5 Contrast of the construction sequence: (*a*) 'top down' in soil nailing; and (*b*) 'bottom up' for reinforced soil.[1]

(iii) Grouting techniques are usually employed to bond the reinforcement to the surrounding ground: load is transferred along the grout to soil interface. In reinforced earth, friction is generated directly along the strip to soil interface itself.

11.2.6 Benefits and limitations of soil nailing

Several factors have contributed to the growing popularity of soil nailing as a construction technique, and these include:

(i) Economic advantage—it would seem that the cost saving for excavations of the order of 10 m deep can be 10 to 30% relative to conventional underpinning, or an anchored diaphragm or Berlin wall alternative in appropriate conditions. This is supported by a claimed saving of 30% on a soil-nailed excavation in Portland, Oregon,[22] and 35% in a more recent San Francisco excavation[23] featuring an underpinning application.

(ii) Construction equipment—drilling rigs for reinforcement installation, and guns for shotcrete application are relatively small-scale, mobile and quiet. This is highly advantageous in urban environments where noise, vibration or access restraints may pose problems. Equally, in remote rural areas it may prove impossible to deploy large-scale equipment such as is needed for piling or diaphragm walling.

(iii) Construction flexibility—soil nailing can proceed rapidly (150–200 m^2 per shift) and the excavation can be shaped easily. It is a flexible technique, readily accommodating variations in soil conditions and work programs as excavation progresses.

(iv) Performance—field measurements indicate that the overall movements required to mobilize the reinforcement forces are surprisingly small. These generally correspond to the movements to be expected for well-braced systems (Category I) in Peck's[24] classification (Figure 11.6). Furthermore, nailing is applied at the earliest possible time after excavation, and in intimate contact with the cut soil surface. This minimizes the disturbance to the ground, and so the possibility of damage being caused to adjacent structures.

Naturally, the technique has certain practical limitations to its application:

(i) Soil nail construction requires the formation of cuts generally 1–2 m high in the soil. These may then have to remain unsupported for at least a few hours, prior to shotcreting and nailing. The soil must therefore have some natural degree of 'cohesion' or cementing. Otherwise, a pretreatment such as grouting may be necessary to stabilize the face, but this will add both complication and cost.

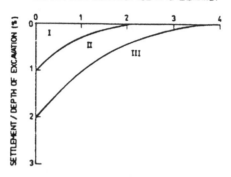

Figure 11.6 Field performance of open excavation systems.[24] I: well-braced systems; II: intermediate performance, e.g. temporary berms and raker support; III: systems permitting ground loss.

(ii) A dry face in the excavation is desirable for soil nailing. If groundwater percolates through the face, the unreinforced soil will slump locally on initial excavation, making it very difficult to establish a satisfactory shotcrete skin.

(iii) Excavations in soft clay are also unsuited to stabilization by soil nailing. The low frictional resistance of soft clay would require a very high density of in-situ reinforcement, of considerable length to ensure adequate levels of stability. Bored pile or diaphragm walls with anchorages are more suited to these conditions.

11.3 History and evolution

The principles and techniques of stabilizing excavations in rock by in-situ reinforcement have long been applied by mining engineers. Beveredge[25] noted that the use of mechanical rock bolts grew immediately after World War II, whilst by 1959 the first fully bonded (resin) reinforcements were being installed in German mines. The New Austrian Tunnelling Method,[26] evolved in the early 1960s primarily as a hard rock tunnelling system using a combination of shotcrete and fully bonded steel inclusions to provide early, efficient excavation stability with minimal movement. It was later adapted successfully to less-competent formations comprising graphitic shales—as in the Massenberg Tunnel[27]—and Keuper Marl—as in the Schwaikhem Tunnel.[28]

This latter project confirmed the viability of the technique in less-competent materials, and soon trials were conducted in soil such as silts, gravels and sands, and in very weak rocks. The earliest applications were in small cross-section metro tunnels in Frankfurt in 1970. Soon after, in Nürnberg, the technique again proved successful in the construction of a double tube of

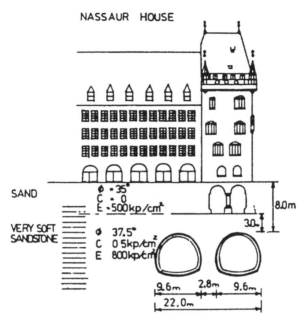

Figure 11.7 Subway station in Nürnberg, Germany formed using the New Austrian Tunnelling Method.[28]

a subway station with cross-passages adjacent to delicate and historic buildings (Figure 11.7).

By this time, the use of dowels and bolts to stabilize rock slopes was also well established. For example, Bonazzi and Colombet[29] described the stabilization, by 'ancrages passifs,' of a rock slope in schists at the Notre Dame de Commier Dam, France in 1961 as being one of the first major rock slopes stabilized in that way. They also reviewed applications in other civil engineering projects such as the 45 m high slope on the A9 Autoroute, in France.

The French contractor Bouygues had gained experience with the New Austrian Tunnelling Method and saw that similar techniques could be applied for the temporary support of soft rock and soil slopes. In 1972, in joint venture with the specialist contractor Soletanche, they started work on a 70° cut slope in heavily cemented Fountainbleau Sand for a railway widening scheme near Versailles. A total of 12 000 m² of face was stabilized by over 25 000 nails grouted into predrilled holes up to 6 m long. This appears to be the first published case history of a true soil nailing project, although it would seem that similar techniques were by then being used for tunnel portal support, and deep excavation stability in Western Canada. Indeed, Shen *et al.*[30] refer to the execution by soil nailing of 'several hundred thousand square feet of excavation, to depths of up to 60 ft,' in a variety of ground conditions in that region prior to 1976. It is likely that nailing also began to be used in California and West Germany about the same time for similar applications.

However, engineers in each of these three distinct areas appear to have proceeded independently—especially with respect to design methodology—until a Paris conference on soil reinforcement in 1979 provided an international forum for the exchange of information. Despite—or even because of—these often heated debates, soil nailing has continued to expand and is still one of the fastest growing specialty geotechnical techniques, described recently[31] as a "solution looking for problems."

First applications continue to be reported from different countries, as already noted, while the scale of the market in older established areas was eloquently described by Condon[32]: over 100 000 square meters of nailed excavations in over 200 different sites—some seismic and as deep as 20 m—over the last 10 years or so in Southern California alone. Nicholson[33] described the growth of the technique from 1982 in the Eastern states, where the current total of completed projects is now approaching Californian levels. Worldwide, about 15–20% of all applications are for permanent structures.

Fundamental research programs have been conducted since the late 1970s with federal and private funds in France,[34] Germany[5] and the United States,[35,36] designed to improve understanding of nail and structure performance, and to evolve design manuals. Demonstration projects have also been sponsored by interested parties, including one of the few contracts so far completed in Britain.[1]

11.4 Applications

Soil nailing has been used successfully in temporary and permanent application, in new and remedial construction, and in rural and urban settings. The following categories of applications can be identified, and selected references are given for each.

11.4.1 New construction

11.4.1.1 Retaining walls. For excavations associated with foundations of buildings, underground car parks, and cut and cover constructions for transportation systems (Figure 11.8).[5,35,37,38]

11.4.1.2 Slope stabilization. For abutment and embankment cuts required for new or widened railway lines or roads (Figure 11.9).[33,39,40]

11.4.1.3 Stabilizing tunnel portals. To provide excavation stability to tunnel portals and adjacent slopes (Figure 11.10).

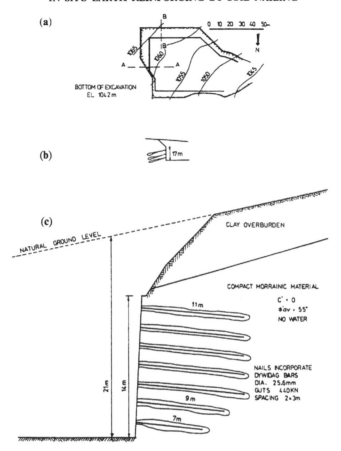

(a)

(b)

(c)

NATURAL GROUND LEVEL

CLAY OVERBURDEN

COMPACT MORRAINIC MATERIAL

c' · 0
ø'av · 55°
NO WATER

NAILS INCORPORATE
DYWIDAG BARS
DIA. 25.6mm
GUTS 440KN
SPACING 2×3m

BOTTOM OF EXCAVATION
EL 1042m

Figure 11.8 Retaining wall for an underground car park at La Clusaz, France: (a) plan; (b) cross-section A (anchored Berlin wall); (c) cross-section B. After Guilloux et al.[37]

11.4.2 Remedial works

11.4.2.1 Repair of reinforced earth walls. To replace the effect of the re-inforcing strips or fasteners damaged by overloading or corrosion (Figure 11.11).[42, 43]

11.4.2.2 Repair of masonry gravity retaining walls. After or just before failure caused by long-term decay of wall, or movements behind (Figure 11.12).

11.4.2.3 Stabilization of failed soil slopes. After collapse of slope due to failure or inadequacy of pre-existing support methods, or catastrophic movements due to changed hydrogeological factors (Figure 11.13).

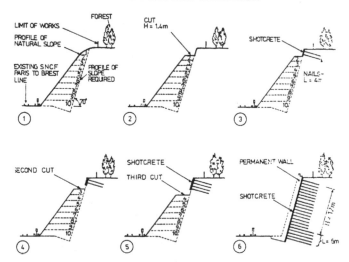

Figure 11.9 Slope stabilization construction sequence at Versailles, France.[39] Fontainbleau Sand (cemented). $\phi = 33$–$40°$; $c' = 20\,\text{kN/m}^2$. Soil nails: 2×10 mm diameter bars grouted into 100 mm diameter holes, spacing 0.70×0.70 m (two per square meter of face).

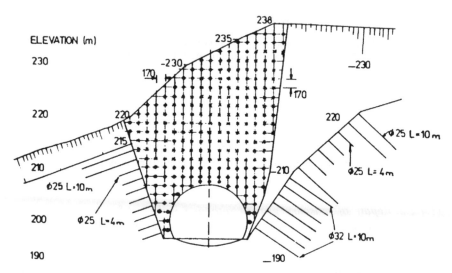

Figure 11.10 Stabilization of tunnel portal and adjacent cut slopes.[41] Nails: (●) $L = 10$ m, $\phi = 25°$; ($\times$) $L = 4$ m, $\phi = 25°$; (○) $L = 10$ m, $\phi = 32°$.

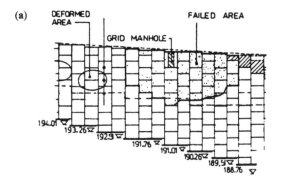

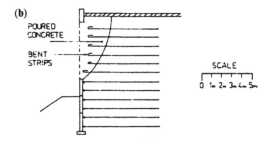

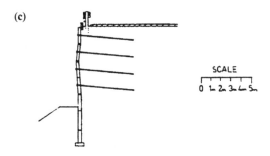

Figure 11.11 Repair of reinforced earth wall at Fréjus, France. After Long *et al.*[43] (*a*) Front face view of wall; (*b*) rebuilding of facing of collapsed area; (*c*) nailing of repaired zone.

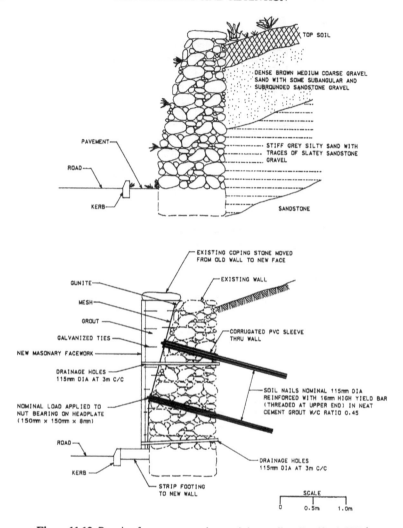

Figure 11.12 Repair of masonry gravity retaining wall at Bradford, UK.[1]

11.4.4.4 Repair of anchored walls. After failure of the prestressed rock anchorages by structural overloading or by corrosion of tendon (Figure 11.14).

11.5 Construction

The purpose of this section is to highlight aspects of soil nailing construction that may be considered as being good practice, or are regarded as having potential for future application.

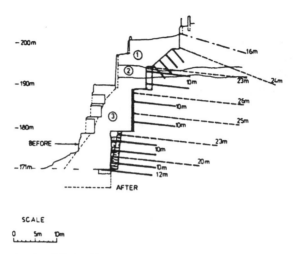

SCALE

0   5m   10m

Figure 11.13 Repair of a failed soil slope at Herbouville, Lyons, France. After Gausset.[44]
① = Sandy silts, $C_u = 0$; ② = silty sands (molasse), $\phi = 25°$; ③ = sandstone, $c = 0.05$ MPa,
$\phi = 35°$. (—) nail; (———) prestressed anchors (----) passive anchors.

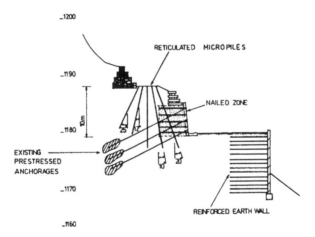

Figure 11.14 Repair of an anchored wall at Fréjus, France. After Corte and Garnier.[45]

11.5.1 Excavation and facing

The maximum cut depth at each level of excavation is dictated by the ability
of the exposed face to 'stand up'. In addition, where deformation must be
minimized, the cut depth may be reduced to the smallest value consistent with
site practicalities and commercial considerations. Cut depths of more than

2.0 m or less than 0.5 m are rare in granular soils. Greater, single cut depths have been used in heavily overconsolidated clays.

A level working bench at least 5 m wide should be provided for the nailing equipment. Usually the length of a single cut is dictated by the area of face that can be stabilized in the course of a working shift. Where deformations must be minimized, and especially in wet or very sensitive soils, the nailing may be executed in alternative primary and secondary panels, typically 10 m long, but occasionally as short as 2 m.

The excavation equipment must minimize the disturbance of the ground to be retained, and must provide a reasonably smooth and regular slope profile. Any loosened areas on the face should be removed prior to the facing support being applied. Pretreatment in the form of grouting may be necessary in loose or dry soils without natural cohesion, especially where the face is subjected to external vibrations. In this context the possible effects of blasting in adjacent areas must be evaluated. Grout columns may also be necessary to provide early support where the soil is cohesive and will not 'take' shotcrete.[46]

As a rule, the face support must be placed at the earliest time to prevent relaxation or ravelling of the ground. Typically this involves pinning a reinforcement mesh to the face and spraying a shotcrete cover before drilling the nail holes. The final face thickness varies from 50–150 mm for temporary applications, to 150–250 mm for permanent projects. The face may be built up in one, two or more layers, depending on the nail type, the construction and stressing sequence, and the longevity of the structure. Short bars may be driven into the face before spraying to serve as a depth gauge for the sprayed concrete, and a final screeding can be achieved using a piano wire as a guide. Architectural finishes may be applied with a final layer of sprayed concrete—say 50 mm thick—to blend color, or with larger aggregate to give a rugged finish.

Both 'wet' and 'dry' sprayed concrete may be used depending on the scale of the project and the availability of equipment and materials. Maximum aggregate sizes of 10–15 mm are usually specified, and admixtures are often incorporated to accelerate set in very wet conditions or, less commonly, to reduce early creep of the hardened concrete. Minimum cement contents of 300 kg/m^3 are typical, and 400 kg/m^3 is common. Control 'panels' or boxes are recommended for on-site quality assurance, at frequencies of about one per 100 m^2. Accelerated shotcrete should give an unconfined compressive strength of around 5 N/mm^2 in 8 h, whilst it is best to let it cure for 24 h prior to further works. The proper curing of the sprayed concrete face is important if surface cracking is to be avoided. Steel or plastic fibers can also be added to the shotcrete to enhance performance during and after spraying.

Spraying is often discontinued about 300 mm above the bottom of the cut. This facilitates both the fastening of the mesh for the next lower cut, and an overlapped construction joint for the sprayed concrete, which is further aided by chamfering at 45°. Careful screeding can eliminate, visually, this interface.

Final pointers on good construction practice were provided by Condon.[32] Special care is warranted with the top row of nails—'the key to a good wall'— while the soil above this row should be sloped back at about 2:1 wherever possible to reduce nailed area, and avoid near-surface utilities.

11.5.2 Drainage

An early aid is to excavate a drainage ditch parallel to, and along the crest of the excavation, to lead away surface water. The ditch may be lined with concrete during the spraying of the first cut. Thereafter, three types of drainage can be applied to the retained soil mass:

1. Shallow drains: pipes 300–400 mm long, to release water immediately behind the facing. These drains are usually about 100 mm in diameter. Their spacing depends on the groundwater conditions and the likelihood of frost damage.
2. Deep drains: slotted tubes, usually longer than the nails, about 50 mm in diameter and inclined upwards at 5 to 10°. Their spacing depends on the soil and groundwater conditions but is typically less than one per $3\,m^2$ of face.
3. Face drains: these are placed vertically against the cut slope at regular horizontal intervals before spraying the concrete face. The spacing depends on groundwater conditions and the threat of frost or ice action, but may typically be between 1 and 5 m. These drains are extended continually over the full height of the excavation and are connected by overlaps at the bottom of each successive cut. At the base they discharge into a collector system with weep holes. The drains may be prefabricated from geotextiles and need protection against impregnation by the sprayed concrete with, for example, a polyethylene sheet backing. Face drains are an alternative to shallow drains.

11.5.3 Installation of nails

In many respects the 'good practice' recommendations or stipulations of Codes of Practice covering ground anchorages would be applicable for soil nails.

Drilling techniques and methods vary with the ground conditions, the geometry of the installation, and the resources and experience of the contractor. The most common systems (excluding simple 'open hole' methods such as uncased rotary, or down-the-hole hammer—suitable for strong, cohesive soils, rock or concrete) are:

1. Duplex drilling. This rotary or rotary percussive method involves the simultaneous advancement of a temporary outer casing and an inner drill

rod.[47] Water or air flush is usually employed, although care is needed with air flush in urban environments.

2. Auger drilling. This rotary method is commonly used in clay soils without boulders, or in cemented sands. In unstable conditions the reinforcement and grout can be introduced through a 'hollow stem' auger during withdrawal of the string, although this system does mean that drill hole diameters are considerably larger than average. This method is particularly common in California.

Based on the experience from ground anchorages, the temporary support of boreholes by bentonite or other mud suspensions is not recommended, as 'smear' on the borehole walls may reduce the subsequent grout-to-ground bond.

Most recently Louis[48, 49] has reported patented systems of nail installation for which very high rates of production are claimed. In the 'jet bolting' system (Figure 11.15) very high pressures—over 200 bar—are used to inject cement grout through small apertures at the tip of the nail whilst it is being installed or percussed into the soil. This jet grout lubricates the penetration of the nail and, on setting, is claimed to provide an enhanced bond capacity for the nail. An improvement of the performance of loose sand or soft clay between nail locations is also claimed, but is as yet unquantified. To the author's knowledge, jet bolting systems have not yet had significant commercial application outside southern France.

In general, borehole diameters range from 76 to 150 mm for drilled and grouted nails, although the use of the hollow stem auger usually raises the upper limit to 200 mm. Such diameters usually permit a grout annulus of at least 20 mm thickness around the reinforcement, so providing a degree of

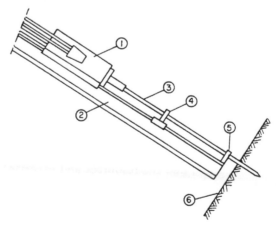

Figure 11.15 The 'jet bolting' technique for soil installation—combines vibropercussion and high grout pressure at the nail tip.[49] ① = Vibropercussion hammer; ② = sliding support; ③ = reinforcement to be inserted; ④ = sliding guide; ⑤ = fixed guide; ⑥ = soil to be treated.

corrosion protection. As nails are relatively quite short and close together, the drilling tolerance does not have to be as precise as it is for gound anchorages, and this allows higher production rates. Holes inclined downwards (even as little as 10–15°) are easier to grout effectively than those that are horizontal or inclined upwards. However, jet bolting can operate equally well over a range of inclinations.

Grouting is usually carried out with stable, cement-based grouts ($w = 0.4$–0.5) under gravity or very low excess (less than 5 bar) pressure. The use of higher pressure is often restricted by the risk of hydrofracture or leakage. Also, the potentially beneficial effects on bond of higher grout pressures do not justify the higher grouting costs for most soil nailing applications. The reinforcement should be placed and the grouting completed with the minimum delay after drilling.

11.5.4 Reinforcement and corrosion protection

Although polymer-based reinforcements such as plastic rods or fiberglass, becoming common in underground works, are being promoted for soil nailing, it would seem that steel bars are still used universally. High yield bars of 25–32 mm diameter are the most typical choice.

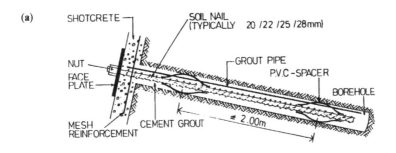

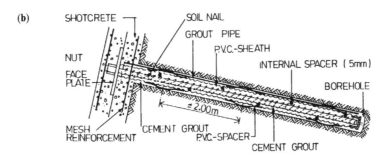

Figure 11.16 Examples of good practice for drilled and grouted nail installation: (a) for temporary applications; and (b) for permanent applications. (Based on (formerly) West German experience, and reproduced with permission of Bauer, AG.)

For temporary applications in standard environments, corrosion protection is usually provided only by the grout, but occasionally also with an epoxy coating to the steel surface (Figure 11.16(a)). For permanent works, the degree of protection may be increased by providing an outer sheath of plastic material, ensuring an inner grout annulus of at least 5 mm thickness (Figure 11.16(b)). Other proprietary systems have also been developed to overcome the potential problems arising from microfissuring of the grout under tension (e.g. the 'Intrapac' nail of Intrafor-Cofor).[7] In all cases, centralizers are placed at regular intervals (say 2 m) along the reinforcement to encourage concentricity with the borehole.

It is interesting to compare the approach in Codes of Practice dealing with ground anchorages and with reinforced earth. All international codes on ground anchorages require protection by at least one sheathing over the tendon. Conversely, in codes for reinforced earth the galvanized steel strips for permanent installations are allowed to remain in direct contact with the soil.[50,51]

One of the most recent studies on corrosion in reinforced earth[52] has demonstrated once more that the understanding of corrosion mechanisms for metals is incomplete and that long-term problems can occur. This would suggest that good practice for permanent soil nailing installations should require direct protection by at least one sheathing along the lines developed, and codified, for ground anchors (e.g. Figure 11.16(b)). A thorough summary of corrosion mechanisms and protection has been provided by the FIP state-of-the-art review *Corrosion and Corrosion Protection of Prestressed Ground Anchorages*, prepared by a working group under the chairmanship of Professor G.S. Littlejohn[53].

The way in which soil nails work—virtually their whole length is bonded to the soil and available for load transfer—means that it is unnecessary and impractical to apply significant degrees of post-tension after installation. However, a load of about 5–10% of the working load is usually locked in, with a torque wrench and lock nut arrangement. This tension is applied to 'seat' the soil/ facing/nail system so that it acts in immediate response to subsequent soil deformation . Since the 'lock off' loads are relatively low, the steel bearing plates are quite light ($150 \times 150 \times 10$ mm or $200 \times 200 \times 10$ mm), and stiff wales are generally not required—although some US contractors provide 4×12.7 mm bar wales. The nominal post-tensioning is normally applied during or just after the installation of nails in the cut immediately below.

11.5.5 Slope claddings

To date most applications of soil nailing have been temporary and so the appearance of the nailed structure has not been a significant consideration. However, there is an increasing number of applications where precast or prefabricated facing units are being used to facilitate construction, improve appearance, provide better long-term durability, or enhance noise absorp- on. Such panels may be placed directly in contact with the slope face during construction and the nails placed at the centers or corners of each panel. Alternatively, the excavation may be completed with a normal sprayed concrete facing and then covered later by precast panels. Drainage arrangements are often attached directly to the back of facing panels, or the gap between shotcrete and panel filled with suitable filter material.

Exactly as for reinforced earth, the benefits of a prefabricated facing include fabrication under controlled conditions to ensure high quality, and the wide range of shapes and materials that may be used to give an attractive, individual finish. The combination of vegetation with an open or terraced structural facing is also used, and this has great potential for providing an environmentally sensitive finish to a permanent face.

11.5.6 Instrumentation and monitoring

In contrast to ground anchorages, it is not routine or necessary to test each individual nail. This reflects the fact that it is the overall performance of the soil-nailed mass which is paramount. Selected nails should, however, be subject to pull-out tests during each level of excavation, to verify the design assumptions on bond capacity. Louis[49] recommends that, for good practice, four or five short bars should be installed and tested for pull-out capacity in each type of soil to be excavated at a site, before the main contract starts, while Condon[32] tests 5% to some test load.

By strain gauging individual nails at regular intervals along the bar, the development and distribution of the nail forces may be measured, and this can provide vital feedback to designers, assuming the data are consistent and truly reflective of in-situ conditions. Load cells at the nail head also provide useful data, particularly where near-surface effects, such as freezing, may be significant. Pressure cells placed under the shotcrete cladding can help in determining soil pressures on the face of the excavation.

Arguably the most useful measurement of overall performance of the system is the deformation of the wall or slope during and after construction. Slope inclinometers at various distances back from the face provide the most comprehensive data on ground deformations. The face movements can be measured directly by surveying, and prisms attached to selected nails permit electronic distance measurements to be made.

Continual monitoring of the ground during the progress of the works allows the actual performance to be checked against the continuous record of performance, thereby allowing modification of the construction details in response to changed conditions—most importantly if poorer soils are encountered. Readings maintained after construction can be equally informative in gauging time-dependent movements.

11.6 Design

11.6.1 Background

The most controversial aspect of soil nail technology is undoubtedly their design and, in particular, the choice of method to compute stability and the role of shear in nail capacity. Recent publications by the respective antagonists have reached a level of stridency not commonly observed in the technical press.[54,55] As in all such matters there is probably no one 'best' method, as illustrated in several papers in a Technical Session at the recent ASCE Conference at Cornell University.[56]

To put the issue into perspective, it is instructive to review the published data to fundamentally determine the major factors influencing the behavior of soil-nailed structures. It would appear that the key design-related elements are:

1. the ultimate capacities of the individual nails;
2. the forces mobilized in the nails under service
3. the various equilibrium methods used to compute stability.

A fourth design consideration, namely the deformation performance of the walls, is discussed in section 11.7.

11.6.2 Ultimate nail capacities

A major conclusion from the wealth of studies, tests and observations is that the ultimate pull-out capacities of nails are strongly influenced by their method of installation. It also seems likely that the noted variations in capacity are due in large part to variations in the normal pressure (σ_o) between the nail and the surrounding ground, as well as the geotechnical categorization of the soil itself (Figure 11.17), and the depth of cover. Another possible source of difference in capacity is the angle of interfacial friction between the nails and the surrounding ground (δ), which is larger for grouted nails than for driven nails.

Schlosser[6] indicated that, in the case of driven nails, the stress (σ_o) should be equal to the overburden pressure. In the case of grouted nails (presumably gravity grouted) he indicated that the normal stress may be very low and approximately constant with depth.

Nicholson[58] suggested that, for grouted nails, the pull-out capacities can be estimated using the same techniques used for grouted anchorages. This suggests that the value of (σ_o) would be assumed equal to the grout pressure in the case of pressure grouted nails.

11.6.3 Mobilized nail forces

Juran and Elias[59] summarized nail forces measured by others in walls constructed in silty sand, poorly graded sands, and clayey sand and residual

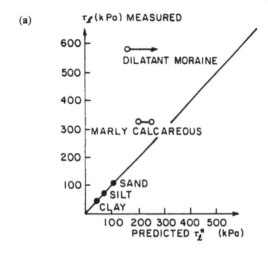

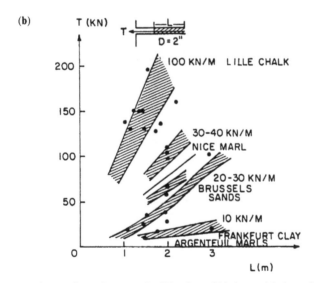

Figure 11.17 (a) Comparison of measured soil bar lateral friction and design guidelines for friction piles using pressuremeter test results;[57] (b) variation of pull-out resistance of reinforcing elements with depth of embedment (cement or resin grout) for different soils (C. Louis, unpublished results). (●) driven bars; (○) grouted bars in borehole.

soils. They converted the measured nail forces to apparent pressure diagrams— as Terzaghi and Peck[60] had done for measured strut forces in braced excavations. The distributions of apparent pressure (shown in Figure 11.18) are very similar to those developed by Terzaghi and Peck, increasing with depth in the upper half of the wall, reaching a maximum near mid-height, and decreasing again below mid-height. It seems likely that the similarity in apparent pressures results from the fact that nailed walls—like conventional braced excava-

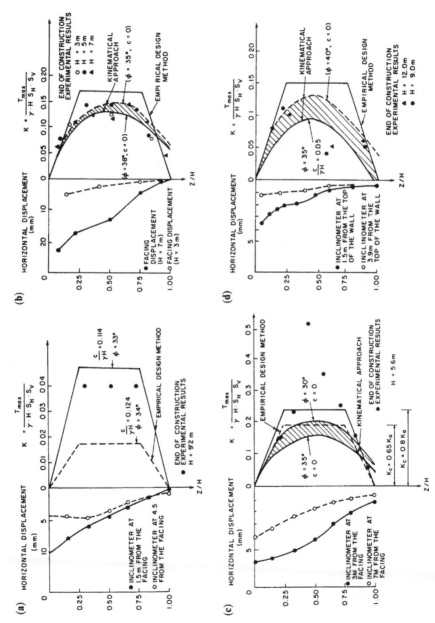

Figure 11.18 Apparent lateral earth pressure for soil-nailed walls—experimental data and theoretical predictions of tension forces. After Juran and Elias.[59] (a) Davis wall; (b) full-scale experiment CEBTP; (c) Parisian wall; (d) Cumberland Gap wall.

tions—are constructed from the top down, with the nails being installed at the top first. The stress distributions are therefore likewise different from those for reinforced earth.

Juran and Elias[59] also found that nail forces increased by 15 to 30% with time after construction in a wall in poorly graded sand. The nail forces measured in the Cumberland Gap wall in residual soils increased by 50 to 70% after construction. During the first winter the nail forces in the Cumberland Gap wall reached values as large as 100% more than the values at the end of construction, and did not decrease afterward. These findings indicate that some allowance should be made for increase in nail forces after the end of construction (assuming these data were not overly influenced by ice build-up at the face).

Experimental studies have shown that the displacement required to mobilize the ultimate pull-out capacities of soil nails are small, typically of the order of 4 to 7 mm.[61] Although the displacements required to mobilize tension in the nails are small, considerably larger deformations are required to mobilize shear forces in the nails.[62]

Shewbridge and Sitar[63] found that displacements of 20 to 40 mm were required to mobilize shear capacities in model tests, and it would be expected that the displacements required to mobilize shear resistance would be considerably larger than 40 mm for full-scale nails in the field.

Since larger displacements are required to mobilize shear resistance, it appears that the principal mode of behavior of soil nails in practice is tensile reinforcement and that the stabilization they impart results mostly from tension forces[6,54,56,64]. It seems reasonable, therefore, to ignore shear forces in soil nails as a stabilizing mechanism and to rely only on their tensile action in design analyses although it must be noted that this conclusion is strongly disputed in certain quarters.

11.6.4 Stability computation procedures

Review of the literature shows that most analyses of soil-nailed walls have been performed using computational procedures based on static equilibrium approaches, similar to those used for conventional slope stability analyses. At least five distinctly different equilibrium stability analysis procedures have been developed for nailed walls. The features of each of these can be summarized as follows.

Gassler and Gudehus.[40] This procedure has been called a 'kinematical' procedure. It is based on a two-part wedge mechanism with a bilinear shear surface. Nail forces are assumed to be known. The equations of horizontal and vertical force equilibrium are used to compute the magnitude of the forces in the soil on the bases of the wedges, and between the two wedges. The factor of safety is then defined as a *ratio of work* done by external and internal forces rather than as a ratio of forces or shear strengths, as is done in other procedures. A simple bilinear shear surface and only two wedges ('slices') are considered. Accordingly, the procedure is limited with regard to the types of soil conditions that can be considered, since the base of each slice must be entirely within one type of soil.

Beech and Juran.[65]   This procedure assumes a curved shear surface that appears to be a logarithmic spiral. The shear surface is assumed to be perpendicular to a line extending back into the slope from the crest of the slope, parallel to the nails, i.e. the shear surface intersects the uppermost nail at approximately 90°. Kotter's equation is integrated to obtain the distribution of normal stresses along the shear surface, and the shear strength is assumed to be fully mobilized along the shear surface. Thus, the shear stresses can be expressed in terms of the normal stress and shear strength parameters (c and ϕ). Forces are resolved in the horizontal direction for segments of the shear surface—corresponding to the bases of inclined 'slices'—to compute both shear and axial forces in the nails where they cross the assumed shear surface. Internal forces are ignored when forces are resolved in the horizontal direction. Unlike other equilibrium analysis procedures, Beech and Juran's procedure is intended for computation of forces in individual nails, rather than as a procedure for computing overall stability and safety factors. It appears that the nail forces calculated by this procedure are used subsequently to compute a factor of safety against failure of the nails using an estimated ultimate capacity for the nails. Juran and Elias[59] have referred to Beech and Juran's procedure as also being a 'kinematical' method.

Schlosser.[6,66]   Schlosser's procedure is based on what appears to be conventional limit equilibrium approaches employing circular shear surfaces. Reference is made to Fellenius's (Ordinary Method of Slices) procedure and Bishop's procedure; however, the specific limit equilibrium procedure employed by Schlosser is not specified. Considerable attention is paid to the nail forces, including particularly the shear forces in the nails. Four failure criteria are considered in arriving at suitable nail forces, and these criteria consider:

1. Structural failure of the nailed inclusion (nail and grout) under combined shear and tension at the point where the nail crosses the shear surface.
2. Structural failure of the nail by bending at the point of maximum moment, which occurs some distance away from the shear surface.
3. Pull-out failure of the nails.
4. Failure due to 'flowing' of the soil around the nail, much like what occurs around a laterally loaded pile as it undergoes excessive deflection.

The lowest of the nail forces found by considering these failure criteria is used in the stability calculations. Separate factors of safety may be applied to each of the failure criteria to compute the nail forces. The governing nail forces may vary along any given nail and from nail to nail. Failure due to shear in the soil itself is accounted for in the normal way by the factor of safety with respect to shear strength, which is computed from the limit equilibrium analyses. A computer program called TALREN performs the computations. The procedure is applicable to non-homogeneous soil conditions, provided that the critical shear surface can be approximated accurately by a circle.

Bangratz and Gigan.[67]   Bangratz and Gigan described a procedure for computing the factor of safety based on the Ordinary Method of Slices. Accordingly, the procedure employs a circular shear surface. Nail forces are estimated and are included in the equilibrium equations. The component of the nail forces perpendicular to the base of the slice contributes to the normal force on the base of the slice, and thus the forces in the nails are considered to contribute to the available shear strength. The component of the nail forces acting tangential to the circular shear surface contributes to the resisting moment. The factor of safety is defined with respect to shear strength in the conventional manner used in limit equilibrium slope stability analyses.

Shen et al.[30]   This method, sometimes referred to as the 'Davis' method, is based on the assumption that the shear surface is parabolic. However, only two slices are used, and failure mechanism is therefore actually a two-part wedge with a bilinear shear surface. The requirements of equilibrium in the horizontal and vertical directions are satisfied by the equations used to compute the factor of safety. Nail forces are estimated independently of the limit equilibrium computations. Only the axial components of the nail forces are considered and included in the equilibrium equations. The factor of safety is defined with respect to the shear strength of the soil. Unlike most limit equilibrium procedures, this procedure may actually correspond to different values of the factor of safety for each slice. However, only an 'average' factor of safety is actually computed. If a factor of safety is applied to the nail forces, it must be included in the values of nail force used in the analyses. Since this procedure employs only two slices, it is limited with regard to the types of soil conditions that can be considered, as the base of each slice must be entirely within one type of soil. Some engineers (e.g. reference 68) state that this method 'overpredicts the width of the active zone.'

In summary, different methods exist because: (i) different shapes of shear surface are assumed; (ii) different equilibrium conditions are satisfied; and (iii) different definitions of the factor of safety are used. None of these procedures of analysis satisfies all of the conditions of static equilibrium, and none is capable of considering non-circular shear surfaces, as might be appropriate if a layer of distinctly weaker soil exists in the cross-section. In addition, three of the procedures (Gassler and Gudehus, Beech and Juran,

Shen *et al.*) use so few slices that, for practical purposes, they are restricted to homogeneous soil conditions.

There does not appear to be any good reason why procedures that: (i) satisfy complete static equilibrium; (ii) are applicable to non-homogenous soils; and (iii) can be used to analyze any shape of shear surface, should not be used for analysis of soil-nailed walls. In fact, considerable advantages would result from their use. Such procedures can be used to consider stability along shear surfaces passing through the nails (internal stability). They can also be used to examine the possibility of bearing and sliding failures of the nail-stabilized soil mass (external stability). Efforts to develop these procedures are currently underway, albeit often on a proprietary basis, in different academic and commercial centers; a detailed example is provided by Bridle.[69]

11.6.5 Development of design methodology

The studies described in the previous section can provide a basis for understanding the behavior of soil-nailed walls, and for deciding what factors should be considered in their design. A reliable design methodology can, simplistically, follow the following steps.

1. Estimate the forces that will be mobilized in the nails. The nail forces can be estimated using apparent pressure diagrams of the type that have been used for many years for estimating loads on excavation bracing systems. Apparent pressure diagrams suitable for nailed walls will need to be developed using the results of experiments and field measurements.
2. Calculate the minimum nail lengths required to develop these forces, with suitable factors of safety against pull-out. The minimum length is equal to the required nail force divided by the allowable load per unit length that can be applied to the nail. Procedures for estimating these allowable loads need to be developed and checked against the available information regarding results of pull-out tests on nails.
3. Perform stability analyses using a method that satisfies all conditions of equilibrium, and that can be used to analyze slip surfaces of any shape. Analyses should be performed for all stages during construction, as well as at the end of construction. The wall should be stable for the conditions existing prior to installation of each row of nails, as well as after. The soil strength parameters used in the analyses should be either undrained or drained, consistent with the permeabilities of the soils and the time available for drainage. The long-term condition should be analyzed using drained strength parameters and the most adverse groundwater conditions that could develop during the life of the wall.
4. If necessary, increase the lengths of the nails, in order to increase the value of the factor of safety to an acceptable value.
5. Estimate the deformations that will develop during and following construction. These deformations can best be estimated based on the results

of previous experimental and analytical studies described in the literature. If the estimated displacements are larger than those that can be tolerated, the design can be modified to reduce their magnitudes.

11.7 Data from published case histories

Published information on case histories, containing construction details for soil-nailed structures in a variety of soil conditions, was tabulated by Bruce and Jewell.[1]

11.7.1 Derived parameters

Four derived parameters or ratios were also calculated for each project:

1. The overall geometry of the structure.

$$\text{Length ratio} = \frac{\text{Maximum nail length}}{\text{Excavation height}} = \frac{L}{H}$$

2. The nail surface area available to bond with the soil.

$$\text{Bond ratio} = \frac{\text{Hole diameter} \times \text{Nail length}}{\text{Nail spacing}} = \frac{(d_{hole})L}{\text{spacing}}$$

where the spacing is the nominal vertical area of face supported by each nail.
3. The strength of the nail arrangement. For steel reinforcement this can be expressed as the ratio of the area of steel to the area of soil. For bar reinforcement, this may be represented by the parameter:

$$\text{Strength ratio} = \frac{(\text{Nail diameter})^2}{\text{Nail spacing}} = \frac{(d)^2_{bar}}{\text{spacing}}$$

4. The performance of the nailed structure. The most frequently made measurement is the outward movement of the top of the excavation, leading to:

$$\text{Performance ratio} = \frac{\text{Outward movement}}{\text{Excavation height}} = \frac{\delta_{horizontal}}{H}$$

Where the nail is not a circular bar, equivalent values for the nail diameter and the hole diameter were calculated and entered in the tables. The equivalent nail diameter gives an equal steel area, and the equivalent hole diameter gives an equal surface area for bonding with the soil.

There are many other references to soil nailing projects in the literature, and some of these contain interesting information. For this chapter, however,

370   UNDERPINNING AND RETENTION

case histories have only been tabulated where the soil and nailing cross-sections have been fully described.

11.7.2 Observations on case histories

A few general observations may be made based on the tabulated data:

11.7.2.1 Steep granular slopes. For steep slope (80° or more) projects in granular soils there is a reasonable correlation of the derived parameters as shown in Table 11.1. Overall, for projects in granular soils, the driven nails are slightly shorter than those that have been drilled and grouted. Probably to compensate for the relatively smooth surface of driven nails, about twice as much surface area is provided for bonding with the soil than is the case with the drilled and grouted nails.

The most striking difference, however, is in the strength ratio, which shows that about three times as much cross-sectional area of steel is used with driven nails compared to drilled and grouted nails. At least part of this, however,

Table 11.1 Comparison of drilled and grouted, and driven nails for steep slope case histories in granular soil.[1]

	Drilled and grouted	Driven
Length ratio	0.5–0.8	0.5–0.6
Bond ratio	0.3–0.6	0.6–1.1
Strength ratio (10^{-3})	0.4–0.8	1.3–1.9
Performance ratio	0.001–0.003	No data

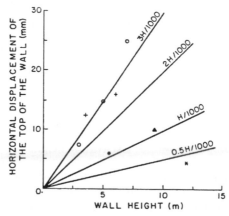

Figure 11.19 Horizontal displacement of nailed walls.[7] (+) medium sand, driven nails (Gassler et al., 1981); (▲) silty sand (SM), grouted nails (Shen et al., 1981); (●) fine sand (SP) to clayey sand (SC), driven nails (Cartier and Gigan, 1983); (*) residual clayey silt weathered shale, sandstone, grouted nails (Juran and Elias, 1986); (○) Fontainbleau Sand (SP), grouted nails (Plumelle, 1986).

Table 11.2 Comparison for drilled and grouted nails for
steep slope case histories in granular soils and Moraine or
Marls.[1]

	Granular soils	Moraine and Marl
Length ratio	0.5–0.8	0.5–1.0
Bond ratio	0.3–0.6	0.15–0.20
Strength ratio (10^{-3})	0.4–0.8	0.1–0.25

must be caused by providing more surface area for bonding with the driven
nails.

The performance ratios for drilled and grouted nails show consistently an
outward movement of up to 0.3% of the excavation depth (Figure 11.19).
Similar excellent performance would be expected for the driven nails,
although no measurements were reported on the commercial projects.

11.7.2.2 Comparison between projects in granular soils and stiff clays. For
drilled and grouted nail projects, less bond and less strength are provided for
the excavations in Moraine or Marl than for the excavations in granular soil.
The results are shown in Table 11.2. Although the length ratio is similar for
projects in the two types of soil, about two or three times less surface area for
bonding is provided in the Moraine and Marl projects. The cross-sectional
area of steel used to stabilize the Moraine and Marl excavations is about
four times less than was the case for granular soils.

By comparison, the one project in Moraine using driven nails had a similar
bond ratio and strength ratio to the typical values for driven nails in granular
soil.

11.7.2.3 Comments on reported failures. The failure at Les Eparris (Table
11.3) is well documented and the slip was due to lack of available bond
between the reinforcement and the clay. This is reflected in the repair cross-
section, where the bond ratio was increased by a factor of three but the
strength ratio is little changed.

Much less information is available for the Gard du Nord failure but both
the bond ratio and the strength ratio were increased in the repair cross-
section by a factor of two to three.

Participants are normally reticent to discuss failures, but various personal
communications confirm that several unpublished failures have occurred
throughout the world. Lack of bond, leading to nail pull-out, is the most
common cause, but it is clear that other construction-related factors have
often contributed. These include—individually or in combinations:

(i) inefficient and late excavation face stability prior to nailing, including the
exposure of too-large face areas in one pass;

(ii) use of too-high pressure air during drilling, causing fracturing of the soil
mass into vertical slices;

Table 11.3 Case histories of failures of drilled and grouted soil nail systems.[1]

Project name / Description and scale	Date	Main references	Ground conditions			Slope angle (degrees)	Height (m)	Nail length (m)	Nail diameter (mm)	Nail hole diameter (mm)	Nail spacing			Nail angle (Horiz.) degs
			Cohesion (kN/m²)	Friction (degrees)	Unit weight (kN/m³)						Horizontal (m)	Vertical (m)	Area (1 per m²)	
Les Epparis, France 70 degrees clay cutting 4.2 m high in ground sloping at 15 degrees	1981	6, 70	Plastic Clay (18 > PI > 15) 0	28	20	70	4.2	4.5	28.0 (eq.)	100 (approx.)	3.00	1.40	4.20	20
Repaired to a 60 degree slope 5.2 m high	1982	70, 71	0	28	20	60	5.2	10.0	26.0	105	2.00	1.73	3.46	30
Paris Gard du Nord, France Excavation of 10 m vertical in Marls	1979	70	Fill overlying 0 Heterogeneous Marls 50	30 20	–	75 overall	10.0 overall	6.5	32.0	100 (est.)	2.50	1.60	4.00	20 overall
Repaired as a vertical wall 8.5 m high	1979	70	Assumed strength 0	25	–	90	8.5	10.0	32.0	100	1.50	1.25	1.88	15

(iii) lines of pre-installed dewatering or instrumentation holes, parallel to, and at some distance back from the face. These act as a vertical 'presplit' line in the soil mass; and

(iv) insufficient attention paid to the shape of the wall in plan (especially where it includes a corner or bend), or the profile of the slope in section (major backslope giving additional surcharge).

11.8 Two recent case histories

The following two North American case histories have been selected to illustrate many of the points raised in the chapter to date. The first—from Edmonton, Canada—focuses on nail application, construction and performance. The second—from Seattle, Washington—provides one view of the design process, as supported by on-site measurements of nail strains.

11.8.1 Excavation for tunnel portal, Edmonton, Alberta, Canada[72]

11.8.1.1 Background.   The city of Edmonton recently commissioned Phase II of the South Light Rail Tansit (SLRT) Extension. This involved the construction of the South Tunnel and Portal, from the south bank of the North Saskatchewan River to the University of Alberta campus (Figure 11.20). As shown in Figure 11.21, the initial 80 m of the tunnel alignment was marked by unfavorable ground conditions—created by slump debris from an ancient landslide comprising saturated sand deposits. For economic and technical reasons, this initial portion was designed as a cut-and-cover section.

The original concept featured soldier piles, prestressed ground anchorages, and struts to secure the excavation (160 m long, 18 m wide to a maximum depth of 19 m), in concert with peripheral dewatering wells. The successful contractor proposed an alternate shoring system featuring soil nails—known locally as the 'Ground Control Method.' This was approved, but with the

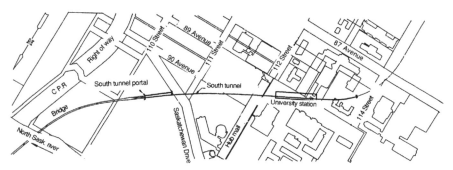

Figure 11.20 Excavation for tunnel portal at Edmonton, Canada—general site plan.[72]

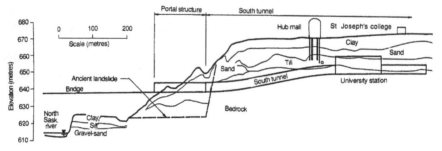

Figure 11.21 Excavation for tunnel portal at Edmonton, Canada—simplified geological cross-section.[72]

caveat that the whole system was appropriately monitored to confirm acceptable performance.

11.8.1.2 Site and geological conditions. The portal was located at about mid-height in the valley slope, immediately below an intermediate terrace at an elevation of 650 m. From this terrace, the ground surface rose at a slope of about 20° to the prairie level at an elevation of 670 m.

The detailed stratigraphy is shown in Figure 11.22. The site investigation confirmed a complex and variable stratigraphy—a direct result of the landslide processes. The general stratigraphy consisted of clay over sand over Cretaceous bedrock. The superficial lacustrine clay was stiff and highly plastic. The sand was fine-grained but variable in silt/clay content (7–46%, average 23%). It was generally medium-dense with SPT values ranging from 5 to 30 (per 300 mm penetration). Additionally, layers of clay and clay till were found in the sand. The bedrock comprised claystones, siltstones and sandstones, with occasional bentonitic and coal seams.

Piezometers confirmed the water level as varying from an elevation of 646.5 m (north) to 654 m (south), about 12-20 m above the base elevation, while pump tests indicated a sand permeability of 8×10^{-5} m/s. Dewatering

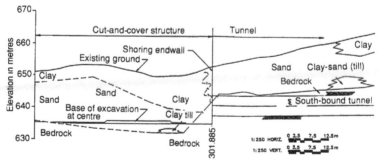

Figure 11.22 Excavation for tunnel portal at Edmonton, Canada—detailed stratigraphy of the portal area.[72]

was therefore a prerequisite for construction, and a contractual obligation was for the piezometric surface to be 2 m below the excavation elevation.

11.8.1.3 Design considerations.   The inherent variability of the stratigraphy and groundwater conditions caused serious concerns for the project engineers. In addition, this alternate and locally untried technique could in no way be allowed to impact the overall slope stability. No precedent could be found in the literature for a 19 m high, soil-nailed wall in cohesionless landslide debris. The design was therefore conservatively derived and meticulously reviewed. The following major criteria were employed:

(i)   A triangular soil pressure distribution, using $K_a = 0.3$, plus an allowance of 9.6 kPa for live load, was adopted to determine 'the total driving force' for the design of the nails. Practical experience was used to determine the level by level distributions of nail loads.
(ii) The safety factor for stability (both overturning and shear movement) varied from 6 to 9, considering shearing through the sand only.
(iii) The safety factor for designing the nails 'and the ground confinement' ranged from 2 to 3.

The excavation was broken down into eight different zones, based on typical wall height, nail spacing and stratigraphy. Details for each zone are shown in Table 11.4. An example of the shoring design for Zone No. 2, including shotcrete requirements, nail spacings, loads and lengths, is shown in Figure 11.23. The wall is shown at a slight batter angle in this figure but, in reality, was installed near vertical for the most part.

Weep holes, which comprised slotted 50 mm diameter PVC pipes, were designed through the shotcrete wall, on a regular pattern. These drains were a back-up measure to ensure that no water pressure built up on the wall, were the dewatering pumps to fail for any reason.

11.8.1.4 Construction.   Installation of dewatering wells around the peri-

Table 11.4 Details of the shoring zones for the excavation for tunnel portal at Edmonton, Canada.[72]

Zone No.	Description	Length (m)	Average height (m)	Vertical nail spacing (m)	Horizontal nail spacing (m)
1	E and W walls	47.0	15.0	2.0	1.8
2	E and W walls	16.0	14.5	2.0	2.0
3	E and W walls	21.0	11.5	2.0	2.1
4	E and W walls	26.0	7.5	2.1	2.4
5	E and W walls	23.0	6.5	2.1	2.4
6	E and W walls	12.0	5.5	2.1	2.1
7	Parking lot	31.0	6.5	2.1	2.0
8	South wall	28.4	19.0	2.0	1.8

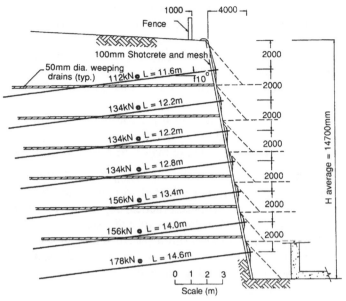

Figure 11.23 Excavation for tunnel portal at Edmonton, Canada—typical shoring design for Zone 2.[72]

phery preceded the excavation and the drilling of nails by one month. The construction procedure then included the following steps:

1. Small panels of soil, approximately 2 m high by 2 to 6 m long, were excavated with intact panels of ground in between. The excavated panels were cut with a sloping face, as illustrated by the dashed line in Figure 11.23.
2. Nail holes, 90 mm in diameter, were drilled with an air-track drill, and Dywidag bars were installed.
3. These bars were tremie-grouted with High Early strength cement at a water/cement ratio of 0.45. No pressure grouting was done.
4. After approximately 24 h of set time, the sloping face of the panel was excavated back to vertical. If located in sand, plywood was quickly installed around the nail to limit soil movement. Wire mesh and horizontal walers of rebar were placed; shotcrete was then sprayed over the entire surface of the panel. The shotcrete consisted of a low slump concrete, which was to develop a compressive strength of 30 MPa in 28 days.
5. After another 24 h, an anchor plate was installed on the end of the bar. A hydraulic jack was used to bring the nail to 133% of its working load for at least 1 min before being unloaded back to its working load. If the nail did not hold its required load after several loading attempts, another hole would be drilled adjacent to it.

Excavation, in alternating panels, proceeded around the site until a complete level was done. Before proceeding to the next lower excavation level, the piezometers inside and outside the site were read to ensure that the water was at least 2 m below the required elevation.

The construction of the shoring system was completed in less than five months.

11.8.1.5 Performance.   Slope inclinometers and piezometers were installed around the portal excavation works prior to construction, at the locations shown in Figure 11.24. The purpose of the instrumentation was twofold:

(i) to monitor the performance of the temporary retaining system; and
(ii) to provide early warning of potential deep seated slide movements in the weak bedrock layers below the base of the excavation, as there was a potential concern that the construction works would reactivate the ancient landslide.

The slope inclinometers were installed in vertical boreholes and extended to well below the old bedrock failure planes. The standpipes and the pneumatic piezometers were also installed in boreholes, located within the sand deposit and the bedrock units, respectively, to monitor the effects of the groundwater dewatering program.

The instrumentation was monitored throughout the portal excavation and backfilling, and for several months after completion. The instruments were

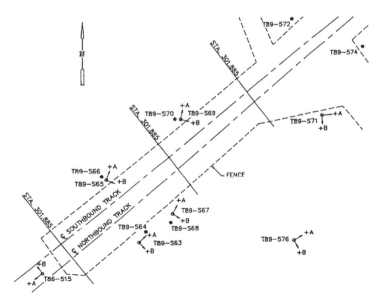

Figure 11.24 Excavation for tunnel portal at Edmonton, Canada—location of slope monitoring instruments.[72] ⭑= slope inclinometer; ⊕ = piezometer.

initially read as the excavation reached each level of nails. This frequency was increased as the excavation neared the final depth.

Slope inclinometers.   The inclinometer monitoring results provided profiles of lateral deflection in both the downslope ('A') direction and the cross-slope ('B') direction i.e. towards the excavation. The A and B directions for each instrument are shown in Figure 11.24.

A typical profile of lateral deflection towards the shotcrete wall is shown for inclinometer T89-S63 in Figure 11.25. The lateral deflection is shown at several stages of excavation. The maximum lateral movement at ground surface was approximately 84 mm towards the excavation, corresponding to a final depth of excavation of 15 m. The lateral movement increased relatively uniformly over the height of the excavation. Small, distinct shear movements of approximately 13 and 8 mm were also evident at depths of 19 m (elevation 632.0 m) and 27.4 m (elevation 623.5 m), respectively. When these shear movements were subtracted, the lateral movement over the excavation depth was approximately 64 mm.

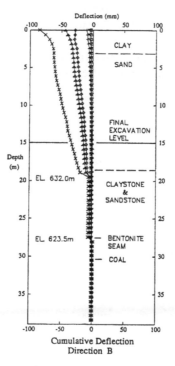

Figure 11.25 Excavation for tunnel portal at Edmonton, Canada—plot for slope inclinometer T89-S63 initiated on 6 March 1989 at an excavation depth of 0 m.[72] (□) 14 March 1989, depth 5 m; (▽) 5 April 1989, depth 7 m; (+) 17 April 1989, depth 9 m; (◇) 16 May 1989, depth 12.8 m; (△) 7 June 1989, depth 13.3 m; (×) 8 November 1989, depth 15 m.

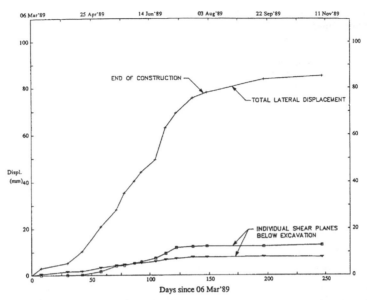

Figure 11.26 Excavation for tunnel portal at Edmonton, Canada—lateral displacement v. time for slope inclinometer T89-S63.[72] Displacements are shown in the negative (B) direction. (□) 18.9/19.5 m; (▼) 27.4/28 m; (+) 0/37.8 m.

The lateral deflections are also plotted against time in Figure 11.26. The upper line represents the lateral movement measured over the entire depth of monitoring, while the lower two lines represent the shear movements on the individual shear planes at 19 m and 27 m depth, respectively. As noted, the rate of movement—over both the excavation depth and along the weak bedrock slip planes—subsided immediately after excavation was completed. The lateral displacement over the entire depth of monitoring is also plotted against depth of excavation in Figure 11.27.

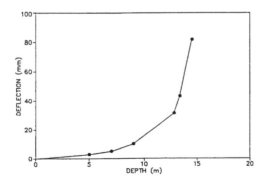

Figure 11.27 Excavation for tunnel portal Edmonton, Canada—lateral displacement v. depth of excavation for slope inclinometer T89-S63.[72]

Table 11.5 Summary of lateral displacements towards the excavation for tunnel portal at Edmonton, Canada.[72]

Slope inclinometer number	Maximum depth of excavation H (m)	Total lateral displacement* (mm)	Distinct shear displacement below excavation (mm)	Lateral displacement over excavation depth y (mm)	Angular distortion y/H (%)
T89-S15	20	60	5	55	0.28
T89-S63	15	84	24	60	0.40
T89-S67	13	60	15	45	0.35
T89-S69	8	82	45	37	0.46

* Measured between the top and the bottom of the slope inclinometer

Results of the slope inclinometer monitoring are summarized in Table 11.5. The lateral movement, y, recorded over the approximate excavation depth, was obtained by subtracting distinct shear movements below the base of the excavation from the total recorded movement. The resulting angular distortions (lateral displacement, y, divided by excavation depth, H) ranged from 0.28 to 0.46%. The resulting lateral movements were well within tolerable limits and the observed slope performance was considered to be acceptable, in terms both of angular distortion, and of overall slope stability.

Other monitoring methods. Two other methods were used to monitor the deformations of the shotcrete retaining walls.

(i) Electronic distance measurement monitoring prisms were attached to the south headwall and two monitoring stations were located at the north end of the site. Deformation of the headwall could thus be monitored from a remote location. Early readings indicated that the deformation of the wall could be measured with this system. Unfortunately, the construction sequence required the removal of the monitoring stations at an early stage, and therefore no further readings were recorded.

(ii) Graduated measuring tapes were attached, perpendicular to the wall, to several of the upper anchor heads. A theodolite was then used to sight between two points located parallel to the wall. In this way, it should have been possible to directly measure the movement of the anchor heads (and the shotcrete wall) in towards the excavation.

The limited readings fluctuated in terms of direction and could not be correlated with the slope inclinometer readings. Consequently, only the slope inclinometers were relied on for accurate deformation measurements.

Comparison with other case histories. Table 11.6 provides a comparison of the results from case histories for granular soils presented in Bruce and Jewell,[1] and for two locations on the project site. The length ratios were at the upper

Table 11.6 Comparison of other case histories with SLRT results.[72]

Derived parameters	Results from drilled and grouted nails in granular soils	Results from SLRT project T89-S15	T89-S63
Length ratio	0.5–0.8	0.6	1.0
Bond ratio	0.3–0.6	0.3	0.4
Strength ratio (10^{-3})	0.4–0.8	0.1	0.1
Performance ratio (%)	0.1–0.3	0.3	0.4

*Bruce and Jewell[1]

limit of the other case histories, perhaps reflecting the high factor of safety (6 to 9) used against shear and overturning. The bond ratio values were within the expected range. The most divergent result was the low strength ratio values: the published examples had four to eight times as much cross-sectional area of steel to soil as did the SLRT project.

The most important parameter is the performance ratio. In terms of this parameter, the SLRT shoring was at the upper values of other case histories, reflecting the influence of the variable stratigraphy and the greater height of the walls.

Problems encountered. Three serious problems were encountered with this shoring system over the course of the project. These included a relatively rapid 40 mm translation of the soil over the bedrock on a portion of the west wall, and a small localized shoring failure on the south wall.

The translational shearing observed on the west wall occurred in an area where the bedrock was at relatively shallow depth below the base of the excavation. This was indicative of one of the major concerns of the entire temporary shoring, namely mobilization of weak bedrock shear planes. The recommended remedial measure in such instances was to increase the number and length of soil nails in such areas to provide additional close reinforcement.

Late in construction, while attempting to read slope indicator T89-S65, the probe encountered a blockage in the casing at a depth of 18.0 m. The immediate reaction was that severe wall movement had sheared off the casing. Subsequent investigation of the blockage indicated that the shoring contractor had installed a nail through the casing.

Four weeks later, a 3 m high by 7 m wide panel of shotcrete failed at the bottom of the south wall. This particular section of the shotcrete wall had not been anchored, since it corresponded in location with the northbound tunnel, and the shoring contractor felt that no nails would be required here because the shoring was up against the bedrock. Unfortunately, the blocky nature of the bedrock, and resultant lateral pressure, caused this panel of the wall to be pushed out.

Remedial measures included building a soil buttress up against the failed mass, rebuilding the wall, and pressure grouting behind the wall.

11.8.1.6 Conclusions. The soil nailing shoring proposal proved to be a viable and relatively well-performing system, even in the variable and relatively poor ground conditions of landslide debris. It allowed the contractor to form and pour the portal structure without interference from bracing struts—measures that saved significant time and money.

The performance of this soil-nailed shoring system was similar to other published case histories with wall deflections 0.28 to 0.46% of the height.

The system was not, however, trouble-free. Constant monitoring had to be a prerequisite to evaluate the performance of the walls on a continual basis, provide early warning of potential hazards and enable timely implementation of slope remedial measures.

11.8.2 Design, construction and performance of a soil-nailed wall in Seattle, Washington[68]

11.8.2.1 Background. The soils in the Seattle area are generally heavily overconsolidated glacials, many of which are ideal for soil nailing. After a long history of the local use of prestressed ground anchorages, nails were first used in 1987, for temporary support of a building excavation. The present project was the first, however, to be designed and constructed by local firms. It was part of a temporary shoring system for a building excavation just east of the downtown area. The project consisted of two nailed walls, with heights of 10.7 m and 16.8 m adjacent to city streets, and two soldier pile and tieback walls adjacent to existing buildings. Soil nailing was not used to support the existing buildings because of the lack of experience with the system in local soil conditions, and an understandable initial concern with the potential for excessive movements associated with an unstressed shoring system.

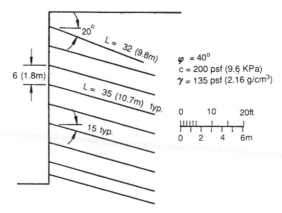

Figure 11.28 Design, construction and performance of a soil-nailed wall in Seattle, Washington—high-wall section.[68]

The soil conditions consisted of fill to a depth of 2.4 m, underlain by very dense glacial outwash sand and gravel, and very dense lacustrine fine sand and silt. The contact between the outwash and lacustrine deposits was encountered at the base of the excavation on the high wall, and at about mid-height on the low wall. Groundwater was below the base of the excavation. Soil properties used in the design analysis are shown in Figure 11.28.

Nails were generally installed on 1.8 m centers horizontally and vertically to a maximum length of 10.7 m. Holes were drilled with a 203 mm diameter, continuous flight, hollow stem auger. Nail bars consisted of Grade 150 Dywidag bars ranging from 25 mm to 32 mm in diameter. Nails were typically installed at an inclination of 15°, although the first row on the high wall was installed at 20° to avoid utilities. A typical section of the high wall is shown in Figure 11.28.

11.8.2.2 Design considerations. A force limit equilibrium analysis was chosen for the design of this wall. It was based on the 'Davis method'[30], but had been extensively modified to include more versatile modeling of a variety of wall and backslope geometries, variable nail lengths and spacings, and a treatment of nail capacities more consistent with local tieback design practices. The authors acknowledged that the limit equilibrium method has a number of limitations as a design tool. Perhaps the most significant was that limit equilibrium analyses can only provide an indication of the total nail force required to maintain a given factor of safety. The distribution of the total nail force among the individual nails cannot be solved for intrinsically, but must be assumed. The major shortcomings with existing limit equilibrium formulations for soil nailing design, including the Davis method, had been poor assumptions with regard to the nail force distribution.

In addition to difficulties with nail load distribution, limit equilibrium models do not address displacements of the reinforced soil mass. For applications such as an excavation shoring system in urban environments, such displacements are critical.

11.8.2.3 Instrumentation program. Instrumentation was installed to provide a better understanding of the aspects of soil nailing performance not well predicted by the limit equilibrium analysis, and to provide baseline information regarding the performance of soil nailing in specific local soils.

Vibrating wire strain gauges were installed on five nails, in a vertical section on the high wall (Figure 11.29). Four to six gauges were spot welded to the steel along the length of each nail. Economic constraint forced the use of a single gauge at each location. Thus, bending movements could not be measured and the effect of bending on the measured strains could not be explicitly determined. All gauges were located at the 3 o'clock position on the bar to minimize the potential for bending moment interference. All

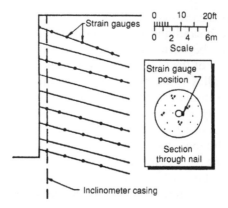

Figure 11.29 Design, construction and performance of a soil-nailed wall in Seattle, Washington—instrumented section.[68]

strain gauges were read daily during construction of the shoring wall, and monthly thereafter until the permanent basement wall was completed.

Two inclinometers were installed at a distance of 0.9 m behind the face of the wall. One was installed on the high wall, 3 m away from the strain-gauged section, and one was installed on the low wall. Inclinometers were read weekly during construction of the shoring wall, and monthly thereafter until the permanent basement wall was completed.

11.8.2.4 Data interpretation

Inclinometer data.   Deflections measured on the high wall are shown in Figure 11.30. The total deflection at the top of the wall after the last lift of

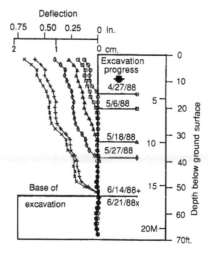

Figure 11.30 Design, construction and performance of a soil-nailed wall in Seattle, Washington—wall deflections.[68]

soil was excavated was 15 mm. One week later, after the nails and shotcrete facing had been installed on the last lift, the deflection at the top of the wall had increased to 18 mm. Monitoring of the inclinometer casings continued for another 10 weeks. No additional movements, within the accuracy of the instrument, were measured during that time.

The deflections were understandably slightly higher than might have been expected for a typical soldier pile and tieback wall in similar soil conditions. The maximum movement at the top of the wall was close to 0.1% of the height of the wall, or about what would be required to mobilize active earth pressure in dense granular soil.[73]

Strain gauge data.   The strain gauge data provided considerable insight into the behavior of the nailed wall, including the short-term development of load on the nails, the change in nail loads with time, and indications of the combined influence of the steel and concrete portions of the nail section.

Figure 11.31 shows early strain measurements from nail level 6, plotted as a function of time. The excavation of each successive lift below nail level 6 can be seen as an initial rapid increase in strain related to the actual excavation of the next lift, followed by a slow increase in strain with time as the nails and shotcrete facing were installed. The construction of each of the three lifts below nail level 6 can be seen clearly in the strain data. This type of behaviour was seen on all nail levels. The influence of the excavation was most significant when the bottom of the excavation was within about 6.1 m of the nail. Following this the excavation effect decreased progressively but was still noticeable during every lift.

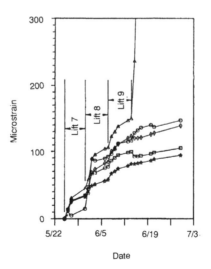

Figure 11.31 Design, construction and performance of a soil-nailed wall in Seattle, Washington—short-term strain from nail level 6.[68] (○) Gauge 1; (□) Gauge 2; (△) Gauge 3; (◇) Gauge 4; (∗) Gauge 5.

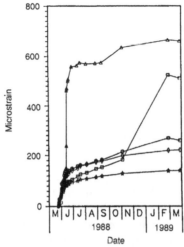

Figure 11.32 Design, construction and performance of a soil-nailed wall in Seattle, Washington—long-term strain from nail level 6.[68] (○) Gauge 1; (□) Gauge 2; (△) Gauge 3; (◇) Gauge 4; (∗) Gauge 5.

Figure 11.32 shows the complete record of strain measurements from nail level 6. The strain increased rapidly at first, as active excavation took place. On completion of the excavation, the rate of strain increase slowed down in a pattern consistent with long-term creep of the nails. At the time monitoring was completed (nine months later), strains had increased by a factor of more than two over their values at the end of construction. This behavior is consistent with other cases where long-term measurements have been obtained.[59] It was ascribed to soil creep, and concrete creep and cracking (e.g. Figure 11.32 'jumps').

The determination of nail loads from strain data is not simple. It requires an assessment of the time-dependent effect of the grout on the overall stiffness of the nails. Concrete has a low modulus in comparison to steel, but the area of concrete in the composite nail section is greater than the area of steel by a factor of 50. Thus, the stiffness of the nail is strongly affected by the modulus of the concrete as long as the concrete remains effective in tension. When the concrete cracks, the local stiffness of the nail decreases substantially.

Two approaches were taken to calculate loads from the strain data. The first approach was to calculate a composite stiffness for the steel/concrete section based on reasonable assumptions regarding the section geometry and concrete modulus. A time-dependent concrete modulus was used, based on 3-, 7-, and 28-day strengths from 17 cylinders of the nail grout. The resulting maximum nail loads at each level are shown in Figure 11.33 as indicated by the curve titled 'Indirect method.'

The disadvantage of this approach is that the area and modulus of the concrete are based on assumptions that may not reflect actual conditions.

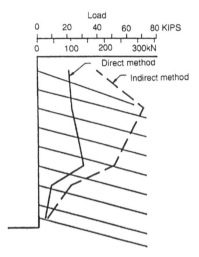

Figure 11.33 Design, construction and performance of a soil-nailed wall in Seattle, Washington—measured load v. depth.[68]

Furthermore, it is assumed that the load is shared in direct proportion to the stiffness ratio of the concrete and steel, with no allowance made for creep in the concrete. As indicated previously, creep may account for half or more of the deformation, particularly early on when the strength of the concrete is low and the nail loads are increasing most rapidly. This will result in less of the load being carried by the concrete than the stiffness ratio suggests and actual nail loads will be lower than predicted. Thus, loads determined by this method are considered to be limiting upper bound values.

The second approach was to assume that the jumps in strain caused by cracking of the concrete represented the full release of tensile stress in the concrete. Thus, the strain on the low side of the jump was related to the stiffness of the composite section, while the strain on the high side of the jump was related to the stiffness of the steel alone. By assuming that the total load remained constant over the strain jump, the stiffness of the uncracked composite section could be back-calculated. The resulting maximum nail loads are shown in Figure 11.33, as indicated by the curve titled 'Direct method'.

The advantages of this approach are that no assumptions need be made regarding the variation of concrete modulus with time, and any transfer of load to the steel which may have occurred by creep in the concrete is accounted for directly. The disadvantage is in the assumption that the strain jump results from the full release of the tensile stress previously carried by the concrete. This is only true if the gauge is located at the crack. If the strain gauge were located some distance away from the crack, the increase in strain would reflect only a portion of the tensile stress released by the concrete. The jumps in strain do provide a lower bound indication of the tensile stress in the concrete and thus a lower bound indication of the total nail load.

Loads determined by the indirect method were substantially in excess of those required under active conditions. However, the magnitude of the wall deflections suggests that active conditions had been established. Consequently, loads determined by the indirect method may not represent the best estimate of actual nail loads. Nail loads calculated by the direct method were consistent with active conditions, and thus may be more realistic.

The location of the maximum bar force is not affected by the assumptions made in converting from strain to force. The distribution of load along the length of each bar is shown in Figure 11.34, using the loads calculated by the direct method. Two maxima were seen on some bars. The peak at the front of the bar may be related to bending forces caused by the weight of the shotcrete facing hanging from the nails during excavation of underlying lifts. The peaks farther away from the face map the locus of maximum strain in the soil mass. This should, in theory, coincide with the critical failure surface predicted by the limit equilibrium analysis used to design the wall. The surface predicted by the analysis is shown in Figure 11.34. The maximum tensile force line typically used in the design of reinforced earth walls[74] is also shown and is actually in better agreement with the measurements than the limit equilibrium prediction.

11.8.2.5 Critique of design methodology. Limit equilibrium analyses are currently common tools for the design of soil-nailed walls in the USA,

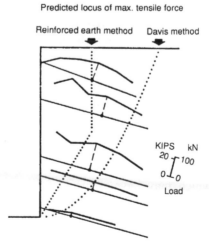

Figure 11.34 Design, construction and performance of a soil-nailed wall in Seattle, Washington—load distribution on nails.[68]

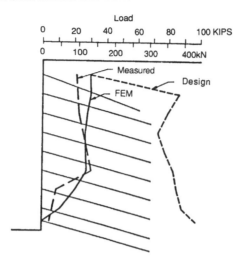

Figure 11.35 Design, construction and performance of a soil-nailed wall in Seattle, Washington—maximum nail load *v.* depth.[68] FEM = finite element method.

although it is well known that such analyses do not provide good estimates of the magnitude and location of the maximum nail forces.[6,7,75] However, working stress analyses can be cumbersome as design tools. Hence, limit equilibrium analyses are commonly used to estimate nail forces, and to design wall systems accordingly. Experience with the Davis method indicates that the predicted nail forces are conservative.

Figure 11.35 presents a compilation of the maximum nail force at each nail level as determined by the original design methodology (Davis method), the field measurements (only the direct method interpretation is shown), and finite element analysis. Clearly, the nail loads determined using the Davis method are excessive in comparison with loads obtained from the field measurements and the working stress analysis.

A number of factors contribute to the overestimation of nail loads. In the first place, the design loads calculated using the Davis method incorporate a factor of safety of 1.5 on soil shear, resulting in more load being carried by the nails. The other loads do not include factors of safety. However, the factor of safety alone does not explain the difference in nail loads.

The primary shortcoming of the Davis method is in the assumptions made regarding the distribution of the total nail forces among the individual nails. The method assumes that the force carried by each nail is proportional to the length of the nail behind the critical failure surface. This bears little resemblance to the distribution of nail forces predicted by working stress analyses or measured in case histories.

During the early stages of the excavation—when the nail length may be considerably greater than the height of the excavation—this assumption leads to excessively high nail forces, and high overall factors of safety. As the excavation is lowered, the failure surface extends deeper within the soil mass and the effective anchor length of the nails decreases. This often results in predicted nail loads that decrease with increasing depth of excavation. The load distribution also results in higher nail forces being carried by progressively lower nails.

This is not a good model of soil-nailed wall behavior. Field data[59] indicate that the soil strength is mobilized fairly quickly. In other words, the factor of safety with respect to soil shear is close to 1 during the entire excavation process. The forces mobilized in the nails are just sufficient to maintain the system at a factor of safety of 1. As conditions change, for example the excavation is lowered, additional forces are mobilized in the nails as needed. As long as sufficient excess capacity is available in the nails the wall will remain stable. The force mobilized in a nail is related predominantly to successive lowering of the excavation. Thus, loads in the lower nails are typically small.

A more rational approach to modeling the mobilization of nail forces would be to compute the total nail force required to maintain a factor of safety of 1 at each stage of the excavation. Limit equilibrium methods are capable of this analysis, since it is essentially an ultimate condition. For design purposes, suitable factors of safety could be applied reflecting the degree of uncertainty in the various strength parameters, which results in a somewhat higher total nail force than for the factor of safety of 1 condition.

This total nail force must then be divided between the individual nails. Limit equilibrium methods are not able to distribute nail loads explicitly, but assumptions can be made on the basis of empirical data—as has been done

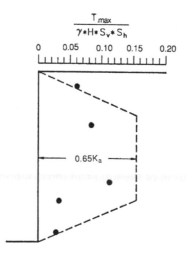

Figure 11.36 Design, construction and performance of a soil-nailed wall in Seattle, Washington—normalized nail force v. braced cut pressure distribution.[68]

for years in the design of other types of braced excavations. In fact, field data[59] indicate that the pressure diagram for a conventional braced excavation[60] may provide a reasonable approximation of the nail load distribution. As shown in Figure 11.36, the project data also fit this model. By accepting the limitations of the limit equilibrium method and applying an empirical understanding of load distributions, more reasonable designs may be achieved.

11.8.2.6 Conclusions.   Considerable insight has been gained into the mechanics of the performance of soil-nailed walls through the instrumentation program described in section 11.8.2.3. In particular, a better appreciation has been developed of the creep behavior of grout, and its effect on nail loads calculated from strain data.

Appropriately calibrated finite element analyses were able to provide a close approximation of the behavior of the wall system. Properly characterized joint elements between the soil and nail elements proved to be critical to the accurate calculation of nail loads.

Incorporating a more rational approach to the modeling of nail force mobilization and distribution, based on the body of knowledge that has been developed in recent years regarding the actual behavior of soil-nailed walls, may significantly improve the ability of limit equilibrium analyses to predict nail forces.

11.9 Final remarks

Soil nailing is a technique of great potential, proven cost-effectiveness and excellent technical performance. It has been used routinely—in appropriate ground conditions—in North America and Western Europe for over 20 years, while each year there are reports of first applications in other countries. Soil nailing is used on both new and remedial projects, and for both temporary and permanent purposes.

Soil-nailed structures are relatively easy and fast to construct, and lend themselves well to informative monitoring and test programs. The ongoing debate over design methods has ensured that the technique has continued to attract the attention of experienced and innovative geotechnical engineers. The main drawback of current design methods is that they do not provide an estimate of either the structural displacement, or the nail forces mobilized at the expected working loads. Indeed, the development of the technology has been essentially empirical, and field experience has markedly preceded theory and fundamental research. The ongoing programs of full-scale experiments, and laboratory and theoretical studies are, however, leading to significant improvements in design methods and predictive capabilities.

Consequently, there is every reason to believe that the number of applications will continue to grow, to the benefit of owner and engineer alike, on a world scale.

Acknowledgements

Much of this chapter is drawn from previous research conducted with Dr. R.A. Jewell of Oxford University, and more recent work by Professors Duncan and Wright at Virginia Polytechnic Institute and State University. Contributions have also been made (often unwittingly) by others, including Dr. Ilan Juran, Mr. M. Condon and the Nicholson brothers. To each, the author extends grateful thanks and full acknowledgment of the contribution.

References

1. Bruce, D.A. and Jewell, R.A. (1986, 1987). Soil nailing: application and practice. *Ground Engineering* 19 (8), 10–15; 20 (1), 12–38.
2. Frankipile (South Africa) (1988). First local application for soil nailing technique. *S.A. Construction World*, December 1987–January 1988, 2 pp.
3. Woodward, S. (1988). Soil nailing. *Report on Auckland Branch Joint Structural and Geomechanics Group Technical Session*. 10 May, 7 pp.
4. Banyai, M. (1984). Stabilisation of earth walls by soil nailing. *Proc. 6th Danube Conf. Soil Mech. Found. Eng.*, Budapest, 459–465.
5. Stocker, M.D., Korber, G.W., Gassler, G. and Gudehus, G. (1979). Soil nailing. *Int. Conf. on Soil Reinforcement*, Paris, March, Vol. 2, 469–474.
6. Schlosser, F. (1982). Behaviour and design of soil nailing. *Int. Symp. on Recent Developments in Ground Improvement Techniques*, November 29 to December 3, 1982, AIT, Bangkok, 399–413.
7. Juran, I. (1987). Nailed-soil retaining structures: design and practice. *Trans. Research Record 1119* Trans. Research Board, Washington, D.C., 139–150.
8. Boley, D.L. and Crayne, L.M. (1985). INSERT wall saves roadway. *Highway Builder*, Fall, 14–16.
9. Lizzi, F. (1970). Reticoli di pali radice per il miglioramento delle caratteristiche di resistenza del terreno. *Convegno di Geotecnica*, Bari.
10. Bruce, D.A. (1989). American development in the use of small diameter inserts as piles and in situ reinforcement. *Intl. Conf. on Piling and Deep Foundations*, London, May 15–18, 11–22.
11. Gudehus, G. and Schwarz, W. (1984). Stabilisierung von Kriechhangen durch Pfahldubel. *Vortrage der Baugrundtagung Dusseldorf*, 660–681.
12. Gudehus, G. (1983). Design concept for pile dowels in clay slopes, Discussion, Special Session 5. *Proc. 8th Eur. Conf. Soil Mech. and Found. Eng.*, Helsinki, Vol. 3, 1182.
13. Jewell, R.A. (1980). Some effects of reinforcement on the mechanical behaviour of soils. *PhD Thesis*, University of Cambridge.
14. Bruce, D.A. and Bianco, B. (1991). Large landslide stabilization by deep drainage wells. *Proc. Intl. Conf. on Slope Stability Engineering*, Shanklin, Isle of Wight, April 15–19, 8 pp.
15. Lizzi, F. (1982). *The Static Restoration of Monuments*. Sagep Publisher.
16. Dash, U. and Jovino, P.L. (1980). Construction of a root-pile wall at Monessen, Pennsylvania. *Transportation Research Record No. 749*.
17. Berardi, G. and La Magna, A. (1984). Le project du reseau de pieux. *Proc. Int. Conf. on In situ Soil and Rock Reinforcement*, Paris, October, 33–38.
18. Baker, R.F. and Yoder, E.J. (1958). Stability analysis and design of control works in landslides and engineering practice. *HRB Special Report 29*, E.B. Eckel (ed.), 189–216.
19. Verrier, G. and Merlette, P. (1981). Confortement des remblais ferroviaires—une technique particulière: le clouage. *Special Revue Travaux*, March, 76–81.
20. Winter, H., Schwarz, W. and Gudehus, G. (1983). Stabilisation of clay slopes by piles. *Proc. 8th Eur. Conf. Soil Mech. and Found. Eng.* Helsinki, May, Vol. 2, 545–550.
21. Vidal, H. (1966). La terre armée. *Annales ITBTP*, Paris, Nos. 223–229, 888–938.
22. *Engineering News Record* (1976). Sprayed concrete wall cuts overall cost by 30% in underpinning, shoring. August 19, 26.
23. Engineering News Record (1990). Soil nailing used again. December 17, 2 pp.
24. Peck, R.B. (1969). Deep excavations and tunnelling in soft ground. *Proc. 7th Int. Conf. Soil Mech. Found. Eng.*, Mexico City.

25. Beveredge, R.L.W. (1973). Repairs and extensions to concrete structures using resin-anchored bars. *Civil Engineering and Public Works Review*, July, 7 pp.

26. Rabcewicz, L.V. (1964–1965). The New Austrian Tunnelling Method. Parts 1 to 3. *Water Power*, London, December 1964 and January 1965.

27. Settler, K. (1965). Die neue Osterreichische Tunnelbauweise, statischei Wirkungsweise und Bemessung. *Der Bauingenieur* 40, 8.

28. Bauernfeind, P., Muller, F. and Muller-Salzburg, L. (1977). Tunnelbau unter historischen Gebauden in Nürnberg. *Rock Mechanics*, Suppl 6, Springer-Verlag, Wien.

29. Bonazzi, D. and Colombet, G. (1984). Reajustement et entretien des ancrages de talus. *Proc. Int. Conf. on In Situ Soil and Rock Reinforcement*, Paris, October, 225–230.

30. Shen, C.K., Herrmann, L.R., Romstad, K.M., Bang, S., Kim, Y.S. and De Natale, J.S. (1981). An in situ earth reinforcement lateral support system. *Report No. 81–03*, Dept. of Civil Engng., U.C. Davis, March, 187 pp.

31. Anon (1991). Soil nailing–a solution looking for problems. *Ground Engineering* 24(1), 42–43.

32. Condon, M. (1991). Soil nailing application and construction. *Short Course on Specialist Construction Techniques for Slope and Excavation Stability*, University of Wisconsin, Milwaukee, March 21–22.

33. Nicholson, P.J. (1986). In situ ground reinforcement techniques. *Int. Conf. on Deep Foundations*. DFI and CIGIS, Beijing, China, September, 9 pp.

34. Plumelle, C. (1986). Le clouage des sols—experimentation en vraie grandeur d'une paroi clouée. *CEBTP Information Note*, 3 pp.

35. Shen, C.K., Bang, S., Romstad, K.M., Kulchin, L. and DeNatale, J.S. (1981). Field measurements of an earth support system. *ASCE J. Geot. Eng. Div.*, Vol. 107, ST 12, 1609–1624.

36. Elias, V. and Juran, I. (1988). *Preliminary Draft—Manual of Practice for Soil Nailing.* Prepared for US Dept. of Transportation, FHA, Contract DTFH-61-85-C-00142.

37. Guilloux, A., Notte, G. and Gonin, H. (1983). Experiences on retaining structures by nailing in Moraine soils. *Proc. 8th Eur. Conf. Soil Mech. and Found. Eng.*, Helsinki, 499–502.

38. Goulesco, N. and Medio, J.M. (1981). Soutènement des sols en déblais à l'aided'un paroi hurpinoise. *Tunnels et Ouvrages Souterrains*, Nr 47, Sept.–Oct., 9–17.

39. Hovart, C. and Rami, R. (1975). Elargissement de l'emprise SNCF pour la desserte de Saint-Quentin-en-Yvelines. *Revue Travaux*, January, 44–49.

40. Gassler, G. and Gudehus, G. (1981). Soil nailing—some aspects of a new technique. *Proc. 10th Int. Conf. Soil Mech. and Found. Eng.*, Stockholm, June, Vol. 3, 665–670.

41. Louis, C. (1981). Nouvelle methode de soutènement des sols en déblais. *Travaux 553*, March, 67–75.

42. Garcia Goytia, R. and Guitton, G. (1979). Test to repair a reinforced earth wall with corroded reinforcement. *Int. Conf. on Soil Reinforcement*, Paris, March, Vol. 2, 439–443.

43. Long, Livet, Boutonnet, Marchal, Olivier, Nabonne and Plaut (1984). Repair of a reinforced earth wall. *Int. Conf. on Case Histories in Geotechnical Engineering*, 335–339.

44. Gausset, P. (1985). Confortement de versants urbanies: exemple dans l'agglomeration lyonnaise. *Seminaire sur le Renforcement Des Sols*, ENTPE, Lyons, October 16–18, 11 pp.

45. Corte, J.F. and Garnier, P. (1984). Transformation d'un mur ancré par clouage du sol. *Proc. Int. Conf. on In Situ Soil and Rock Reinforcement*, Paris, October, 327–332.

46. Wolosick, J.R. (1988). Soil nailing at Nashville fault zone. *Proc. Annual Meeting, Deep Foundations Institute*, Atlanta, GA, October, 12 pp.

47. Bruce, D.A. (1989). Overburden drilling: a generic classification. *Ground Engineering* 22 (7), 25–32.

48. Louis, C. (1984). New method of sinking/grouting ground anchors. *French Patent 84–02742*.

49. Louis, C. (1986). Theory and practice in soil nailing—temporary or permanent works. *ASCE Annual Convention, GEOTECH 1986*, Boston, October 27–31.

50. SETRA (1979). *Recommendations et Règles de l'Art—Les Ouvrages en Terre Armée.* Ministère des Transports, Paris.

51. DTP (1978). Reinforced earth retaining walls and bridge abutments. *Technical Memorandum BE 3/78*, Department of Transport, London.

52. FHA (1985). Corrosion susceptibility of internally reinforced soil retaining structures. Federal Highway Administration, *Report No. FHWA/RD 83/105*.

394   UNDERPINNING AND RETENTION

53. FIP (1986). *Corrosion and Corrosion Protection of Prestressed Ground Anchorages. State of the Art Report*, Thomas Telford, London.
54. Bridle, R.J., Barr B.I.G., Jewell, R.A. and Pedley, M.J. (1990). Soil nailing—discussion. *Ground Engineering* 23 (6) 30–33.
55. Pedley, M.J., Jewell, R.A. and Milligan, G.W.E. (1990). A large scale experimental study of soil-reinforcement interaction. *Oxford Univ. Internal Report No. OUEL 1848/90*, 17 pp, plus figs.
56. Lambe, P.C. and Hansen, L.A. (eds) (1990). Design and performance of earth retaining structures. *ASCE Geot. Spec. Pub. 25*, Session I—Soil nailing, 612–691.
57. Guilloux, A. and Schlosser, F. (1984) Soil nailing—practical applications. *Symp. on Soil and Rock Improvement Techniques*, A.I.T., Bangkok, Thailand.
58. Nicholson, P.J. (1987). Insert walls—an in situ earth reinforcement technique. *Paper prepared for Central PA Section ASCE/Penn DOT Conference*, Harrisburg, PA, April.
59. Juran, I. and Elias, V. (1987). Soil nailed retaining structures: analysis of case histories. Soil improvement—a ten year update. Proceedings of a Symposium sponsored by the Committee on Placement and Improvement of Soils, *Geotechnical Special Publication No. 12*, ASCE, 232–244.
60. Terzaghi, K. and Peck, R.B. (1967). *Soil Mechanics in Engineering Practice*. 2nd edn, Wiley, New York, 729 pp.
61. Guilloux, A. (1982). Quelques observations et syntheses sur des ouvrages de soutenement par clouage. *ENPC Course on Improvement of Soils and Rocks by Reinforcement*, Paris, October.
62. Juran, I., Shaffiee, S., Schlosser, F., Humbert, P. and Guvenot, A. (1983). Study of soil–bar interaction in the technique of soil nailing. *Proc. 8th Eur. Conf. Soil Mech. and Found. Eng.*, Helsinki, 513–516.
63. Shewbridge, S.E. and Sitar, N. (1989). Deformation characteristics of reinforced sand in direct shear. *J. Geotechnical Engineering, ASCE*, Vol. 115 (8), 1134–1147.
64. Dyer, N.R. and Milligan, G.W.E. (1984). A photoelastic investigation of the interaction of a cohesionless soil with reinforcement placed at different orientations. *In Situ Soil and Rock Reinforcement, Proceedings of the International Conference*, Paris, October, 257–262.
65. Beech, J. and Juran, I. (1984). Analyse théorique du comportement d'un soutenement en sol cloué. *In Situ Soil Rock Reinforcement, Proceedings of the International Conference*, Paris, October, 301–307.
66. Schlosser, F. (1983). Analogies et differences dans le comportement et le calcul des ouvrages de soutenement en terre armée et par clouage de sol. *Annales ITBTP*, No. 418, *Sols et Fondations 184*, October, 8–23.
67. Bangratz, J.L., and Gigan, J.P. (1984). Methode rapide de calcul des massifs cloués. *In Situ Soil and Rock Reinforcement, Proceedings of the International Conference*, Paris, October, 293–299.
68. Thompson, S.R. and Miller, I.R. (1990). Design, construction and performance of a soil nailed wall in Seattle, Washington. *ASCE Spec. Conf. on Design and Performance of Earth Retaining Structures*, Cornell University, New York, June, 15 pp.
69. Bridle, R.J. (1989). Soil nailing—analysis and design. *Ground Engineering* 22 (6) 52–56.
70. Gigan, J.P. (1986). Applications du clouage en soutènement—paramètres de conception et de dimensionnement des ouvrages. *Bull. Lias. LCPC*, 143, May, June, 51–64.
71. Schlosser, F. (1986). Private communication.
72. Cassie, J.W. and Tweedie, R.W. (1990). Performance of an excavation supported by the ground control method. *Proc. Conf. Tunnelling Assoc. of Canada*, Vancouver, November 1–2, 16 pp.
73. Peck, R.B., Hanson, W.E. and Thornburn, T.H. (1974). *Foundation Engineering*. 2nd ed., Wiley, New York, 514 pp.
74. Juran, I. (1985). Reinforced soil systems—application in retaining structures. *Geotechnical Engineering* 16, 39–82.
75. Mitchell, J.K. and Villet, W.C.B. (1987). Reinforcement of earth slopes and embankments. *NCHRP Rpt. No. 290*, Trans. Research Board, Washington, D.C., 323 pp.

Index